LAWS OF THE LAND

Laws of the Land

FENGSHUI AND THE STATE IN
QING DYNASTY CHINA

TRISTAN G. BROWN

PRINCETON UNIVERSITY PRESS
PRINCETON & OXFORD

Copyright © 2023 by Princeton University Press

Princeton University Press is committed to the protection of copyright and the intellectual property our authors entrust to us. Copyright promotes the progress and integrity of knowledge created by humans. By engaging with an authorized copy of this work, you are supporting creators and the global exchange of ideas. As this work is protected by copyright, any reproduction or distribution of it in any form for any purpose requires permission; permission requests should be sent to permissions@press.princeton.edu. Ingestion of any IP for any AI purposes is strictly prohibited.

Published by Princeton University Press
41 William Street, Princeton, New Jersey 08540
99 Banbury Road, Oxford OX2 6JX

press.princeton.edu

GPSR Authorized Representative: Easy Access System Europe - Mustamäe tee 50, 10621 Tallinn, Estonia, gpsr.requests@easproject.com

All Rights Reserved

First paperback printing, 2025
Paperback ISBN 9780691247175

The Library of Congress has cataloged the cloth edition of this book as follows:

Names: Brown, Tristan G, 1988- author.
Title: Laws of the land : Fengshui and the state in Qing China / Tristan G Brown.
Description: Princeton : Princeton University Press, 2023. | Includes bibliographical references and index.
Identifiers: LCCN 2023024993 (print) | LCCN 2023024994 (ebook) | ISBN 9780691246734 (hardback) | ISBN 9780691246727 (ebook)
Subjects: LCSH: Law—China—History. | Feng shui—China—History. | China—History—Qing dynasty, 1644–1912.
Classification: LCC KNN129.F47 B768 2023 (print) | LCC KNN129.F47 (ebook) | DDC 349.51—dc23/eng/20230601
LC record available at https://lccn.loc.gov/2023024993
LC ebook record available at https://lccn.loc.gov/2023024994

British Library Cataloging-in-Publication Data is available

Editorial: Priya Nelson & Emma Wagh
Production Editorial: Jaden Young
Jacket/Cover Design: Heather Hansen
Production: Lauren Reese
Publicity: William Pagdatoon
Copyeditor: Leah Caldwell

Jacket/Cover image: Illustration of the trees at Yongling, the mausoleum of Nurhaci's (1559–1626) ancestors. Reproduced here with permission from Taiwan's National Palace Museum / Palace Memorial and Grand Council Archive.

This book has been composed in Arno

To my parents, Alice and Gary,
who taught me to cherish the moments that never come back,
and the questions that never get old.

CONTENTS

Acknowledgements ix
Dates of Qing Era Names xv
Weights, Measures, and Exchange Rates xvii
*Qing Administrative Units for the
Interior Lands ("China")* xix
Conventions in the Text xxi

	Introduction	1
1	Litigating Graves	20
2	Mapping Fengshui	61
3	Examining Fortune	108
4	Mining Sichuan	158
5	Breaking the Land	193
	Concluding Remarks	223

Appendices 231
List of Chinese Terms 237
Notes 241
Bibliography 285
Index 315

ACKNOWLEDGEMENTS

I AM GLAD FINALLY to have the opportunity to express appreciation to all those who contributed to the making of this book. A few brief paragraphs cannot do justice to the many scholars and friends who offered inspiration, help, corrections, and guidance over the years, but I hope they can capture how grateful I am for having known and learned from the people mentioned here.

I began studying Chinese at Harvard, where I benefitted from the mentorship of Mark Elliot, who inspired my early interest in Qing history, Islam in China, and Manchu; Peter Bol, who introduced me to the poems and paintings of Song literati; and Wai-yee Li, with whom I read *A Dream of Red Mansions* for the first time. Then and even in later years, I cherished the mentorship of Henrietta Harrison, Michael Puett, Michael Szonyi, Jonathan Lipman, Sayed Ilsisi, Robert Gimello, Bill Granara, Roderick MacFarquhar, Luis Girón-Negrón, Ali Asani, Shahab Ahmed, James Russell, Nathan Hill, David Brophy, and Max Oidtmann. Their formative influence is found across the chapters that follow.

I pursued my doctorate at Columbia University under the wise direction of Madeleine Zelin, who introduced me to Chinese law and brought my attention to Sichuan Province's archives. I thank Eugenia Lean for her inspiring Colloquium on Modern Chinese History, which piqued my early curiosity for the history of science. Zvi Ben-Dor Benite and Zhaohua Yang each offered reading seminars during my time in New York that proved invaluable as this project took shape. I also thank Myron Cohen, Jonathan Ocko, Benjamin Liebman, Haruo Shirane, Chün-fang Yü, Jonathan Ocko, Deborah Coen, Gray Tuttle, Shang Wei, Robert Hymes, Feng Li, Guo Jue, Christine Philliou, Richard Bulliet, Michael Como, Adam McKeown, Chengzhi Wang, Pierre-Étienne Will, and Chuck Wooldridge for their feedback and encouragement during my doctoral training.

These years of research and writing were sustained by the friendship of Guy St Amant, who helped think through key ideas in a formative stage of research, and Or Porath, whose timely humor and astute observations were greatly appreciated. I was also fortunate to know and learn from Brian Lander, Ling-Wei

Kung, Kevin Buckelew, Colin Jones, Arunabh Ghosh, Ziyun Liu, Weiwei Luo, Sungoh Yoon, Guangtian Ha, Anatoly Detwyler, Jing Zhang, Myra Sun, Josh Batts, Glenda Chao, Yanjie Huang, Chelsea Wang, Gal Gvili, Ling-wei Kung, Dongxin Zou, Gavin Healy, Elizabeth Reynolds, Ulug Kuzuoglu, John Chen, Qiaomei Tang, Eric Schluessel, Covell Meyskens, John Lee, Devin Fitzgerald, Jomo Smith, Noriko Unno-Yamazaki, and Maura Dykstra. At the Weatherhead East Asian Institute, Ross Yelsey helped demystify the publication process and always pointed me in productive directions.

Colleagues from China extended support far exceeding what I could have ever hoped. Wu Peilin provided me with opportunities to visit Nanchong's West China Normal University and offered his keen insights into the Nanbu Collection, helping me understand what was typical and what was rare to find in the archive. At Sichuan University, Xu Yue generously shared insights into popular religion during the late Qing. I also thank Liu Yonghua, Yang Der-Ruey, Xiangyu Hu, Chen Tingxiang, Li Zan, Li Deying, Wu Yan, Liu Xinjie, Li Xiaoyu, Lü Yi, Pan Jiade, Liang Yongjia, Zhang Jiajing, Qiu Jingjia, Zhu Haohao, Wang Ning, and Liu Mingqin. As I finished work on this study, Hua Ye corresponded with me extensively about the nuances of geomantic terminology and brought an important fengshui manual to my attention.

I thank Matthew Sommer for seeing potential in the project when it was still only a dissertation, for bringing me to Stanford University for a precious year of research, and for later commenting on the manuscript at a workshop. In Palo Alto, I had the pleasure of meeting and befriending Alexander Statman and Koji Hirata, both of whom provided essential feedback at various stages of writing. Kären Wigen was very kind to share her knowledge on the history of geographical thought. I also appreciated the guidance of Thomas Mullaney, Jun Uchida, Mark Lewis, John Kieschnick, Wen-Hsin Yeh, Sixiang Wang, Nicolas Tackett, David Felt, Yanshuo Zhang, George Zhijian Qiao, Emily Mokros, Jinyi Chu, Wenfei Zhou, Henry Xu, and Kelly Min.

This book was greatly improved during a residential fellowship at St John's College, Cambridge. Joseph McDermott spent hours talking with me about imperial Chinese history and lent me several key sources for this study that I regret not returning prior to his passing. Adam Chau always asked the best questions about the exact phrasing and translation of core concepts. I also enjoyed conversations with David McMullen, Noga Ganany, Michael Loewe, Hans van de Ven, Imre Galambos, Stephan Feuchtwang, Donald Sutton, Lars Laaman, Hannah Theaker, Junqing Wu, Mickey Adolphson, Dror Weil, Cheng Yang, Avital Rom, and Christoph Hess. I sincerely thank Flavia Fang Xi, Ria Roy, Huiyi Wu, Shirley Ye, Bill Moriarty, and Rando Künnapuu for reading the manuscript during my stay in Britain. Over those years, college dinners with John Weisweiler, Sylvana Tomaselli, Howard Hughes, Morgan Ng, Paul Wood, Joana Meier, Andrew Chen, Clara Manco, Dhruv Ranganathan, Robert Tombs, and Peter Johnstone were the perfect distractions from writing.

At MIT, colleagues in History and beyond have provided me with an ideal space for research and collaboration. I am grateful for the support of my chairs, Jeff Ravel, Chris Capozzola, and Craig Wilder, who went out of their way to make me feel at home as a new faculty member arriving during the pandemic. Emma Teng influenced this book long before I had the chance to meet her, and I feel lucky to have her as a colleague. Harriet Ritvo, Megan Black, Kate Brown, and Deborah Fitzgerald provided vibrant community and conversation around the university's Seminar on Agricultural and Environmental History. I also thank Sana Aiyar, Hiromu Nagahara, Peter Perdue, Kenda Mutongi, Eric Goldberg, Elizabeth Wood, David Kaiser, Will Broadhead, Lerna Ekmekcioglu, Catherine Clark, Caley Horan, Anne McCants, Tanalís Padilla, Malick Ghachem, Graham Jones, Wiebke Denecke, Alexander Forte, Marjoleine Kars, Ece Turnator, and Victor Seow for sharing their wisdom and experience as I wrapped up writing. I also thank Patty Alves, Meghan Pepin, and Kathleen Lopes for their kind patience in answering my questions about the Institute.

I finished writing this monograph at the Max Planck Institute for the History of Science in Berlin, where I have relished working alongside Dagmar Schäfer and Shih-Pei Chen. During my stay in Germany, I also had the pleasure of befriending Sarah Schneewind, who became my lunch buddy. After walking to one of the dining options available in our corner of Dahlem, Sarah offered critical insights into the late imperial novel *The Plum in the Golden Vase*, shared her profound erudition about Ming history, and provided generous feedback on chapter 3. Marta Hanson drew my attention to historical parallels between fengshui and medicine that made their way into chapter 2. I also appreciated conversations with Lisa Onaga, Siyen Fei, Joachim Kurtz, Qiao Yang, Damián Fernández, Di Luo, Ying Zhang, Melissa Charenko, Jeffrey Kotyk, Martina Siebert, Yijie Huang, and Chun Xu. Nanny Kim and Hailian Chen generously shared their expertise on Chinese mining with me and provided feedback on chapter 4.

Workshops in France allowed me to share my work with Vincent Goossaert, whose deep knowledge of Chinese religion is inspiring a new generation of scholars in the field. He drew my attention to imperial-era daybooks, calendars, and ritual time as I made sense of the arguments in imperial-era legal cases. In Paris, Poitiers, and Aussois, I also benefitted from the feedback of Christian Lamouroux, Michela Bussotti, Alain Arrault, Cynthia Brokaw, Marcus Bingenheimer, Huayan Wang, James Robson, Daniela Campo, Katherine Alexander, Gregory Adam Scott, Jinhui Wu, and Jiechen Hu.

People previously unmentioned who generously read and commented on part or all of this manuscript over the years include Fei-Hsien Wang, Paul Katz, Bony Schachter, Ian Miller, Daniel Burton-Rose, Judd Kinzley, Lillian Li, Gilbert Zhe Chen, Xin Yu, Ole Bruun, Robert Weller, Chris Coggins, Claire Yi Yang, Ronald Knapp, Richard Von Glahn, Praesenit Duara, Qiaomei Tang, Donia Zhang, Michael Paton, Celine Sui, Min Tao, and Volker Olles. At talks and conferences, I garnered helpful feedback from Robin Yates, Kenneth Dean, David Bello, Daniel Kane, Pär Cassel, Jenny Huangfu Day, Shellen Wu, Meng Zhang, Sonya Lee, Huaiyu Chen, Ari Levine, David Ambaras, Peter Golas, Johan Elverskog, Margaret Kuo, John Lee, Larissa Pitts, Robert Campany, Philippe Forêt, Beth Harper, Yue Du, TJ Hinrichs and members of the Cornell University Classical Chinese Consortium; Li Chen, Taisu Zhang and members of the International Society for Chinese Law and History; as well as Kalyanakrishnan Sivaramakrishnan, Mark Frank, and members of the Yale University Topics in the Environmental Humanities workshop.

This book would never have been possible without the encouragement and friendship of Richard Smith and Danni Cai. Rich took an interest in my work and, specifically, the development of this book. Over five years of countless emails and even a few long phone calls, Rich helped me think through difficult sources on fengshui and yin-yang cosmology. Danni proved to be an extraordinary careful and critical reader of this study while sharing her encyclopedic knowledge of Chinese and foreign scholarship. Rich and Danni were important companions over this long journey, always reminding me why this subject was worth the time and energy whenever I lost sight of the big picture.

Support for the research and writing of the book came from the Social Science Research Council, the Henry Luce Foundation, the Andrew W. Mellon Foundation, the Daniel and Marianne Spiegel Fund Grant, the American Council of Learned Societies, the British Academy, the James P. Geiss and Margaret Y. Hsu Foundation, the Max Planck Society, as well as a MIT research leave. I am grateful for the knowledge and help of the staff at the Archives of Academia Sinica's Modern Historical Institute, Taipei; Bayerische Staatsbibliothek, Munich; C. V. Starr East Asian Library, University of California at Berkeley; Hatcher Library, University of Michigan; the First Historical Archives of China, Beijing; Starr East Asian Library, Columbia University; Harvard-Yenching Library, Harvard University; the Langzhong Municipal Archive, Langzhong; the Library of Congress, Washington D.C. (accessed online); the Nanbu County Archive, Nanbu; the Nanchong Municipal Archive, Nanchong; the Sichuan Provincial Archive, Chengdu; Sichuan University China Southwest Bibliography Research Center, Chengdu; the Shanghai Library, Shanghai; Staatsbibliothek Berlin, Berlin; Taiwan National University,

Taipei; National Palace Museum Palace Memorial and Grand Council Archives, Taipei (accessed online); and Tōyō Bunko, Tokyo.

At Princeton University Press, I thank Priya Nelson, Emma Wagh, Jaden Young, Heather Hansen, Leah Caldwell, Lauren Reese, and three anonymous readers for their feedback and support in bringing this project to publication. The book required many maps and images—and I acknowledge the generosity of archives and individuals on that front especially. The Nanchong Municipal Archive, the Langzhong Municipal Archive, and the National Palace Museum each provided images in their original form. Kazumasa Yamashita and Wei Yuangang allowed the reproduction of images in their possession. Rob McCaleb and Dimitri Karetnikov created the detailed map found in the book's front material.

Last but certainly not least, the travel, research, and writing for this book required many sacrifices from my cherished family due to missed holidays, birthdays, and many celebrations. I thank my parents, Gerard and Alice, and my grandparents of the Brown and Mahfouz families for their love and support over the past thirty-four years. I thank Emily for looking out for me with her medical training and advice of all kinds. And I thank Sophie for our walks together, where she reminded me in tough times that fengshui is about the future, and the best is yet to come.

DATES OF QING ERA NAMES

Qing era names do not correspond exactly with the dates of imperial enthronements. The Kangxi Emperor for instance was enthroned in 1661, but his era name was only used for year identification and numbering upon the following New Year in 1662. The list below provides the dates for each era name.

Shunzhi (1644–62)

Kangxi (1662–1723)

Yongzheng (1723–36)

Qianlong (1736–96)

Jiaqing (1796–1821)

Daoguang (1821–51)

Xianfeng (1851–62)

Tongzhi (1862–75)

Guangxu (1875–1909)

Xuantong (1909–12)

WEIGHTS, MEASURES, AND EXCHANGE RATES

Area

1 *mu* = c. 0.1647 acre; c. 0.0666 hectare

Length

1 *cun* (inch) = c. 1.4 English inches

10 *cun* = 1 *chi* (foot, c. 14.1 English inches; c. 35.6 centimeters)

10 *chi* = 1 *zhang* (c. 11.7 English feet; c. 3.56 meters)

180 *zhang* = 1 *li* (c. 0.333 English mile; c. 0.5 kilometer)

Weight

1 *liang* (tael) = c. 1.333 English ounces

16 *liang* = 1 *jin* (catty, c. 1.333 pounds; c. 0.6 kilograms)

100 *jin* = 1 *shi* (picul; c. 133 pounds)

QING ADMINISTRATIVE UNITS FOR THE INTERIOR LANDS ("CHINA")

Eighteen provinces (*sheng*) were divided into circuits (*dao*).

Circuits were divided into prefectures (*fu*), independent departments (*zhili zhou*), and independent subprefectures (*zhili ting*).

Prefectures were divided into counties (*xian*), departments (*zhou*), and subprefectures (*ting*).

CONVENTIONS IN THE TEXT

ALL INDEXING NUMBERS from the Nanchong Municipal Archive (hereafter Nanbu County Qing Archive) are written in the following tripartite convention: "Section Number. File Number. Document Number." For example, in "7.720.01," 7 refers to the Guangxu reign, 720 is the file number, and 01 is the first document in the case file. Indexing numbers for the Langzhong Municipal Archive, the Nanbu County Archive, Sichuan University's China Southwest Bibliography Research Center (hereafter SUCSBRC), the Sichuan Provincial Archives (hereafter Ba County Qing Archive), the First Historical Archives of China, Taiwan's National Palace Museum Palace Memorial and Grand Council Archive, Taiwan's Academia Sinica's Archive of the Grand Secretariat, and the Institute of Modern History Archives, Academia Sinica follow the labeling conventions of each institution.

I transcribe Chinese names and terms in Hanyu pinyin according to Standard Mandarin pronunciation, except for words and phrases from Taiwan. In Chinese, family names (*xing*) are not surnames since they precede the personal name, but I retain surname for clarity. I provide the birth and death dates for eminent persons whenever possible and use "d.u." to denote "dates unknown." Northern Sichuan is capitalized throughout the text since the term refers to a Qing administrative unit: the Northern Sichuan Circuit (*Chuanbei dao*).

While alternatives like "topomancy" or "topographic siting" have their merits as translations for the word fengshui, I have retained "[Chinese] geomancy" as it has been the most common rendering in English. As will be made clear in the following pages, "fengshui" was just one Chinese term for geomantic beliefs and practices, and it was not always the most preferred in the imperial era. Other terms include *dili, kanyu, budi, dizhan, dixue, zhanzhai, zedi, xiangzhai*, and *yinyang jia yan*. For readability, I generally use "fengshui" or "geomancy" throughout the chapters but draw attention to the alternatives mentioned above when relevant.

The imperial lunisolar calendar divided the year into twelve months, with the new moon falling on the first of each lunar month and the full moon appearing on the fifteenth. For ease of reading, all dates in the main text are provided in the Gregorian solar calendar. References in the endnotes follow the Chinese lunisolar calendar with reign name following by year, month, and day (SZ: Shunzhi; KX: Kangxi; YZ: Yongzheng; QL: Qianlong; JQ: Jiaqing; DG: Daoguang; XF: Xianfeng; TZ: Tongzhi; GX: Guangxu; XT: Xuantong). For example, "JQ19.8.7" would be "the nineteenth year of the Jiaqing reign (1814), the eighth month, and the seventh day" in the semilunar calendar and "September 20, 1814" in the Gregorian calendar. Intercalary months are indicated with an asterisk (*).

MAP 1. Sichuan Province and the Qing Empire (China proper), circa 1800. Map created by Rob McCaleb and Dimitri Karetnikov.

Among the classical texts on earthly principles [fengshui], there are those that can be consulted and those that cannot be. Those that can be consulted include *Classic of Qingwu, The Book of Burial, Snow Heart Rhapsody, Inverted Rod Pamphlet, Suspecting the Presence of Dragons Classic, Shaking Dragons Classic, Emanating Profundity Doctrines, Cavern Condition Rhapsody, Nine Star Pamphlet, Eight Forms Lyric, A Precious Mirror of Geomancy, Passing Courtyards Classic,* and *Humble Opinions on Geomancy*. These texts are all authentic books about earthly principles, and they must be read.

地理經書，有可讀者，有不可讀者。可讀者，惟青烏經、葬書、雪心賦、倒杖篇、疑龍經、撼龍經、發微論、穴情賦、九星篇、八式歌、堪輿寶鏡、趨庭經、堪輿管見，此皆地理正宗，不可不讀也。

—*CORRECT DOCTRINES OF FENGSHUI* (1740)
Composed by the Qing Astronomical Bureau

LAWS OF THE LAND

Introduction

THIS BOOK CONCERNS the roles of fengshui in law during the Qing (1644–1912), the last of China's imperial dynasties. Literally wind (*feng*) and water (*shui*), fengshui refers to the practice of analyzing landscapes to determine the most auspicious sites and orientations for houses, graves, temples, and other kinds of structures, based on principles of harmony between humans and their environments. Fengshui has long been misunderstood in the West. Early Western observers dismissed it as superstition, while New Age practitioners enthusiastically adopted it for interior design and home decoration. Scholarly publications have shed much light on fengshui's theories yet have tended to focus on its applications in city planning, architecture, and aesthetics. From today's standpoint, the law court probably would be the last place to go looking for fengshui. Yet for the Qing period, the legal system was one of fengshui's most recognizable arenas of play.

During the Qing dynasty, people across China submitted lawsuits about harm done to fengshui. Not all disputes involved such claims, but many did—especially those involving houses, graves, natural resources, and public spaces. In response, courts inspected fengshui, mapped sites where fengshui was at issue, weighed the effects of harming fengshui, scolded people for being too obsessed with fengshui, and applauded others for taking good care of ancestral sites and common areas. Many case files and trial transcripts from the dynasty have survived, offering a rare window into how people in China analyzed, understood, and recorded land in relation to fengshui prior to the twentieth century.

Drawing on five hundred legal cases and a host of technical manuals, *Laws of the Land* argues that fengshui was interwoven with Qing governance, especially its legal system. The imperial state incorporated fengshui into its administration of gravesites, forests, mining, and the examinations used to recruit scholars into office. Qing officials invoked and even practiced fengshui to resolve local disputes, secure community harmony, and govern rural society. The result was a practical blurring of the line between law and fengshui, with people constantly invoking fengshui in law, as law. In managing intractable conflicts over the land's interpretation, Qing officials clarified when and where fengshui was disturbed and when and where it was not. The legal system's answers to those conflicts produced continuous evidence for fengshui's reality and relevance.

The interplay between law and fengshui sheds light on a host of fundamental issues relating to and transcending the imperial legal system, including land use strategies in an era of dwindling resources, the methods Chinese people mobilized to confront Western imperialism in the nineteenth century, and the history of knowledge production. What follows is a historical portrait of life and death along a Qing landscape—with fengshui in focus.[1]

Divergent Paths in the Study of Chinese Law and Religion

The Westerners present in the final decades of the Qing dynasty encountered fengshui during disputes over churches, concessions, telegraphs, and railways. Their views of it were not complimentary. Anglo-American diplomats and missionaries described fengshui as "a ridiculous caricature of science," "an abyss of insane vagaries," and "a perverse application of physical and meteorological knowledge."[2] Such statements are reflections of Orientalism, that is, Eurocentric attitudes to non-Western peoples of the "Orient" informed by colonialist biases.[3] Although prejudiced, these perceptions cast a wide net and a long shadow. No less a social scientist than Max Weber (1864–1920) presented fengshui objections to railway construction as evidence of the supposed irrationality of the Chinese legal system.[4] Late Qing Western observers concluded two things from their encounters with fengshui: first, that it was a "superstition" devoid of logic, and second, that China lacked a proper legal system rooted in the rule of law.[5]

Historians and anthropologists later challenged these views, albeit independently of each other. Anthropologists such as Maurice Freedman, Emily Ahern, Stephan Feuchtwang, Ole Bruun, and Robert Weller picked up the study of fengshui through painstaking fieldwork across Greater China.[6] Collectively, their works demystified the practice by documenting its significance to kinship practices and popular religion. Anthropologists in China and Japan, including Chen Jinguo and Segawa Masahisa, addressed fengshui through a blend of ethnography and careful analyses of family genealogies, including those from the Ming (1368–1644) and Qing dynasties.[7] Each time Chinese rural society drew attention, fengshui came into view. Yet, with a few exceptions, scholarship on fengshui, rooted in fieldwork conducted in the twentieth century, did not address the place of fengshui in imperial law.[8]

For their part, historians mounted a major revision to blinkered Orientalist impressions of Qing law and governance.[9] Access to local and national archives in China and Taiwan enabled historians to dismantle earlier critiques that China lacked a judicial system that protected property or conformed to the rule of law.[10] Contrary to earlier views of the Chinese legal system as primarily designed to administer punishments, historians have demonstrated that routine conflicts over land and property constituted a significant part of legal practice.[11] For resolving these disputes, judges weighed the 436 statutes (*lü*) and the more than a thousand substatutes (*tiaoli*) of the *Great Qing Code* (*Da Qing lüli*), as well as model cases, established principles, and community reception.[12] Across the board, the legal system offered, in the words of Philip Huang, "consistent guides for the resolution of real problems and disputes," and thus was far more reasoned and pragmatic than earlier Western observers perceived.[13]

Concomitant with the revised scholarly perception of Chinese law, there has prevailed an assumption that the Chinese legal system was "fundamentally secular," with little influence from religion.[14] Pioneers in the study of Chinese law, such as Derk Bodde and Clarence Morris, supported this argument with the observation that China lacked a divine lawgiver or any divine origin tale for law.[15] The secular reading of Chinese law became widely held among historians and legal scholars, both in the West and China.[16] Until recently, it was scarcely challenged.

The notion of legal secularism persisted because it allowed historians to argue that Chinese law was "substantively rational," even if authoritarian.[17] Key issues posed by Orientalism were conveniently solved as legal historians focused their attention on the "secular matters" of the law code, such as "taxes, corvée labor, regulations for officials, monetary obligations, military regulations, forgeries."[18] In revising earlier impressions of the imperial government, legal historians kept the code and model cases while ceding religion to anthropologists, thus producing a great twentieth-century divergence between the study of Chinese law and fengshui.

Rediscovering Religion in Chinese Law

Recent years have witnessed a changing tide in understandings of the ideological foundations of Chinese law. In Taiwan, researchers have become increasingly aware of an unmistakable and pervasive feature of Chinese religion: it is full of law. Temple murals, vernacular tales, and ritual handbooks depict busy law courts of the underworld and lawsuits brought by the unhappy dead against the living or their deceased kin (*zhongsong*).[19] A wide spectrum of rituals performed for judicial deities offers parties an opportunity to first attempt dispute resolution at temples before heading to the state's courts. These examples form part of what Paul Katz calls the "judicial continuum" between Chinese religious and legal practice.[20]

Another finding emerged from a surprising source—the late imperial law codes—which initial impressions held as merely containing a secular blend of Confucian ethics and stern legalism. In his revisionist history of the *Great Ming Code*, Yonglin Jiang makes a robust case that the ruling elite of the late imperial era "viewed law as a concrete embodiment of the cosmic order."[21] The law expressed cosmological unity between heaven and earth, with its statutes intended to bring the institutions governing human action into line with the cosmic order.

Thus, at present, Katz finds law across religious practice, and Jiang finds religion in the official code.[22] But if law pervaded religion on the ground, it seems likely that religion influenced not only the law code but also pervaded law as practiced daily in the courts.

It did. *Laws of the Land* picks up where Katz and Jiang left off and revises prevailing understandings about Chinese law and its underlining sources of legitimacy. This book demonstrates that a sophisticated cosmology was assumed in law, permeating down the bureaucracy to the circuit, prefecture, and county levels.[23] When applied to land, this cosmology manifested as fengshui. Because the imperial state recognized that harming fengshui was a crime, people thought of fengshui-related problems as potential legal problems and sought official intervention when conflicts arose. Through invoking fengshui as laws of the land, litigants appealed to officials in ways that did not undermine the government's authority, but affirmed it.

The key to resolving the fracture between nineteenth-century impressions of Chinese superstition and twentieth-century visions of Chinese secularism lies in disrupting the dichotomy of law as rational and fengshui as irrational. Consider the purported secular rationality of Western legal systems. Western law and religion do not just embrace resonant doctrines—penance and penalty, covenant and contract, sin and crime—but law and religion in the West also are performed in similar fashions.[24] Affirmations with holy scriptures remain a standard procedure for individuals testifying in courts. Oaths for newly inducted judges include phrases such as "so help me God." While adjudicating cases, judges don flowing black robes and, in some regions, powdered wigs (perukes). After assuming the bench, many of them work in courthouses displaying grand murals or monuments of the Ten Commandments. In the United States specifically, the overt use of religious references in judicial decision-making remains "not rare" through the present time.[25] These theologically derived rituals, symbols, and features all bear witness to the notions that "we have never been modern" and that Western law is not completely disenchanted.[26]

Consider in turn the logics of fengshui in Qing China. The imperial state issued technical texts about geomantic positioning and time selection, and various official and private specialists mastered fengshui's intricacies. Rules and regulations dictated when and where fengshui could be invoked in legal settings. Qing authorities ordered lands to be precisely mapped to appraise the merits of claims presented in a fengshui-related lawsuit. With these maps, officials could identify and dismiss specious geomantic claims while accepting and enforcing claims they understood as compelling.

This arrangement is not so difficult to understand. The chapters that follow reveal communities drawing on fengshui to regulate building heights, cacophonous activities, and flammable industries. These deployments bear some resemblance to zoning laws, which in the twentieth century came to represent smart development—not backwardness.[27] Further resonances are found today in instances where jurists are tasked with weighing unpredictable future consequences, like climate change. Some US-based legal experts have proposed invoking the constitutional protection of religious freedom to compel courts to recognize indigenous claims over territories through which crude oil pipelines are planned and constructed.[28] In the Qing, people would have pursued a similar strategy through fengshui.

The Western Orientalists of the nineteenth century were wrong not because they identified fengshui as important to Qing society, law, and governance. They were wrong because they condemned fengshui as irrational and the antithesis of proper law. That distinction offered a false choice then, and it remains one now. By taking law and fengshui together not as an oxymoron but as a shared nexus of principles that supported imperial governance and administration, this book fundamentally revises understandings of both.

Practicing Fengshui in Qing China

People did three things with fengshui during the Qing dynasty. First, they studied it. Scholars voraciously collected, read, and cited manuals about fengshui, approaching it as a rigorous academic subject. Literate commoners also acquired information about it from technical tracts, almanacs, and popular encyclopedias. Even illiterate farmers had access to useful geomantic information provided verbally by ritual masters in towns and villages.[29] Second, people practiced fengshui, often for a price. Geomancers sold their knowledge to people needing to construct a house, grave, temple, shop, school, or bridge, offering insurance to investors and merchants who sought to limit the risk of litigation. They also instructed people how to present compelling lawsuits in court. Finally, people narrated fengshui. People mobilized it to explain phenomena in the material world, creating tales of success and loss that formed the scaffolding of social status, family histories, and property claims.

Although there was no universal system of fengshui across China, specific works provided a measure of state-sponsored guidance. Such texts included the *Imperially Endorsed Treatise on Harmonizing Times and Distinguishing Directions* (*Qinding xieji bianfang shu*) and the *Imperially Endorsed Almanac for Time Selection* (*Qinding xuanze lishu*), the former of which was printed at the imperial palace's Hall of Martial Valor (*Wuying dian*) for official distribution. Another source, *Correct Doctrines of Fengshui* (*Qintianjian fengshui zhenglun*), cited in the epigraph to this book, was composed by the Astronomical Bureau in Beijing under the auspices of the Erudite Scholar (*boshi*) Gao Dabin in 1740. Not printed for general circulation, *Correct Doctrines of Fengshui* was reserved for use by central government authorities. The text carefully lays out the government's position on fengshui by inveighing against the prevalence of inadequate knowledge and listing "false" fengshui texts alongside those deemed reliable.[30]

Like all other geomantic works in China, *Correct Doctrines of Fengshui* draws on cosmological concepts dating from pre-imperial antiquity (before 221 BCE) that found expression in many realms and activities, including astrology, music, and medicine.[31] Canonized Confucian texts, most notably the hallowed *Yijing* (*Classic of Changes*) and its commentaries, validated these notions.[32] In brief, the foundational generative forces of *yin* (female, moon, passive, etc.) and *yang* (male, sun, active, etc.) constantly interacted in the material world to produce the so-called five agents. These five agents possessed the qualities of fire (heat), water (coolness), metal (sharpness), wood (flourishing), and earth (stability).[33] These agents became manifest as the constituent elements of *qi*, translated as ether, energy, pneuma, vital essence, or material force. All things, visible and invisible, were composed of *qi*.[34]

When properly understood, the operations of *yin* and *yang*, the five agents, and other cosmic forces were harnessed for analyzing gravesites, termed houses of *yin* (*yinzhai*), and residences, called houses of *yang* (*yangzhai*).[35] *Correct Doctrines of Fengshui* covered principles for both structures. Reflecting the gendered dimensions of fengshui's cosmology, houses of *yin* energy were the most anxiety-inducing relative to other dwellings since these "passive" sites required active upkeep to avoid misfortune.[36] The importance of the "house of *yin*" in Qing society is hard to overstate. Acquiring a house or a field did not render someone a recognized member of most village communities. An ancestor's grave did.[37]

There were several schools of fengshui in Qing times, but two were dominant. The Compass School (*liqipai*, lit. "principles and energy" school) paid close attention to the orientations of graves and houses as directed by a geomantic compass (see figure 2.7 for an example). This device, typically about four to eight inches in diameter with a magnetic needle pointing south, displayed a series of concentric circles representing the abovementioned cosmic variables, with compass points identified in large part by an ordering system known as the ten heavenly stems and twelve earthly branches.[38] For instance, a gravesite identified as *mao shan you xiang* meant that it was situated on an eastern (*mao*) mountain facing west (*you*).

The Form School (*xingshipai*, "forms and configurations" school) emphasized the balance of *qi* in landforms naturally shaped by mountains and water. Although scholarly and popular literature has often discussed these schools, the distinction between the two was rare in legal cases. *Correct Doctrines of Fengshui* elevated the Form School above the Compass School but minimized their differences by drawing on both techniques.[39] And, in practice, exponents of fengshui, whether professionals or amateurs, drew upon both schools by orienting graves with the compass and identifying relevant landforms.

Knowledge of fengshui was diffused across the general population through handbooks and almanacs, but there were specialists. Many families preferred to hire geomancers to take on the responsibility of site selection and, if necessary, to assume the blame if results were poor. Practitioners of fengshui could be called by several titles, with geomancers (*dilijia*) or geomancy specialists (*kanyujia*) two of the more common titles.[40] Geomancers and diviners were usually male, but female fortune tellers were not unknown.[41] The imperial state, for its part, established official positions for the study of *yin* and *yang* throughout China and hired individuals who demonstrated knowledge of officially sanctioned astronomical and geomantic texts. These functionaries were known as officers of Yin-yang schools (*yinyangxue guan*), with "schools" referring to local institutions responsible for propagating correct applications of Yin-yang cosmology. Their institutional roles were important, but because there was only one such officer at any given time in a jurisdiction, the market for private geomantic and astrological consulting remained vast and largely unregulated over the dynasty.

A core concept in *Correct Doctrines of Fengshui* was earth veins (*dimai*), which referred to invisible subterranean channels thought to direct *qi* through the land.[42] These veins animated the earth, and their flow across bends in the terrain revealed the land's living body.[43] That body was deemed healthy by the presence of lush vegetation and flowing—not still—water, identified as the blood of the land.[44] Topographical features were related to each other through veins, for instance, with the vein of an "ancestor mountain" flowing into the "parent mountain" surrounding a site of development. Upon burial, the fates of living descendants (*mingmai*) became linked to the well-being of earth veins. Such veins were deemed "dragon veins" (*longmai*) when they aligned with sites of imperial power, major mountain ranges, or large watersheds. All veins could be harmed by cutting down a tree or trees, modifying water flows, or mining a mountain. Protracted legal disputes often focused on the alleged harm done to earth veins by activities of this sort.[45]

The importance ascribed to earth veins draws attention to the fact that most lawsuits invoking fengshui disputed the exploitation of resources, with fengshui almost always cited in law for halting development. As such, fengshui's invocation parallels other cultural practices identified by economic historians as contributing to the limitations on the free disposal or alienability of land during the late imperial era.[46] The chapters that follow share many examples of the ways fengshui shaped economic activity through land sales, agriculture, water control, riverine transportation, mining, construction, timber harvesting, and, in the very late Qing, political decisions about telegraphs, mines, and railroads. It may not be feasible to quantify the economic impact of fengshui, but its impact was real. Those who condemned fengshui as "superstition" after 1900 were frustrated precisely by the ways it interrelated with what they saw as economic development.

Although fengshui's economic relevance is beyond dispute, its environmental applications have been more contentious. Lynn White's 1967 article, "The Historical Roots of Our Ecologic Crisis," which criticized Western religions for legitimizing overexploitation of the land, galvanized general interest in alternative belief systems like fengshui.[47] Not long thereafter, some Westerners started taking fengshui as a form of "Eastern wisdom" designed to harmonize the actions of humans with the forces of nature. Pushing against these notions, Ole Bruun argues that they amount to the invention of tradition by considering fengshui outside its historical and cultural contexts.[48] Bruun is correct that fengshui was historically neither a self-conscious doctrine to protect the environment nor a social practice that maintained pristine nature.[49]

Like other kinds of indigenous knowledge now drawing scholarly attention, fengshui was the major medium of expression to discuss and analyze environments and their changes.[50] When a place lost all tree cover, people did not talk about deforestation, but a loss of healthy *qi*. When contesting mining, people did not speak about the risks of pollution, but of severed dragon veins. Qing documents of all sorts refer to land as a dynamic, "living" thing, but hardly in a romantic way that might be envisioned in the popular imagination today. The land's features were not limited to harmonious sites of reflection. They could also be terrifying, summoning visions of illness, famine, or death. The fretful drive to secure and maintain fortune in some cases could justify controls on the exploitation of natural resources over surprisingly long periods of time. Chris Coggins, for instance, identifies the legacy of such practices in the rural fengshui forests of south China, where botanists have now documented the survival of the threatened Chinese Tulip Tree and other rare plants.[51] This research speaks to consequences, not to motives.

Physical environments in China declined, changed, and sometimes recovered, regardless of what people may have believed about the land and what their intentions may have been toward it.[52] People invoked fengshui to put the environment to use for human benefit.[53] And it was this fundamental reality that gave fengshui standing throughout Chinese society and at all levels of the imperial government.[54]

Mobilizing Fengshui in Law

Fengshui's importance to imperial administration, particularly legal practice, may not be self-evident. The law code did not explicitly invoke the word "fengshui" outside of a couple statutes concerning delayed burials and uncovering graves.[55] To understand why people invoked fengshui in the law, one must consider how local cases were adjudicated within Qing legal culture. As Melissa Macauley defines it, legal culture constitutes "a system of symbols, language, and diffusely shared attitudes that produced a common set of legal assumptions and beliefs."[56] The key word in that sentence is "assumptions," or experiential knowledge that was once so obvious to people that few felt compelled to explain it at length. Because fengshui was part of an "educated person's natural furniture" in the late imperial era, it related to matters involving graves, trees, rivers, fields, and mines in ways that were obvious then, but which may not be apparent or intuitive today.[57]

Although the law code was widely distributed, providing a solid framework for adjudicating serious crimes, officials had a great deal of flexibility in adjudicating "minor matters," especially at the lower levels of imperial administration.[58] County magistrates, the lowest-level officials appointed directly by the emperor, sometimes departed from the written law code in their rulings. For instance, Matthew Sommer, drawing on several of the same archival collections used in this book, has shown that county magistrates tolerated the widespread but illegal practice of wife-selling during local adjudication.[59] Magistrates weighed competing factors, especially economic needs, when issuing verdicts.

This administrative flexibility extended to matters of land. The Qing state did not get involved with most private transactions in property. In theory, local governments were supposed to register land for taxation, and all contracts for exchanging land needed to be stamped with an official seal. However, the government recorded little land after the 1730s, and from the late 1700s onward, few contracts were stamped with official seals.[60] The result was that the empire's land tax revenue in 1850 was not much different from the yield in the mid-1700s even though the population had doubled and land yields had significantly increased over that time.[61] As the population boomed, the size of the formal bureaucracy hardly expanded, and around 1,358 county magistrates managed a population of over four hundred million by the end of the dynasty.[62] Nonetheless, even without extensive land records and a limited formal bureaucracy, Qing courts adjudicated property cases until the fall of the dynasty in 1912. They did so in part by directly engaging bodies of knowledge, such as fengshui, which complemented the official legal system.

How did fengshui become so entangled with the legal system? Officials recognized fengshui as a rising source of legal disputes in the Song dynasty (960–1279).[63] Around the same time, scholars within the ascendant Neo-Confucian movement voiced opposition to cremation in favor of bodily burials.[64] Graves became big business, and by the middle of the Ming dynasty (1368–1644), lawsuits over harmed earth veins began appearing in litigators' manuals.[65] By Qing times, not only was there abundant precedent for invoking fengshui in legal disputes, but hundreds of popular geomantic manuals were also in circulation alongside official ones to aid in doing so.[66] A plethora of official handbooks and judicial collections also appeared. Not a few of these texts touched on fengshui, including some that detailed government uses of prognostication in the law.[67]

These facts suggest that the influence of fengshui flowed in two directions. Litigants drew on situated geomantic knowledge recognized in their local communities while making legal petitions.[68] In turn, officials gauged legal claims against their more formal and theoretical understandings of law and fengshui, not infrequently offering the "correct" interpretation in verdicts. We can identify some widely recognized principles through local cases, handbooks, and edicts. Disturbing a grave never helped fengshui; it only hurt it. Opening a distillery, oil-press shop, or slaughterhouse next to a school or government office never helped fengshui; it only hurt it. Mining coal never helped an area's fengshui; it only hurt it. Cutting down a tree, with some exceptions for houses, never helped fengshui; it only hurt it. By contrast, damming a river, dredging an irrigation channel, erecting a pagoda, building a bridge, adding a wall, or moving a school could help or hurt fengshui, depending on the size and location of the project, as well as the immediate reception and aftermath. Elites and commoners all understood the rules of the game, which is why fengshui could be invoked in court—not as a custom that the Qing legal system tolerated, but as part of a larger framework through which law operated.

Considering fengshui's roles in legal practice, one must ask what the officials responsible for issuing verdicts believed about it. There is no simple answer to that question. Qing officials adjudicated claims about fengshui with different motives: some brought a professed commitment to fengshui's principles, and others brought tidbits of practical knowledge concerning house and grave layouts, irrigation, or mining. Some brought ample doses of skepticism, and others brought an eye for expedient solutions to complex conflicts.[69] These diverse motives often overlapped in court in fascinating ways, mirroring the complex inner lives of individuals tasked to uphold proper ritual and enforce the law.

Qing officials did not need to fully believe in fengshui to engage, analyze, and fairly adjudicate questions about it.[70] In fact, they had reasons to be confident at trial. County magistrates were usually the only person in a courtroom who had passed several levels of the civil examinations—the quintessential signifier of favorable family fengshui in late imperial times.[71] If anyone knew geomantic secrets to success in a county, it was likely to be the sitting magistrate. Aware of the great responsibility that came with this authority, many officials provided genuine advice to commoners from the bench. There are limits to this generalization, of course, as commoners resentful of county-level rulings sometimes appealed to higher courts, insisting that a magistrate's interpretation had been wrong.

Holding considerable power over matters of land and law, officials critiquing fengshui typically did not condemn it categorically but distinguished their refined understandings of it from those of commoners. The eminent statesman Zeng Guofan (1811–72) could cite his grandfather's disbelief in doctors and geomancers as an edifying model yet not condemn recourse to medicine or fengshui. Rather, he directed criticism toward plebian practitioners who profited from their purported expertise in those fields.[72] Zeng even ascribed his grandfather's recovery from an illness and his career promotion to the careful positioning of his grandmother's grave.[73] No person reading Zeng's letters in Qing times would have perceived a contradiction.

Elite criticisms of popular practices could also indirectly acknowledge uncomfortable realities of local administration. As a rule, officials bemoaned litigation of all types even though they adjudicated legal cases throughout the dynasty. To the imperial state, litigation represented worrying cracks in the moral order. Consider also non-degree-holding "runners" employed to do the dirty work of local governance. As revealed by Bradly Reed, these functionaries were rhetorically scorned for their subordinate position to examination-passing officials even as they were essential for many administrative functions, ranging from criminal fact-finding to tax collection.[74] Official rhetoric did not always align with governing realities.

Qing officials took fengshui seriously and typically knew something about it. They also expressed concern that commoners might misinterpret fengshui or be deceived by unscrupulous professionals. The widespread anxiety about reliable knowledge and skill—the pervasive fear of being cheated, framed, or harmed—drove the Qing state to issue the *Imperially Endorsed Treatise on Harmonizing Times and Distinguishing Directions*, to annually distribute the imperial calendar (*Shixian li*, or Timely Modeling Calendar; renamed *Shixian shu*, or Timely Modeling Book, in 1735), to appoint and employ Yin-yang officers with learned expertise, and to engage fengshui directly through the courts.[75] Occasional words of concern and critique mattered, but government actions spoke the loudest.

Sichuan Province and Historical Change

Sichuan Province is home to two of the most extensive county-level government archives from the Qing dynasty: the collections of Ba County, available from the 1980s, and those of Nanbu County, which have risen to prominence since the beginning of this century. Sichuan's geographical isolation spared it from much of the destruction caused by the Taiping Civil War (1850–64) and other definitive conflicts of the past two centuries, allowing government records lost elsewhere to be preserved intact in the province. As geographical and temporal context helps in understanding legal disputes, this section introduces the history of Nanbu County and Sichuan province.

Sichuan was transformed from a frontier province in the early Qing to a breadbasket of the empire by the 1800s. After the province's devastation in the 1600s, the newly established Qing dynasty encouraged migration into the province with the promise of cheap land and low taxation.[76] Migrants arriving from Hubei Province and elsewhere faced several decades of tiger attacks, indicative of Sichuan's resurgent forests during the population decline of the Ming-Qing interregnum.[77] Over the 1700s, the tigers retreated as the forests were cut down, their timber sold for construction or charcoal, and much of the land transformed for intensive agriculture.[78] The population boomed. Land, once plentiful, became scarcer. By the end of the dynasty, Sichuan had more than forty-eight million residents, making it the most populous in the empire. For its part, Nanbu County's population grew from seventy thousand in the early 1700s to over six-hundred thousand by 1912—a nearly tenfold increase compared to a fivefold increase nationally over the same period.

Nanbu shared some features with Ba, though the two counties' fortunes diverged with time. Both were geographically large and had roughly commensurate populations before the rapid commercial expansion of Ba's urban core at Chongqing in the second half of the 1800s. Illegal underground organizations ("secret societies") proliferated in the patchwork of rural migrant communities in both counties and across Sichuan, especially in the 1800s.[79] While both counties lay along the Jialing River, Ba rested at its intersection with the Yangzi, connecting the county to the wealthiest regions of the empire and allowing trade to flourish. Ba's farms yielded more than twice as much land tax as Nanbu's and saw higher rates of landlordism by the 1800s.[80] By contrast, officials described Nanbu with phrases like "mountains are many, and fields are few; the soil is infertile, and the people are poor."[81] Early twentieth-century surveys found that over 75 percent of Nanbu agriculturalists owned some of the land they tilled.[82] Regions with low tenancy had fewer protections for tenants, and officials serving in them devoted considerable attention to the survival and sustainability of family units.[83]

No inauspicious events were recorded in the county prior to 1778, after which floods, droughts, and famines became increasingly commonplace.[84] A conflict often considered the turning point in the dynasty's fortunes, the White Lotus Rebellion (1796–1804), engulfed Northern Sichuan, resulting in the killing of a Nanbu magistrate by the rebels.[85] To pacify the rebellion, the Qing relied on militia groups and the mobilization of the rural elite, who thereafter accumulated considerable power in local society. Through this elite, the government obtained basic information about the county's vast territory, which extended for over 2,200 square kilometers. Plenty of information remained concealed to authorities however, and entire swaths of the region's landscapes were not formally charted until provincial surveyors arrived in the 1930s.

The desire of a newly arrived population to claim land in the face of an increasing scarcity of resources for construction, burial, irrigation, and mining contributed to fierce competition over fengshui. Claimants in court had incentives to insist that their land had cosmic significance, not least because doing so increased the gravity of legal claims. For their part, magistrates governed the county and adjudicated disputes within the confines of available information. Remember that magistrates were almost always outsiders in the counties where they served because the so-called rule of avoidance prohibited them from holding office in their home provinces. In the absence of personally acquired local knowledge, magistrates had to rely inordinately on information provided by yamen underlings, on data contained in any available documents, and on oral testimonies offered by litigants at trial.

This final observation is critical for historicizing the book's argument. People engaged fengshui in law since the beginning of the Qing dynasty and centuries before. However, during the dynasty's second half, fengshui took on heightened salience in legal and political discourse. Officials in Sichuan and beyond expressed concern regarding increased conflicts over land, minerals, and trees. To explain these trends, some highlighted cultural problems, such as the preponderance of greedy diviners said to delude the people; others highlighted material ones, such as population pressures and the diminishing of natural resources.

Regardless of their precise diagnoses, officials found it helpful to invoke, engage, and practice fengshui in court to help resolve disputes and to maintain order during the social, economic, and environmental challenges of the mid-to-late-Qing era. Considering fengshui in contestations over development helped officials maintain stability, which was their most important task in office. This legal and political context must be understood to situate concerns over industrialization late in the dynasty, when a dramatic debate—spanning the nature of law, the governing responsibilities of the Qing state, and the best strategy for responding to Western encroachment into China—unfolded across the bureaucracy. Resonating far beyond the palace walls of Beijing, those debates brought fengshui to the wider world's attention.

Sources and the Scope of the Argument

The Nanbu Collection is the second largest county archive in China. In 1960, the remarkable collection was discovered tucked away on the dusty shelves of the county's public security bureau. Not long after, the government transferred the collection to Nanchong City, where it survived the Cultural Revolution under the protection of military guard. In the ensuing decades, scholars at Southwest China Normal University in Nanchong spearheaded the drive to use the archive as a means to exhaustively analyze local administration during the Qing dynasty.[86]

Methodologically, the archive has been used to overturn received wisdom about the era based on more conventional sources such as gazetteers, which were locally compiled records of a county, prefecture, province, or temple. Wu Peilin has shown that the average magistrate served in Nanbu for far shorter periods than the commonly cited length of three years for county officials.[87] With turnover high due to illness, death, mourning obligations, or dismissal, provincial authorities often made temporary or acting appointments, which outnumbered direct appointments from Beijing throughout the dynasty. As gazetteers did not record all these men, their tenures would be unknown without the archive. Even with the existing range of gazetteers, handbooks, and edicts, understanding what was happening in the Qing remains a challenge without access to raw and unedited records, which government staff assumed might never be reread.

The sixteen months I spent in Nanbu, Nanchong, and Chengdu, in addition to the publication of the archive's index (2010) and the subsequent publication of the collection itself (2016), permit a preliminary analysis of the frequency of fengshui-related litigation during the Qing.[88] A total of 11,071 litigation files are preserved in Nanbu's archive dating from the Yongzheng reign (1722–35) to the end of the dynasty. Because a single file may contain more than one legal case, the aggregate number of all cases remains to be determined. Nonetheless, within the collection at least 1,192 legal cases concern graves, trees, burials, temple properties, and house and ancestral hall layouts (Appendix A).[89] For some of these cases, fengshui was cited in a lawsuit's opening lines, while for others it was implicitly invoked (e.g., "forbidden trees," "illicit burial," "[geomantic] obstruction," etc.). Since the archival preservation rate is higher for the nineteenth century than the eighteenth, it is challenging to estimate change over time. However, inferences are possible.

Clerks belonging to one of seven county yamen departments processed submitted legal petitions: war, personnel, works, punishments, revenue, rites, and salt. Two departments, rites and works, handled most cases involving disputes over land claims arising from tensions over field boundaries, irrigation, inheritance, temples, lineage estates, houses, burials, and graves.[90] For relatively well-preserved files of the Guangxu era (1875–1909), these two departments processed on average twenty-seven cases over graves, trees, and burials per year. Those cases represented a significant percentage of all disputes involving landed property. The sixty-nine surviving grave-related cases from the Daoguang era (1821–51; Appendix A) constitute at least a third of all land disputes surviving from that era.

This rough estimate can be compared to data from the Ba Archive. Wei Shunguang's research on that collection indicates that 26 percent of land disputes in Ba County from 1821 to 1838 concerned graves, burial, and trees, which at face value is resonant with Nanbu's files for roughly the same period.[91] There are reasons for caution, however. The Ba Archive is roughly six times larger than Nanbu's and contains many lawsuits related to commercial activities in Chongqing.[92] A greater percentage of Nanbu's files may well concern routine disputes over land and graves. In any case, comparing data drawn from two archival indexes with differing labeling systems is fraught with potential problems. These caveats aside, graves were legally overrepresented in both counties relative to the space they occupied on land, underscoring their importance for claiming land and contesting territory throughout the dynasty.

Nanbu's percentage of case files directly or indirectly related to fengshui was not high compared to other areas of China. In some regions, fengshui was cited as the subject of most legal disputes. A native of Huizhou (a region today split between Anhui and Jiangxi provinces) and an early Qing official, Zhao Jishi (1628–1706), observed that "people in Huizhou ascribe great importance to the theories of fengshui and among routine disputes that become lawsuits, almost half concern it."[93] Echoing similar sentiments, the judicial administrator of Hunan Province, Zhou Renji (1696–1763), lamented in a public notice that every time local courts opened for lawsuit submissions, people contesting fengshui presented half of the plaints.[94] In Jiangxi Province, the 1782 gazetteer for Ganzhou, a prefecture bordering Fujian and Guangdong provinces, claimed that over half of all local litigation concerned fengshui.[95] Taiwan's Dan-Xin Archive, which echoes practices found along the southeastern coast of China proper, contains "thirteen cases alleged about the fengshui of graves, many more than lawsuits over land boundaries."[96] Thirteen is too small a sample to say much, but the trend is clear.

Some provinces with notably high rates of fengshui litigation belonged to the broader Jiangnan (lit., "South of the [Yangzi] River") region, which had distinct socioeconomic and cultural characteristics. Jiangnan was the wealthiest area of the country. For this reason, its lineages composed detailed genealogies, which often included expansive geomantic maps of ancestral properties and gravesites. With more resources available to prepare, Jiangnan's scholars passed the civil examinations, which were the primary route to office-holding in late imperial times, at rates and absolute numbers higher than anywhere else. There was an assumed link between success in the exams and favorable fengshui. In parts of Jiangnan and Guangdong, reburial, which involved exhuming the bones of the deceased and moving them to a more advantageous spot, was enthusiastically pursued.[97] Reburial made the dead mobile and unsurprisingly produced higher rates of geomantic litigation.[98] Reburial was especially common in places where fengshui allegedly drove around 50 percent of legal caseloads.

While Sichuan's rates of fengshui litigation did not reach those of southeast China, there are reasons to consider it a useful lens into Qing governance. Jiangnan is blessed with many written sources, including gazetteers and genealogies, yet surprisingly lacks extensive county court records. This lacuna has meant that historians have not viewed courts considering geomantic questions in real-time. The archival wealth of Nanbu, a place not known for having especially favorable fengshui, fills that gap. Its archive also reveals what officials needed to maintain the rural order, including things that elites perhaps took for granted in the southeast: healthy family units, hopeful scholars, ample timber, peaceful mining, community consensus, and basic geographic information about the counties they governed. As commoners, scholars, and officials faced off in court, fengshui was often on the negotiating table.

Of the legal cases concerning graves, trees, or fengshui in Nanbu's archive, I draw on the contents of 310 case files in the following chapters, including many cases containing judicial maps (Appendix B). Those files are complemented by a hundred cases from the Ba and other archives from Sichuan and Taiwan. An additional ninety cases are drawn from the palace collections of Beijing and Taipei, precedent cases, judicial handbooks, and local gazetteers. These sources, which reveal fengshui's widespread importance as a social and legal phenomenon, have been complemented by intensive research into genealogies, published writings of officials, and diplomatic and missionary sources.

Perhaps the most unconventional sources of this book, fourteen geomantic manuals are invoked throughout the chapters. The Qing state oversaw the completion of two of these manuals, *Imperially Endorsed Treatise on Harmonizing Times and Distinguishing Directions* and *Correct Doctrines of Fengshui*. An additional two manuals originate from Sichuan Province. A Beijing-based Manchu author compiled another in the late Qing. Finally, commercial presses in southeastern provinces published the remaining nine manuals over the Ming and Qing dynasties. Some concern houses, some are about graves, and several cover temples, pagodas, academies, furnaces, oil-press shops, millstones, bridges, mining, and irrigation matters. All contain information that was as relevant to Sichuan's petitions, verdicts, and official orders as it was to the provinces of the southeast, where most of the manuals originated. Even some legal cases from central and northern China, such as Shaanxi, Shandong, and Zhili, invoked principles in these manuals. Despite certain regional variations, many geomantic principles found expression across China during the Qing period.

Chapter Summaries

Appreciating fengshui's roles in governance requires first establishing its significance to various status groups and sectors of Qing administration on the ground. Accordingly, four chapters are broken down by theme: graves, maps, examinations, and mining. Each chapter pays attention to change over time, which spans from the early 1800s through the 1870s, except for exceptional cases dating from slightly earlier or later. The final fifth chapter tackles a period of crisis in the late Qing—from the 1870s to 1912—which saw sustained political discussions over Western encroachment and industrialization across the empire.

The story begins with a chapter on grave litigation. Qing law strongly protected graves, and everyone knew it. Residents enthusiastically sought out auspicious lands, which resulted in many fengshui disputes over graves, trees, and ancestral properties. Some created fake graves, adopted anonymous graves of uncertain pedigree, and composed contracts filled with geomantic information to claim and manage land. A growing population meant that magistrates needed to resolve the steady stream of trenchant disputes over land, but many had little reliable information with which to work. Knowledge of fengshui enabled well-informed magistrates to express their authority in courts of law while resolving disputes in terms that were relevant and acceptable to rural communities.

The next chapter shifts away from why courts engaged fengshui to how they did so. During the Qing, powerful lineages placed detailed geomantic maps of gravesites into their genealogies to stake ancestral claims to land and hedge their bets in case of a lawsuit. Handbooks for officials, cognizant of the difficulties in adjudicating related cases over land, recommended that properties in dispute because of fengshui be formally mapped for consideration in court. Magistrates ordered such maps drawn up by governmental affiliates and interpreted them in court to resolve ongoing disputes. Some officials referenced geomantic principles in court to prove their interpretations of land were correct. The government's mapping of graves, houses, and temples underscored their importance to the dynasty and broader society. Over the last century of Qing rule, Nanbu's court mapped more landscapes in dispute because of graves and fengshui than any other category of property lawsuit.

Chapter 3 moves from commoners to scholars, exploring deployments of fengshui in gentry petitions concerning the civil examinations. The term "gentry" refers to local elites who held scholarly degrees by passing one or another of the state civil service examinations or by purchase. Although scholars in Northern Sichuan passed exams at fair rates in the early Qing, by the turn of the nineteenth century, the provincial capital at Chengdu had risen to provincial dominance while Nanbu and its surrounding counties produced few successes. With local success hard to find, magistrates actively tried to improve local fengshui by moving educational structures, building pagodas, and protecting culturally sensitive landscapes. Sichuan's scholars recognized the power of leveraging fengshui in court petitions and often did so to exert their influence. Fengshui's ties to the examination system were manifested through many local contentions over public space and even came to involve the area's Muslim community.

The following chapter explores mining in Sichuan. With many mines opening in the 1800s, extraction could be contentious from a legal standpoint. Because Qing law recognized fengshui as a valid reason to ban mining, litigants appealed to it to exert control over mineral extraction in their communities. Miners and quarriers of salt, coal, and stone knew the rules of the game. They sought insurance against litigation, often through employing their knowledge of fengshui or hiring resourceful ritual specialists, particularly Daoists. Anxieties over social unrest, exams, agricultural production, and the arrival of Western business interests became more pronounced during these decades, and officials looked for ways to influence public opinion regarding fengshui—generally to justify leaving minerals in the ground. The chapter's focus is not on whether these mining bans protected environments but on what the government's actions reveal about Qing politics and law at a time of considerable economic and demographic change.

The book concludes with a reevaluation of the infamous debates over industrialization. Ambitious infrastructure projects were not new in China's history, but the late Qing's legal, economic, and geopolitical contexts were. Qing opponents of industrialization were not naïve or technophobic. Instead, they offered coherent arguments based on imperial law and legal precedent. Scholars and officials cited the evidence of natural disasters and social unrest to argue that dynastic and provincial fortunes were seriously declining—and that harmed fengshui was a reason. The evidence was hard even for some proponents of new industrial development to completely dismiss. The context of the previous four chapters helps explain why the stakes of these discussions were so high, not least for establishing legal precedents and expectations across the country, including in Sichuan. After Qing codified law changed to accommodate the infrastructural demands of industrial capitalism in the early twentieth century, reverberations were felt widely—even in the remote county of Nanbu.

1

Litigating Graves

QING LAW REQUIRED people to create graves. Cremation was illegal, even if a senior requested it before passing.[1] Upon burial, the law strictly protected the resting places of the dead. A person who dug up a burial mound to the point of revealing a coffin was to be exiled after receiving a hundred strokes of the heavy bamboo, while a person who opened a coffin of a senior or elder was to be beheaded.[2] Anyone who leveled a grave for farming was to be beaten with a hundred strokes of the heavy bamboo, while a person who stole a grave tree received eighty strokes.[3] Further amendments to the law code in the late eighteenth and nineteenth centuries increased the severity of related punishments.[4] Even if courts did not implement the exact punishments mandated by the code for every infringement, they stayed true to their principles by consistently ruling that upon the burial of a human body, graves could not be disturbed or moved save for extenuating circumstances.

Qing property relations were complex, varying considerably according to the type of land in question. In addition to regional practices involving topsoil (i.e., cultivation) and bottom soil (i.e., rent collection) rights, there were a variety of land sale types, including "revocable sales," "irrevocable sales," and "sales by pledging" or "mortgages."[5] Graves were unique forms of property on that legal landscape since the law prohibited the sale of graves and ancestral lands designated for their upkeep. More than any other property type, the imperial state envisioned graves as permanent acquisitions.

People found ways around these prohibitions. Some illegally purchased the ancestral lands of others that a geomancer deemed auspicious. Some buried animal bones in the ground to create fake graves to claim desirable land. Others adopted ancient graves, or the graves of peoples whose descendants were unknown, as their own ancestors' resting places. A picture begins to emerge from the legal record. The significant protections lent to gravesites, coupled with the law's requirements for bodily burial, coincided with extraordinary popular demand for auspicious burial land. This chapter explores litigation arising from that tension and considers its implications for the dynasty's legal administration.

Qing law was not static about graves and burial. In addition to the increasingly stringent punishments for grave desecration mentioned above, a notable policy adjustment came in 1817, when the government added a substatute to the law code clarifying that the practice of selling the land around graves did not constitute grave destruction and was legal for people in poverty.[6] The practice of "selling the land but keeping the graves" effectively separated a grave from its surrounding land into two bundles of use rights. The rural poor practiced "selling the land but keeping the graves" to protect their ancestral graves while obtaining needed money.

This policy change, which sanctioned already prevalent practices in rural society, posed new questions that codified law did not answer. Custom held that graves possessed a zone of protection varying in size depending on regional land prices, topography, and population density. In Ba County, graves were protected within an eighteen-step (*bu*) diameter; in Shunchang County in Fujian Province, the rule was twenty-four feet (*chi*); and in Guangshan County in Henan Province, it was "five feet."[7] Nanbu County did not have a fixed custom, and magistrates there adjudicated grave territoriality on a case-by-case basis. The problem was, if grave land was sold but the grave itself was retained—sometimes to the point of "not retaining an inch of land"—how could living descendants protect their buried ancestors? Could an irrigation channel be dredged next to such a grave? Were trees included in the sale? If so, could they be cut down?

The *Great Qing Code*'s 1817 amendment to grave law reflected and perhaps even exacerbated growing social anxieties around status, fortune, and fengshui on the ground. With more poor commoners compelled by economic need to sell lands around graves in legal or quasi-legal sales, purchasers of well-placed and well-maintained gravesites had the opportunity to elevate their social status at a minimal cost. Related anxieties translated into rising conflicts over land as appellants sought ways to protect what they owned or reclaim what they had lost.

Sichuan's authorities noted this trend in early 1821, when the provincial judicial administrator issued a public notice excoriating the province's residents for being overly litigious and "obsessed with fengshui."[8] The issuance of this notice followed forty years in which Sichuan saw one of the highest population growth rates in the empire and a subsequent increase in land disputes. At first glance, the government notice appears only to contend that more disputes about graves and illicit burials were arising because greedy residents were "obsessed with fengshui" and "fond of lawsuits." However, a striking administrative reality emerges when one reads between the lines of official rhetoric with the aid of local cases: changing demographic, social, and economic circumstances—coupled with evolving legal standards—compelled officials to engage with fengshui to resolve land disputes more than ever before.

This observation builds upon existing work on Chinese property but adds a significant twist. Weiting Guo has observed that lineages in Taiwan incorporated geomantic language in contracts as "a more efficient means of dispute resolution than official litigation."[9] Paul Katz similarly views the proliferation of private judicial rituals as mechanisms to avoid formal litigation.[10] In Sichuan, people also drew up contracts about fengshui and conducted private rituals. However, many still made the often long and costly trek to court to contest fengshui. Because of the law code's strict protections for graves, their cases called for official adjudication but were not easy to solve. Many graves occupied untaxed lands, leaving magistrates with limited options for fact-checking. If magistrates could establish hard facts about a case, they could apply imperial law exactly and recommend the beheading or exiling of egregious grave disturbers. Officials saw those options as last resorts.

How could the magistrates uphold the law in adjudicating challenging cases over graves? They could interpret fengshui. Officials found different uses for fengshui, with some invoking it to justify practical outcomes and others employing it instrumentally in ways that directly shaped their verdicts. Examples of both exist in archival case files.[11] This judicial strategy was powerful in meeting the exacting demands placed on county law courts. Through invoking fengshui, litigants could access the courts despite whatever documentary evidence they may have lacked. On the flipside, through considering fengshui, officials could weigh otherwise unknowable information while maintaining their judicial authority.[12]

This method might at first seem surprising since the law code excoriated diviners for "speaking absurdly of the country's bad and good fortune" at the homes of civil or military officials.[13] For precisely this reason, the Qing government licensed and set clear standards for professional geomancers; chapter 3 introduces their hiring and functions. The law code also denounced people for excessively delaying burial to wait for an auspicious interment time or to obtain auspicious land.[14] These lines forbid excessive practices related to certain interpretations of fengshui; they did not attack or deny fengshui's relevance in the abstract. Another major compendium of Qing law, the *Collected Statutes of the Great Qing*, shares the government's position but adopts a completely different tone: "Whenever surveying fengshui in the construction of great works (i.e., imperial palaces, mausolea, city gates and walls, etc.), the Astronomical Bureau will commission officials to inspect *yin* and *yang*, set the building's direction, and pick an auspicious day to begin construction."[15]

Put simply, the *Great Qing Code* banned harmful burial practices by commoners while nonetheless requiring them to create graves. The *Collected Statutes of the Great Qing* mandated the practice of accurate fengshui by officials. Read together, any contradiction dissipates. The imperial government held that the common people often misunderstood authentic fengshui's intricate principles. Identifying genuine, ethical, and precise fengshui was the prerogative of the imperial state and the emperor who ruled over it. Hence, the morally ambiguous fengshui disputes of commoners needed to be adjudicated in court.[16]

A pattern emerges in trial verdicts. When a case hinged on the fortunes of a collective group, such as a family or lineage, magistrates usually accepted geomantic arguments and ordered protections for fengshui. When a case hinged on an individual desirous of fengshui for selfish reasons, magistrates were quick with a reprimand. But there were intriguing exceptions, and persuasive arguments needed solid knowledge and clear logic—not hearsay or nonsense. The following sections examine cases over real, fake, and ancient graves, all of which could require officials to inspect fengshui. The chapter then turns to questions of contract, land sales, and taxation. The final section examines precedent cases, routine memorials, and palace memorials—that is, written communications carefully drafted by officials to the emperor—to situate the trends seen throughout the chapter within the broader context of Qing law.

Real Graves

People in Qing society valued gravesites for many reasons. They naturally fostered deep emotional bonds to the places where their parents or kin were laid to rest. Graves also instrumentally established a family's belonging to a locality, expressed social status, and extended claims over adjoining natural features such as water and trees. In David Faure's terms, the gravesite "defined the living community" of a rural settlement.[17] As the cases below make clear, this description applied to Sichuan and undoubtedly elsewhere in a very literal sense.

Graves religiously marked the resting places of one of the deceased's two major souls. The corporeal *po* soul, associated with the qualities of *yin*, was thought to remain with the corpse of the dead in a grave. In contrast, the ethereal *hun* soul, related to the qualities of *yang*, was worshipped in an ancestral hall or a house altar. The management of souls in life and death involved many complexities, and one Daoist tradition conceptualized a total of seven *po* souls to the *hun*'s three.[18] Despite regional variations in some details, people across many parts of China thought that graves could bring good fortune if they were situated correctly and well maintained. Gravesites could also kill children if they were created poorly or disturbed via harm to their surrounding earth veins. During the last century of Qing rule in Nanbu County, litigants brought many tales of ill or deceased children to court.[19]

Magistrates appointed by the central government arrived in Sichuan Province from all corners of the empire. Though some were bannermen, most were Han Chinese holding capital or provincial degrees gained through the civil service examinations. Many hailed from prominent lineages in the wealthy regions of southeastern China, which were infamous for their protracted fengshui disputes.[20] These men knew little about the county they were assigned to govern and usually did not stay for long. Most only began formally studying law after receiving their degrees, as legal knowledge was not part of the regular exam curriculum.[21]

After a person submitted a lawsuit, the court first determined whether the allegation required a formal investigation. If approved for one, clerks and runners gathered information by traveling to the contested site to collect contracts and inscriptions. They scanned landscapes and made notes on trees, water, or shrines around houses, graves, and cemeteries, drawing maps if requested to do so by the court. Clerks seldom made notes about ill or dead children while inspecting fengshui. They refrained from doing so—not because courts deemed those claims theoretically impossible—but because infant mortality was high and allegations were easy to fabricate.[22] Clerks preferred to cite the presence or absence of concrete physical evidence, such as a cut tree or a quarried stone. Nonetheless, clerk reports might mention severed earth veins if an infringement was particularly egregious.

On the day of a trial, litigants filed into the yamen—if they were not already being detained there. The magistrate, having read through the file, started asking questions. Trial transcripts contain heart-wrenching testimonies of men and women who explained why they had come to court. Trials over minor disputes—that is, not homicides or robberies—were public and publicized.[23] People gathered to watch the process for entertainment, education, or perhaps both. The magistrate, facing the litigants and the crowd, pieced together a plausible series of events and rendered a verdict.

An example can illustrate the process. In the summer of 1879, a man named Zhang traveled to court, fearful for his life.[24] Zhang was a farmer but worked as a geomancer for extra money. Two years previously, he had selected the site of a grave for the father of a low-ranking member of the gentry named Jing. The results were not ideal. Jing's son fell ill a year later, causing members of Jing's family to blame the poor fengshui of the new gravesite. They assaulted Zhang, threatening to kill him. Zhang was terrified and felt he had no choice but to come to court for the vindication of his reputation and to seek safety.

The following lunar month, Jing submitted a lawsuit. Jing described the methods Zhang had used to select his father's burial place and positioning.[25]

Zhang had allegedly buried Jing's father at a time that was inauspicious for Jing's family. A year later, after his young son suddenly fell ill and died, Jing became convinced that the burial time, coupled with the grave's poor orientation and location, had inflicted a malady (*sha*) known as the General's Arrow (*jiangjun jian*) on his son. Just as Zhang had alleged, Jing's lawsuit contended that the geomancer had killed his son through shoddy grave divination.

The court's investigation into the matter revealed a barren, marshy landscape around the father's grave.[26] The low-lying area was prone to flooding in the rainy Sichuan autumn, and grass—let alone trees—had failed to take root in the damp soil around the grave despite Jing's efforts to plant new foliage. After reading through the case file and hearing the oral testimonies at trial, the magistrate agreed that Zhang had accidentally killed Jing's son, ruling:

> The court has investigated and found that because Zhang Xichou's skill at site selection is poor, Jing's father has not even been buried for three years and the family has already seen the painful loss of a son. On the grave, not an inch of grass will grow, and within the gravesite, there are now termites; Zhang's lack of skill in earthly principles [fengshui] is evident . . . [27]

The magistrate publicly demolished Zhang's reputation as a geomancer, but his harsh words may have saved his life. The verdict continued by forbidding the Jings from ever harassing Zhang again, threatening them with punishment if they did so. After all, although Zhang was an untalented geomancer now effectively banned from public practice, Jing, a local academy student, had chosen to hire him. Since Jing, who had some education as a member of the county's gentry, should have employed a better geomancer, the magistrate made it clear that some responsibility for the child's death lay with him.[28]

The environmental conditions of the gravesite and the unfortunate burial timing led the magistrate to conclude at trial that Zhang was not talented at fengshui. The knowledge by which he came to that judgment was not local to Nanbu or even Sichuan. Nanbu's gazetteer documented floods and droughts over the nineteenth century but recorded no dramatic termite infestations or suspicious deaths in specific seasons. The knowledge that Jing cited in his lawsuit, and the information that the magistrate drew upon in his verdict, came from geomantic and fate calculation manuals circulating around the empire.

One common manual on children's horoscopes in Qing times took the title *Classic for Hitting the Mark Every Time* (*Baizhong jing*). Manuals bearing this name were easy to produce, cheap to buy, and "sold in considerable quantities" when new editions were printed with updated calendar dates.[29] These manuals paid great attention to a person's birth time, recorded through the "eight characters" (*bazi*), that is, the birth year, birth month, birthday, and birth time (each recorded with two characters). The image below, taken from a manual

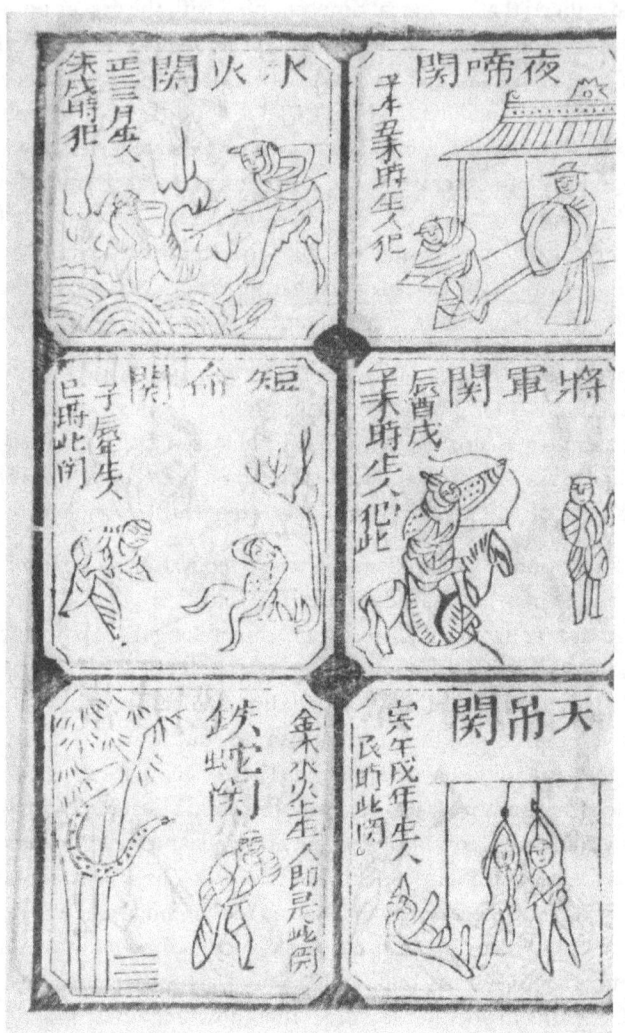

FIGURE 1.1. Image of the General's Arrow Malady Striking Children. The image of the "General's [Arrow] Pass" (*jiangjun guan*) is at the center right. Children with "eight-character" birthtimes containing the earthly branches of *chen, you, xu, zi,* or *wei* times were vulnerable to being struck by the General's Arrow. Image from *Xincan houxu baizhong jing* (d.u./c. 1840). Original manual held at the Bayerische Staatsbibliothek.

containing horoscopes that date from the Daoguang era (1821–51), details the General's Arrow malady, which was one of the two dozen or so dangerous "passes" (*guan*) potentially afflicting children born at certain years and times.[30] Other manuals from the nineteenth century offered precisely the same malady and description.[31] Jing alleged that the burial occurred at an inauspicious time for his son in relation to the General's Arrow, to which a mountain of textual evidence attested.

In turn, the magistrate's insistence that the presence of termites was evidence of poor fengshui finds resonance in fengshui manuals. Consider the "grave afflictions" section of a geomantic manual published in 1831, titled *Two Works on Fengshui, On Forms and Material Forces, Organized by Categories*. It was compiled by Ouyang Chun, a native of Liuyang, Hunan Province. In the diagram of an inauspicious grave cavern seen above, burying a wooden casket in shallow damp soil invites termites, while burying it deeply in damp soil invites mold. Although water is generally auspicious in fengshui, it must flow naturally or at least be allowed to drain away from the dwelling.

For this reason, well-schooled geomancers paid great attention to the direction of water around graves and houses. Where water is allowed to pool without properly draining, as was common near the irrigated wet fields of Sichuan (or Hunan, where Ouyang composed the manual), the subterranean wood of a burial casket or the foundation of a residential house attracts termites, which are drawn to moist soil. Ouyang Chun was on to something: the appearance of termites in poorly drained soil is a recognized problem in agricultural research today.[32]

A final detail—the magistrate's emphasis on "within three years of burial"— was also significant. Geomantic manuals, like Zhao Tingdong's *Five Secrets of Earthly Principles* (1786), reinforced a widespread belief that the three-year period following a burial or a major change to the gravesite revealed a plot's level of auspiciousness.[33] This timespan had special ritual meaning in Chinese society since children mourned their parents' passing for three years, a span that traced back to the first chapter of the Confucian *Analects*. Many corpses were not buried for three years after death despite the law code's mandate for burials to occur within three months.[34] Similarly, for regions that practiced reburial, exhumation might occur three years (or more) after burial. Chapters 3 and 4 will show Qing officials citing this three-years principle as a kind of statute of limitations on alleged inauspicious repercussions, with misfortune reported three years after a major landscape alternation being deemed logically implausible.

These principles allow us to make sense of other cases from Sichuan's archives. Consider the following example from Ba County, dated 1825. Then, the Pengs and the Liaos engaged in a property dispute over a hillside with graves

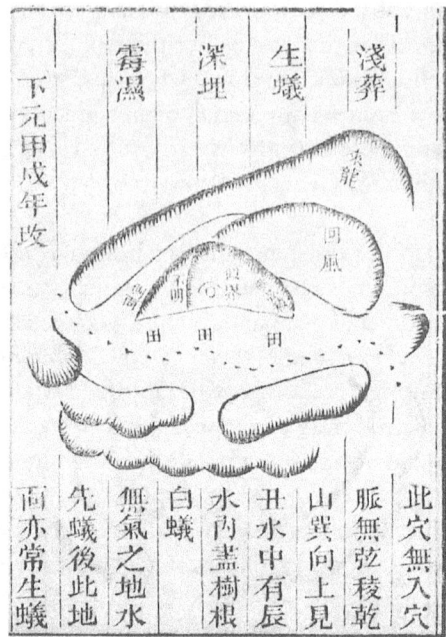

FIGURE 1.2. Geomantic Diagram and Description of a Termite-Afflicted Gravesite. The description atop the diagram reads, "Burying the coffin in shallow soil [with poor drainage] will summon termites; burying it in deep soil [with poor drainage] will invite mold." Diagram from Ouyang Chun, *Fengshui ershu xingqi leize* (1831). Original manual held at the Harvard-Yenching Library.

belonging to both families. The Pengs added an irrigation ditch on land they contractually owned while packing soil onto their ancestor's grave, presumably to protect the site from the drainage problem detailed above. The Liaos were upset by these changes and presented a lawsuit claiming that the irrigation ditch damaged the earth vein of their grave site. After reading through the case file, the magistrate forbade the Pengs from harming the earth vein and demanded that the water channel be closed.[35] In this case, a claim based on fengshui was recognized as valid and upheld in court because one party actively modified the environmental conditions of a site that would affect a second party's grave. People in Sichuan readily invoked graves to control water flows and irrigation, with powerful families who owned large auspicious plots having the upper hand in rural society.[36]

The cases above suggest that magistrates in Sichuan desired to project an image of themselves as fair-minded protectors of graves—not because they necessarily admired every theory of fengshui, but because, in effect, they had no other choice. Of paramount importance was the fact that magistrates wanted to prevent further litigation or, even worse, violent conflict. But they lacked detailed, reliable information about rural property ownership. Magistrates could not know the precise location of every land boundary across a county's vast hinterlands. They also could not keep track of ballooning populations. Yet, admitting those facts would weaken the government's authority, its reputation, and its projection of power. If magistrates could not resolve a case, why would claimants bother seeking out the state?

Thus, in issuing verdicts, magistrates offered interpretations of land that made sense to litigants by reinforcing in clear, understandable terms what the rules of a plot of land needed to be. Fengshui entered the courts as situated rural knowledge and exited as more general information through the state's strategic editing, assemblage, and redaction. The only alternative to fairly and meaningfully protecting the environments of real gravesites was not protecting them at all. And that alternative appears to have been unthinkable for Qing officials serving on the ground.

That said, magistrates serving across Sichuan adopted locally discrete strategies for dealing with conflicts over graves and cemeteries. The phenomenon of illicit burials (*daozang*), which referred to burying a corpse on land one possessed no rightful claim to, is illustrative. The law code's penalty in 1740 for an illicit burial on another's grave land (without physical harm to existing graves) was eighty strokes with the heavy bamboo and removal of the grave to another site.[37] The same 1740 edition's penalty for an illicit burial on farmland was sixty strokes and the removal of the grave to another site.[38] Disturbing another's grave land was the more serious violation of the two since the act of disturbing another's ancestors had potentially significant impact on their living descendants. Following a revision in 1788, the relevant substatute increased the penalties for both offenses while specifying that any "geomancer (*dishi*) or litigation master (*songshi*) who instigates an illicit burial shall be punished as one party with the criminal."[39]

These rules were not always enforced on the ground. Magistrates in Nanbu seldom ordered graves moved, even those proven to have been established illicitly. Consider an individual with no remaining shares in a communal lineage cemetery. If he buried a corpse illicitly in the cemetery, Nanbu magistrates often cited the principle of "one vein" (*yimai*), a concept that envisioned the lineage unit as one collective body sharing a common fortune. They might then mandate sacrificial offerings to appease the souls of the deceased and forbid additional burials. For example, one magistrate ruled, "Since the grave has already been created and, on account of the fact you are of the same vein (*mai*; lineage), the court shall be lenient and not investigate further, but you shall not bury other people in this place in the future. The court orders that you properly arrange and prepare incense, candles, and silk for sacrificial offerings (*jiao*) to appease the ancestral spirits in the cemetery."[40] Mandating sacrificial offerings to the disturbed dead in the wake of grave violations was a common judicial strategy of Nanbu magistrates.

By contrast, in Ba County, magistrates might order graves moved, especially if they harmed the fengshui of other (nonrelated) lineages.[41] In one instance from 1825, a Ba magistrate ordered a grave belonging to a man named Zhang completely dismantled. The grave in question had been created on land aligning the ancestral graves of the Lai lineage that had been illicitly sold by a Lai family member to Zhang.[42] After the new grave was removed, Zhang was ordered to return his purchased land to the Lais for the original sale price of 9,500 cash, or roughly nine and a half ounces of silver. Recall that by 1825, "selling the land but keeping the graves" had been formally legalized, but protests against "illicit" sales of grave land remained common and potentially effective when it could be proved in court that a lineage had not collectively agreed to depart with a plot of communally maintained ancestral land.

The relative difference in judicial practices between Nanbu and Ba may be explained by differing levels of land commercialization, the degree to which reburial was practiced in each place, the ratio of lineage cemeteries to independent family tombs, or the varying degrees to which land disputes involved kin versus non-kin. For instance, when allegedly illicit burials occurred in communally owned lineage cemeteries, as was often the case in Nanbu, officials may have found it expedient to urge the lineage to accept the addition. By contrast, if a grave had been recently created on land belonging to non-kin (of different surnames) and was immediately reported to authorities, officials may have found it wiser to order it moved lest a conflict between non-kin groups escalate. This circumstance may have appeared more frequently in the more commercialized land market of Ba County. Qing officials adopted locally pertinent judicial strategies to keep the peace, but magistrates in both counties considered fengshui when framing their verdicts.[43]

A panorama of Sichuan-based adjudication during the nineteenth century appears in Wang Dingzhu's (c. 1761–1830) *Comments on Cases from Sichuan Litigation*, which contains appeals from thirty-six counties in the province. Wang served as the circuit intendant of eastern Sichuan in Chongqing, an official serving within the imperial hierarchy between the county magistrate and the provincial governor. His collection of legal comments contains hundreds of cases processed during his tenure from 1821 to 1827, dozens of which concerned graves. These cases derived from two major sources. First, county magistrates or other local officials grappling with a complex local conflict could pass the file up the bureaucratic ladder to Wang. Second, residents dissatisfied with county rulings appealed to Wang for a second or third hearing at his circuit yamen. Wang then had to decide if cases were severe enough to be passed upward to the provincial judicial administrator for further review or if case files needed to be sent back down to county courts for further consideration.[44]

Reflecting the concerns of the 1788 amendment to the law code mentioned above, Wang's collection illustrates the lengths commoners would go to contest fengshui upon the instigation of geomancers and litigation masters. One man named Cao traveled to the circuit yamen to protest the rejection of his plaint about an illicitly added grave in Jiangbei subprefecture. Wang was not impressed at the appeal, commenting, "There is no evidence that the earthly principles of that spot are most profound and subtle; thus, you are never permitted to heed an unskilled geomancer's instigation and presumptuously bring litigation saying that an earth vein has been severed and the grave has been harmed."[45] Wang rejected Cao's argument that the new grave affected fengshui by stressing he did not have any witnesses supporting that claim from the rest of the lineage. A sole geomancer's testimony was not enough and likely raised the official's suspicion. But even in this case, Wang did not dismiss fengshui as irrelevant; rather, he concluded the evidence was simply not sufficient to make a claim.

Qing officials also ruled that some allegations concerning fengshui were false or faulty.[46] False accusations were a real problem in the Qing legal system. Nonetheless, remember that bringing a lawsuit involved considerable time and expense, and in rural Sichuan, it often required a long journey to the county seat. Most litigants did not wish to waste their time and money with entirely baseless claims. Consider the following two cases concerning roadbuilding. In 1895, a court ruled that an allegation that the opening of a road disturbed fengshui was false because it discovered that natural rainfall—and not the human-constructed road—had caused the grave in question to move.[47] Yet, during a case adjudicated a decade earlier, the court ordered a road and water channel moved and compensation of six thousand cash, or roughly six ounces of silver, paid for alleged harm to a gravesite.[48] Here, the magistrate found the claim valid and ordered compensation for the potential misfortune brought upon the descendants of the buried dead.

What was the approximate balance between "true" and "false" allegations? Based on a sample of 110 legal cases about graves from Ba County, Wei Shunguang found that the local court ultimately deemed around a fifth of the accusations to be false or faulty.[49] Evidently, the risks of having one's allegation dismissed by the courts were not high enough to stop people from presenting cases about fengshui through the end of the dynasty. A reason for that trend was that courts judged most accusations about gravesites as falling between true and false. Litigants who presented lawsuits about fengshui believed they had compelling points of argument and good evidence. Court verdicts prove that some did.

Fengshui Groves

Legal battles over fengshui were not limited to graves alone. They also involved disputes over trees that stood in the proximity of graves and other physical structures like ancestral halls, houses, and temples. Kinship and religious groups in Sichuan identified these community groves as "fengshui" trees.[50] Everyone related to the interred dead or enshrined ancestors claimed a cosmic connection to the trees. In theory, all male descendants, who were typically members of a communal Qingming (Tomb Sweeping Day) lineage association, had to agree on the timing of their transformation by axe from protective fengshui into monetary wealth or another material resource.[51] Before that time, designated kin could gather bark, brush, and branches.[52] These trees accounted for a large portion of the litigation recorded in Nanbu's archive, constituting around half of the thousand-plus surviving disputes related to graves, burials, or temples from the Qing period. Details from case files reveal at least two reasons for their importance, breaking down by class status.

First, it was a point of pride for members of the county's elite to own sizeable groves of trees. Evergreens such as pines and cypresses were preferred for planting around burial grounds since their year-round foliage symbolized the continued presence of the spirits of the deceased and the continuity of the family lineage.[53] Regardless of species, the highest prized trees were ancient. Elites claiming affiliation with ancient trees could craft elaborate narratives of ancestral merit and examination success across time. The presence of older trees on ancestral land indicated longtime residence in Sichuan, which served as tangible evidence that the family had migrated to the locality during the early Qing, late Ming, or even earlier, unlike the newer migrants of later years.[54] As we will see in the next chapter, one enterprising lineage, the Chens, seemingly convinced everyone—including the county government—that they had resided in Sichuan for over nine hundred years by identifying and mapping ancient trees and graves as the resting places of their alleged ancestors.

Second, for people of more modest means, a connection to fengshui trees helped ensure the survival of their family units. A cemetery grove, even one that had already filled up, still permitted descendants to gather "timber, firewood, or even leaves," thus extending the practical import of the cemetery across time.[55] People guarded these cemeteries closely. Timber was needed for cooking, heating, coffin-making, and building. Sichuan's officials were aware of this reality: a provincial gazetteer published in 1816 noted the fengshui woodlands that dotted the countryside for family and community subsistence.⁵⁶ Resource geography added urgency for certain counties. Coal was not locally mined in Nanbu, meaning that merchants had to import coal from the far north, much of it going to the county's salt furnaces and local sericulture industry.⁵⁷ Although coal was used, timber retained local importance as a fuel source into the early twentieth century.

In legal records, litigants spoke of their lives as connected to fengshui trees so that if a person cut one improperly, an earth vein could be severed (*baimai*), causing children to fall ill or die.⁵⁸ Why was this argument potentially compelling in court? Fuel anxiety was a genuine concern for many families, and Qing officials knew it. As Kenneth Pomeranz observes, the cost of fuel in the form of timber or coal, coupled with sustained fuel shortages in poor hinterlands, resulted "in enormous suffering" during the nineteenth century.⁵⁹ With pressures rising in Nanbu, to have good fortune was to be connected to trees: the older, the better; the more numerous, the better; the shadier (due to an abundance of leaves and *qi*), the better. Accordingly, in Sichuanese idiom, the word "fengshui" became closely linked and sometimes interchangeable with the word "tree" (*shumu*) in the sense of being scarce, valuable, and essential for a family's future.⁶⁰ Magistrates arriving in the county from outside the province never seemed to struggle understanding what it meant.

The photograph seen in figure 1.3, taken around 1920, was described by John Lossing Buck (1890–1975) as "a typical graveyard of a well-to-do family in North China. It occupies the equivalent of almost a whole field."⁶¹ Although not an image from Sichuan, many commissioned court illustrations of burial grounds from Nanbu showcase similar groves of trees (a family or lineage's "fengshui") surrounding graves in large lineage cemeteries.

Trees were also important because, like tombstones, they announced where the dead "lived." Tombstones existed in abundance, but not to the extent one might expect because they required additional expense for their carving and placement. In Sichuan, they often served as status markers. The inscribed tombstones of landed elites typically stood in physical isolation from other graves (figure 1.4), showcasing the independent fortunes of a prosperous or ambitious family from the main ancestral line, represented by the lineage cemetery.⁶² Most lineages tended to plant trees around the otherwise unmarked graves in a communal cemetery, sometimes adding an inscribed tombstone for the burial place of a common ancestor. Figures 1.5 and 1.6 showcase mapped examples of Nanbu's forested cemeteries.

Magistrates sought to protect fengshui trees from being cut by reinforcing the idea of their cosmic powers to the people through the courts. This posture applied to both lineage cemeteries and independent gravesites. If a conflict arose from cutting trees to pay for expenses benefitting an entire community, such as repairs to a temple, ancestral hall, or community millstone, magistrates

FIGURE 1.3. Photograph of Lineage Cemetery with Trees in North China. The cemeteries of prosperous lineages in the North China Plain occupied much space on the land. Part of the value of these cemeteries was that by the turn of the twentieth century they held some of the only remaining local sources of timber. Photograph from John Lossing Buck, *Chinese Farm Economy* (1930).

typically "condoned" the action retroactively. They then usually ordered the writing of a new contract protecting fengshui in the future, the holding of a lineage banquet, or the public performance of a temple sacrifice.[63] In at least one instance, a magistrate ordered an opera performed after the cutting of fengshui trees—despite the fact the law code banned the performance of operas during burial rites:

> The court rules: we have learned that the felling of Su Qixian's ancestral grave trees by Su Jigui's workers is true, and therefore order that Jigui and others, along with Qixian, perform the *jiao* ritual before the graves, offer the ancestral sacrifices, and have an opera performed as a warning to future offenders.[64]

However, when individuals privately spent money from the timber of fengshui trees for their exclusive benefit, the defendant was typically ordered to repay the lineage.[65] When a tree or grove's status was in question, magistrates gauged its distance from a house, grave, or temple, sometimes issuing orders for the protection of fengshui over the entire area to remove any future ambiguity.[66] In a case from 1832 concerning a threat to the fengshui of a mountain, the court issued a public notice that any timber merchants who violated the prohibition on cutting trees would face punishment.[67]

FIGURE 1.4. Photograph of an Inscribed Tombstone and Woodland (Nanjiang, Sichuan). This photograph of an inscribed tomb and surrounding grove is from Nanjiang County, (Qing-era) Baoning Prefecture. Photograph from Huang, You, and Li, *Chuan dongbei Qingdai mubei jicheng* (2016).

Temple groves, planted and maintained by clerics to financially support religious estates, were also identified by surrounding residents and patrons as connected to fengshui. Because temple grounds also kept cemeteries containing the graves of past patrons, or because a surrounding community felt a solid connection to a particular estate, residents tended to call those trees "fengshui," thus tying them to their communities.[68] Clerics had to tread carefully. In one instance, residents around the Propagating Buddhism Temple accused two monks of cutting down thirty cypresses and planting grain in their place, thus harming fengshui and threatening their children's lives.[69] The county court dispatched a clerk to inspect and map the temple, where he observed the monks using the timber to repair the temple's Meditation Hall.[70] The clerk's commissioned map depicted the necessary repairs to the temple from the wood of the felled trees, which the court deemed a valid use of fengshui. Had the trees been sold for profit on private timber markets, the outcome might have been different.

Trees ascribed with cosmological significance were not separate from the broader timber economy. In practice, they were an integral part of it and could be cut down under certain conditions. The planting and protection of fengshui trees were designed to provide a family with a potential source of communal income ("good fortune") for a rainy day down the road. This very practical and material fortune served as a kind of social security for families. Accordingly, when a lineage collectively agreed to sell trees once deemed connected to fengshui while still retaining the cemetery, magistrates could reject retroactive claims to the trees if presented with a reliable contract documenting previous family consensus over the sale.[71] Still, if protests arose, the county government became involved in determining whether a grove's utilization had been lawful.

While magistrates could retroactively condone the felling of fengshui trees, they were cautious about giving advance permission to cut them down. Although the timber could, in theory, find good use in aid to the poor or local charities, corrupt intermediaries or greedy family members could steal the funds all too easily. People took the matter of fengshui trees so seriously that they petitioned the court in advance to permit the felling of a shrine's trees. In 1877, a member of the gentry petitioned the court to cut sixty trees around a Buddhist temple—a grove that he claimed was "not connected to fengshui" (*wuguan fengshui*)—to aid hungry people during a drought.[72] The scholar emphasized in his petition that the trees were unconnected to fengshui because he knew that this was the only condition under which the court could grant permission for their sale. The magistrate replied to the request by writing that the petitioner had presented no proof that the trees were unconnected to fengshui and that he could only permit the request if leaders of all families affiliated with the temple came to the court to unanimously verify the claim.[73]

The picture that emerges from nineteenth-century tree disputes is one of social precarity, resource scarcity, and sustained government involvement through the courts. In Nanbu, kinship units were relatively weak, burial lands were scarce, and property conflicts were common. With a growing population requiring more gravesites and the county facing periodic spikes in fuel prices, good fengshui became harder to find and more essential for families. Facing the people at public trials, Qing officials considered claims over fengshui for many reasons, not least of which was to maintain order amid significant demographic and economic changes. For officials, that meant rejecting frivolous accusations, selectively permitting trees to be felled in extenuating circumstances, and, most of all, unambiguously protecting authentic grave and temple groves.

Most striking is the consistency with which Sichuan's officials attempted to sustain community groves until very late in the dynasty. It is difficult to find legal verdicts, judicial comments, or county orders issued before the 1900s that encouraged cutting down grave or temple trees under most circumstances. Official motivations were related to maintaining family unity and community harmony, but one suspects that magistrates may have been aware of broader economic transformations occurring across Sichuan's rural landscapes during the nineteenth century. With hundreds of legal disputes over these groves recorded in the county over these decades, one may conclude that these sites held some of the region's only tree cover left.[74]

Fake Graves

When corpses were not available for burial, people created fake graves by packing the earth into mounds and sometimes burying bones in the soil. These bones shocked potential grave-levelers into paying monetary compensation for disturbing what could be, in fact, the final resting place of a pig or cow. People created these mounds to claim a burial spot in advance of a death in the family or to steal farmland.

Here again, courts consulted contracts and other written records. These documents might be sufficient for resolving a case, but there were times when contextual geomantic information was helpful for judicial interventions. Because imperial law forbade the desecration of graves, courts could not order the mounds excavated, meaning that interpreting fengshui might be the only means available for officials to gauge whether the grave was real or fake. The sudden appearance of a new "grave" without any apparent consideration of its geomantic placement or orientation arose suspicion and could be useful in trial arguments.

Consider the following case, which saw a magistrate conclude that the Xu lineage had created five fake graves on their neighbor's land. The Xus owned a cemetery filled with fengshui trees, but a century earlier, they had sold land south of the cemetery to the neighboring He family for tilling. For four generations, the Xus resented the sale, and when the Hes tilled a bit too close to the cemetery one spring, the Xus decided to strike back. They breached the limits of the lineage cemetery and created five fake grave mounds on the Hes' farmland, claiming the area was part of their ancestral cemetery. The court dispatched a clerk to analyze and map the site in contention. This map revealed that the five mounds in question had no tree cover and thus lay outside the cemetery's boundaries, confirming their inauthenticity.[75]

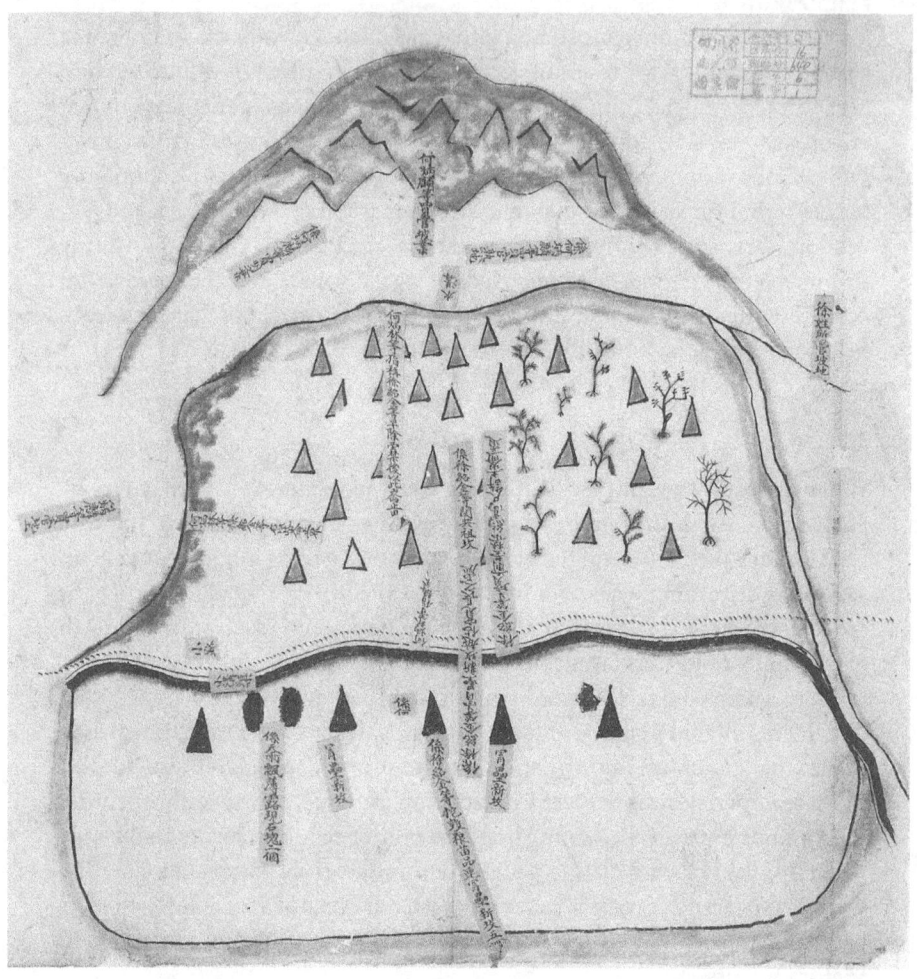

FIGURE 1.5. Commissioned Court Map of Fake Graves. This judicial map has nineteen annotations written on crimson red strips of paper. The three long annotations aligning the mountaintop as well as the single annotation on the far left confirm that these lands had been purchased by the Hes. Only the cemetery and the plot directly to its right belonged to the Xus. The dividing line between the top of the Xu cemetery and the Hes' property is marked by an irrigation ditch, while the dividing lines between the left and bottom of the Xu cemetery and the Hes' property were marked by stones. The bottom of the map contains annotations for the five fake graves, drawn in a darker tone for emphasis.
Map from Nanbu County Qing Archive.

During the subsequent trial, the magistrate never asked whether the Hes paid tax on the purchased land. There was, however, discussion about the cemetery's trees. Nine trees of different species enlivened the Xu cemetery area, a detail the clerk responsible for designing the map emphasized by finely drawing the discrete leaves of cypresses and pines. The map depicts individual trees adorned with dozens of brushstrokes representing their precise leaf cover—a measure of the life force, or qi, that the graveside trees held. The clerk had also noted the sizes of the cemetery's trees by the circumference of their trunks. This information was relevant for gauging the cemetery's age and alerting the magistrate how filial the Xus had been in the upkeep of their cemetery in the years before this dispute.[76] Since the dead could not speak, trees often had to do the talking.

After analyzing the map, the magistrate ordered the Xus to return the stolen land to the Hes immediately. Drawing attention to the tree cover of the Xu cemetery, the magistrate emphasized that if the Hes planted trees on their side of the border in the future, they would be unconnected to the fengshui of the cemetery. The Xus could not claim them.[77] In this case, trees ("fengshui"), which, unlike graves, people could not easily fabricate, provided the physical evidence for historical property ownership necessary to resolve the dispute. But in his final verdict, the magistrate also had to preempt the possibility that the Xus would again attempt to claim the land should the Hes plant trees near the site in the future.

Likewise, officials might inspect fengshui in cases where people created fake graves to claim a spot for future burial. A dispute of the Zhong lineage profiles one such instance. The Zhongs maintained a communal cemetery that was nearly full, making competition between the family branches for remaining spaces fierce. Trouble ensued when men from one side of the lineage snuck into the cemetery and constructed a fake grave. In court, this side claimed the allegedly fake grave was the foundation for a future grave of one of the elderly lineage members.[78]

Another side of the lineage claimed that the fake grave had severed an earth vein and that, as a direct result, their youngest son had succumbed to a wind disease (*fengbing*) and died.[79] This serious allegation needed to be investigated. As in the previous case, the court dispatched a clerk to map the cemetery in question, and both parties presented contracts.[80] One deed supported the idea that one side privately owned part of the mountain, while the other deed suggested that the entire mountain was common lineage property. The magistrate dismissed the contracts at trial, observing that the submitted deed and depositions stood in contradiction.[81]

To reach a compromise, the magistrate referenced the clerk's map of the cemetery and ruled that the foundation for the new grave should be moved several meters from its present location "so that it does not sever the earth vein or block the geomantic positioning of the existing graves."[82] This ruling allowed the future burial to proceed while contending that an earth vein would not be damaged if the family moved the grave foundation. Following the trial, the family afflicted by a wind disease remained unsatisfied and presented another plaint claiming that by then, two daughters and a son had died because of the severed earth vein. The court dutifully dispatched a runner to investigate these claims.[83] The case file ends here, but one can speculate as to the eventual outcome. Upon confirming that the grave foundation had been moved as ordered, the court presumably rejected the theory of geomantic harm behind the children's deaths.

Cases involving fake graves reveal much about official methods for adjudicating complex property claims. Because the imperial state did not control the production of knowledge about land, it often had to consider a broad array of information to resolve disputes, including fengshui.[84] In cases concerning fake graves, there was substantial pressure on the courts to render definitive judgments on their authenticity. Since it was difficult to know the identity—or even the species—of the interred dead, understanding the motivations and backgrounds of litigants was essential to making legal judgments. Even then, facts could remain evasive: at the conclusion of a trial in Ba County from 1853, a magistrate ordered uninterred bones immediately reburied just in case, even though he admitted there was no way to know whose bones they were.[85]

Creating a fake grave was audacious, but people knew imperial law strongly protected graves.[86] One intriguing detail shared in both previous cases was that the victims of the property theft did not dare to disturb the fake graves before going to court, even though they were infuriated by their creation. Accusations about grave desecration were common, with many people claiming that graves were injured (*shang*) or destroyed (*hui*). For the fifty-five cases about graves and fengshui preserved in Nanbu's archive from the Tongzhi era (1862–75), nineteen concern grave destruction or injury, and the remaining thirty-six involve grave environments (i.e., fengshui issues related to stone quarrying or tree cutting).[87] When claims turned out to be verifiably true, legal punishments could be considerable.

The statutory hierarchy of punishments between illicit burials and the more serious offense of destroying a grave may have informed people's actions. A memorandum (*shuotie*) composed by legal experts at the Board of Punishments for a case originating in Sichuan (dated 1845) showcases these two crimes within a single incident. The case concerned a grave illegally created on contractually purchased land and subsequently removed (i.e., destroyed) by the new owner of the land. In line with the statutes of the *Great Qing Code*, the creator of the grave had been ordered physically punished with the heavy stick because he conducted an illicit burial and violated a previous local court ruling forbidding the addition of a grave. Yet because the property owner destroyed the newly created grave, he (the owner) was punished for destroying the grave with a sentence lessened by one degree since legally it was his land.[88] This punishment amounted to penal servitude for three years and a hundred strokes of the heavy bamboo. In brief, having someone create an illicit grave on one's private property did not confer the right to remove that grave. Only the imperial state could authorize such an action.

Accordingly, one can surmise that victims of property theft via the creation of fake graves harbored caution toward removing them without a court's explicit consent because, in the slight chance that a grave was authentic, considerable legal trouble could follow. Legal allegations of fake graves were not especially common, but they also were not rare: Nanbu's court processed at least six cases about fake graves from 1877 to 1886, or about one every other year.[89] Fake graves were also identified and disputed in Taiwan around the same era.[90] Considering not all case files from these decades survived, the actual number of fake graves created during these decades was likely higher. One may infer that the practice was more common than the legal record suggests since some people undoubtedly got away with it.

Ancient Graves

The strategic adoption of ancient or orphaned graves shared similarities with the creation of fake graves, with one crucial difference. Ancient graves were real, meaning that even if their genealogical pedigree could not be established, courts were bound by law to protect them. In many cases where the identities of the dead could not be established, courts ordered the graves to be protected and even sanctioned the offering of sacrifices while denying litigants the right to exclusively claim their fengshui.[91] Here, the key question was how the courts could be certain that a grave held the remains of a person from whom a litigant claimed descent.[92]

The strategy of adopting ancient graves also calls our attention to the stark realities of rural poverty in Sichuan. There was immense pressure across rural society to obtain an acceptable burial site for one's parents. Identification with a grave separated a person "from the sorry lot of the homeless who were forced to drift from place to place," and established him or her as a recognized resident of a rural settlement.[93] Throughout the 1800s, people sold children or wives to obtain money for burials, and county courts issued verdicts mandating compensation in the form of burial costs.[94] Adopting ancient graves was not exclusive to the poor, but a good number of desperate people tried to do it. Magistrates often blocked the act when it involved stealing land from other lineages, but in the following case concerning an impoverished young man and his wealthier relative, the presiding official did not. Instead, the magistrate permitted the adoption of an orphaned grave along with its fengshui.[95]

Prior to the outbreak of conflict, the Xus had lived in Sichuan for well over a hundred years. Their history was documented by a large, forested cemetery, an inscribed tomb of an early ancestor, and a thick collection of property deeds spanning the 1700s and 1800s. Troubles began when a well-off member of the lineage discovered a new grave near his privately owned fields. The creator of the new grave, a man named Xu Yingcheng, had buried his mother on land he allegedly did not own, sparking a harsh altercation that culminated in a trial at the county's court.[96]

Upon receipt of the plaint, a clerk from the county's Department of Works mapped the site in contention. The clerk framed his map with the inscribed tomb of the common Xu ancestor on the far left and the densely packed lineage cemetery on the far right. That cemetery was full, leaving no available spaces for Yingcheng to use in burying his mother. Between these resting places of the dead were two rows of farmland separated by a ridge.

Yingcheng, a poor member of the lineage, owned no fields or any land along the ridge. The plaintiff, a man named Xu Longfu, presented a deed documenting his purchase of the relevant land, which was annotated in the center of the map with crimson slips of paper. In court, Longfu emphasized that the lineage never offered sacrifices at the unmarked graves, which he claimed existed prior to the arrival of the Xus to the area. Even though the three unmarked graves had no connection to the Xu lineage, no one apparently dared to disturb them for over a hundred years.

Why did Yingcheng bury his mother on Longfu's land? In court, he pointed to a household division register dating from the mid-1700s. While the register confirmed that the fengshui of the Xu cemetery was communal property for the lineage, the ridge in contention was not close to the cemetery, so the magistrate rejected the claim. Yingcheng also claimed that the unmarked graves held the remains of his grandmother. This allegation drew the magistrate's attention. When his mother had passed away the previous month, Yingcheng had set his sights on the ridge's three unmarked graves, which he assumed had decent fengshui. The magistrate understood Yingcheng's true intention: to find a reasonable place to bury his mother even though he had no money to buy one. Yingcheng was a thief, but a filial one.

The magistrate's verdict did not focus on Yingcheng's theft, but rather on a solution to the new facts on the ground. Acknowledging that he had no way of knowing the true identities of the individuals buried in the three unmarked graves, the magistrate ordered Yingcheng to purchase from Longfu the land around the three ancient graves as well as his mother's newly created grave for

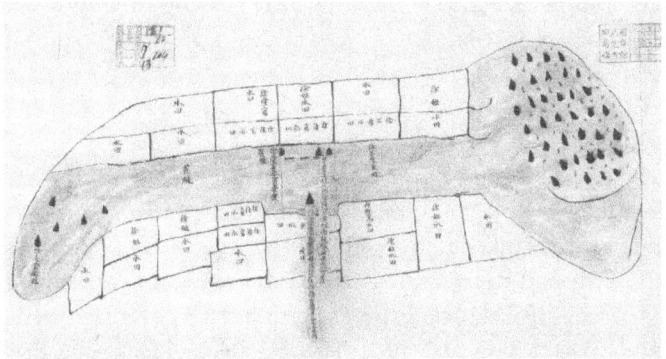

FIGURE 1.6. Commissioned Court Map of Ancient Graves. This judicial map profiles a large area of the Xu settlement. At the far right, the Xu lineage cemetery is drawn with thirty-six gravemounds and over forty trees, here represented by simple lines. At the far left stands an inscribed tombstone for a common Xu ancestor. Within this frame lies rows of agricultural fields, some of which were privately owned, and some of which were commonly held by the entire lineage. The two annotations written on crimson red strips of paper were placed at the center of the map to identify the three ancient graves, with the newly created grave of Yingcheng's mother below them. Map from Nanbu County Qing Archive.

a price of a hundred *chuan* of copper cash.[97] These graves were to be adopted and sacrificed to by Yingcheng as his own ancestral graves, regardless of whether one of them held the remains of his grandmother. Later, when Longfu returned to court complaining that Yingcheng could not afford the mandated price, the magistrate, citing Yingcheng's poverty, lowered the price to forty *chuan*.[98] The magistrate would not budge from his initial ruling.

That ancestral fiction proliferated during the late imperial era is well known among historians of Chinese genealogical writing.[99] But it is quite another matter to witness officials directly sanctioning the adoption of fictional ancestors in courts of law. One may infer that because magistrates were aware that good fengshui was often monopolized by wealthier members of a lineage, they sometimes aided poorer litigants through recognizing their claims to land even if the evidence was feeble. Yingcheng's actions were legally reckless but strategically effective. He evinced the magistrate's sympathy by centering his motives on finding an appropriate spot for his mother's grave rather than on greedily stealing land. Those three ancient graves that were used by Yingcheng to orient his mother's grave constituted the fengshui of the site. The magistrate, noting that the facts of the situation were unknowable, insisted that the three ancient graves be purchased at a below-market rate to avoid further disputes and in recognition of Yingcheng's dignity as a member of the Xu lineage.

Grave Contracts and Land Sales

Fengshui was leveraged by families to contest and control rural landscapes long before they arrived at a county court. Land contracts incorporated geomantic information that empowered lineages to identify fengshui infringements across broad rural spaces. These documents reveal much about fengshui's power in local society, from exerting claims over land well beyond the physical grave, to hiding taxable estates, and to regulating collective lineage property. Lineages undoubtedly employed fengshui as a kinship strategy to ensure their survival and prosperity against competing neighbors. What is also clear, however, is that officials also had reasons to invoke fengshui while adjudicating related disputes.

People drafted contracts for a wide range of matters, but most concerned land. Families wrote these documents to transfer land or establish the rules over the maintenance and ownership over a plot or area.[100] According to the law code, only tax receipts and contracts registered with the government (with an official stamp) were to be accepted as evidence during trials over land, including grave land. A substatute to the law code added in 1768 specified that "old contracts, inscriptions, and genealogies cannot be taken as evidence" during trials over graves and landed property.[101] As seen from the cases above, by the nineteenth century if not earlier, many magistrates found it impossible to uphold those regulatory standards.

Additionally, contracts offered as evidence in legal settings had practical limitations. Over time, rural landscapes changed through natural or human activity, rendering older contracts less useful for understanding the rules of a plot of land. Magistrates dismissed presented contracts to the court for being old, illegible, altered, or forged.[102] For cases about graves, officials often wanted to read contracts and genealogies together. These officials bemoaned the lack of family genealogies among Nanbu's population, and on a couple of occasions ordered them composed in the presence of the court to lessen the potential for future disputes over fengshui trees, burial plots, houses, and water sources.[103] Presumably, these officials hailed from parts of the country where it was unthinkable for a respectable family to lack a decent genealogy.

There are no typical contracts, as the genre spanned almost every feature of social life.[104] Nonetheless, land contracts in Sichuan regularly included geomantic information.[105] One unstamped contract, composed in 1906, observed: "Now, Wei Ziyu has buried his father on new grave land. The land is on Wei's property, but its vein is connected to the dragon vein of the Dus."[106] The Dus were clearly the more powerful lineage in the area and thus demanded the writing out of rules for any new gravesites on land adjoining their property. The result was that the Dus claimed that land contractually owned by outside parties was connected to their fengshui. Contestations of this sort were well known in Sichuan, where contracts were drawn up for the purposes of selling a site's valuable "fengshui."[107]

Land contracts also identified sites located geographically far away from ancestral gravesites with geomantic significance. In a contract composed by the Zhang lineage, failed repairs to a bridge located on the outskirts of their settlement were cited as harming the fengshui of the entire clan.[108] Graves were unharmed in a strictly physical sense by these actions. Rather, money reserved for the ritual upkeep of the graves that had gone toward the bridge repairs was deemed wasted, thus harming fengshui in the eyes of the lineage heads. Was the planned bridge located along the Zhang dragon vein? Were ancient fengshui trees cut down for use in repairing the bridge or to pay for the repairs? The contract did not explicitly say.

The only way to make sense of these contractual deployments of geomantic terms is to understand fengshui as a territorial language of power that expressed connections between people and places, families and land, and social status and environment. This language was often written into contractual records, which in turn entered courts as legal tender. From there, the ready mechanism for officials to understand what these documents meant was to dispatch runners and clerks to interpret the land in light of the available evidence and report back.

The process of judicial interpretation could be extensive for cases involving "selling the land but keeping the graves," which, as mentioned above, accelerated after the law code's 1817 adoption of a substatute sanctioning the practice for the rural poor. One case over sold ancestral land was adjudicated twice at the county level and then appealed at Circuit Intendant Wang Dingzhu's office in Chongqing. There, Circuit Intendant Wang cited fengshui to resolve the dispute at hand, amending the verdicts of the two lower courts. Understanding this case requires a bit of background in Qing regulations for the repurchase of land but provides an excellent demonstration of fengshui's invocation by a relatively high-ranking official.

A common practice in the Qing property market was selling land in priority to kin ("the right of first refusal"), which galvanized relatives to try to purchase property at below-market rates.[109] This custom caused problems for sellers, who may have wished to sell land to members outside of the family to obtain a better price at prevailing market trends. Over the course of the eighteenth century, the state attempted to curtail related practices. Qing law did not recognize the right of first refusal, and an amendment in 1730 sought to restrict its usage in instances of extortionary intent while banning the redemption of land sales explicitly labeled as "irrevocable sales."[110] Further regulations on conditional sales followed in 1740 and 1756.[111] As the population boomed over the eighteenth century, the Qing state sought to limit disputes over land while facilitating the legitimate transfer of landed property between nonrelatives.

Yet, other regulatory forces were at work that complicated attempts to curtail these practices. In 1756, the law code adopted a substatute to the effect that collective lineage lands around a grave or cemetery needed to be protected against illicit sales or purchases. Illicit sales of collective lineage properties that reached fifty *mu* (around eight acres) were to be punished in accordance with the crime of selling a grave mountain, with military exile to a frontier. Illicit sales below that amount were henceforth punished in accordance with the crime of illegally selling government land.[112] In other words, collective lineage properties (i.e., *jitian* or "ritual fields" designated to cover the expenses for ritual ceremonies), "fengshui," and graves were conflated to some degree in legal discourse during the second half of the eighteenth century, precisely when the law also formalized limitations on the redemption of farmland. The aforementioned contracts invoking sprawling geomantic dimensions are local manifestations of that legal understanding.

By the 1750s, imperial law firmly recognized that farmland could be permanently alienated outside of a family or lineage. However, if substantial amounts of sold land were connected to the ancestral graves of a lineage, the transaction could be deemed illicit. The issue was, through fengshui, ancestral land could take on a discursively broad territorial scope, as seen in the above contract concerning failed bridge repairs. How could officials strike a balance between laws that on the one hand contended that "selling the land but keeping the graves" was lawful for people in poverty, and on the other hand mandated that grave lands and collectively owned ritual fields could not be illicitly sold outside a lineage? When, exactly, could fengshui be legally invoked? Magistrates had to answer these questions on a case-by-case basis.

One answer to this question is found in the following appeals case of Circuit Intendant Wang, processed in the 1820s. A member of the Wang lineage (no relation to the circuit intendant) had sold a share of inherited land to a man outside the lineage, surnamed Yang. Other members of the Wang lineage who were not involved in that transaction brought a lawsuit against Yang, saying his purchase of Wang ancestral property was illicit.

A first lower court ruling permitted the sale to Yang, stating that while Wang could not alienate all Wang lineage land, he could sell his share from his family's inheritance. A second lower court ruling voided the land sale on the basis that the plot should have first been offered to members of the Wang lineage. Circuit Intendant Wang disagreed with the logic of the second ruling, writing that the customary principle of "property sold should in priority be returned to the original family and should not be sold outside the family" (i.e., the right of first refusal) was not recognized under Qing law. However, the circuit intendant still permitted the Wangs to repurchase the property from Yang. Considering his rejection of the right of first refusal, how did the circuit intendant justify such a ruling? The circuit intendant found the perfect solution—in fengshui:

> . . . if the property is near an ancestral shrine or tombs, then the land is connected to the fengshui of the ancestral sacrifices (*yinsi*) of the whole lineage and should never have been privately sold.[113] We permit the Wangs to redeem the land back by increasing the price from its original value and affixing the incidentals, such as the broker fee and a new tax deed all together. This land redemption will strengthen the cohesion of the Wang lineage. The statutes of our country clarify the great law and the principles of heaven accord with the sentiments of the people. Now wait for the magistrate of Jiangjin County to investigate the case again and issue the new ruling.[114]

By identifying the sold property as connected to fengshui, the circuit intendant permitted the Wang lineage to repurchase it, albeit at the market price—not at a discount. In this case, a relatively high-ranking provincial official deemed the contested plot to be connected to the fengshui of the Wang family graves because it was allegedly "near an ancestral shrine or tombs."

But can one confidently conclude that the plot of land had been truly part of the Wang lineage's collective estate prior to the sale? Since lineage cemeteries and ancestral shrines were common features of rural landscapes in Sichuan, as they were in most of the rest of China proper, many lands were "near ancestral shrines or tombs." While it is inconclusive whether the plot was understood as part of the Wang collective ancestral estate prior to the trial, a high-ranking provincial official found it convenient to accept that interpretation for resolving this dispute.

The other side of the coin also existed, as county magistrates cited fengshui to allow the repurchase of alienated land in cases where real damage transpired. In one instance from 1834, after a member of the Peng lineage sold mountain land abutting the lineage's ancestral graves, the site became used for cattle slaughter and manure collection. Following an investigation, a Ba magistrate identified the sale as illicit and allowed the Pengs to repurchase it by citing the need to protect fengshui.[115] Put simply, these Qing officials wanted to allow repurchases of alienated lineage land for the same reason they wanted to protect grave trees: keeping families intact over time was a matter of regional security and social stability. Invoking fengshui, whether to proactively arrive at a solution or to retroactively justify a desired one, allowed officials to do that.

Graves and Taxation

Gravesites were also intertwined with taxation, a notoriously complicated topic in Qing history. Here is the gist: by the early 1900s (and probably earlier), Sichuan's antiquated and meager tax records, coupled with the frequent exchange of land, meant that some people paid "taxes" without owning land while some people owned land without paying much in taxes.[116] Many gravesites occupied untaxed lands that were never officially registered with county yamens. This trend did not universally apply because as time went on, people created graves on taxable land as easier alternatives became scarcer. A grave plot's tax status in Sichuan depended on whether the land it occupied had been historically recorded in official registers.

Though tax status varied by plot, people in general saw graves as good, safe monetary investments. Graves and cemeteries occupying tax-exempt land were used by elite families to store and hide assets.[117] During one trial over a lineage cemetery, Nanbu's court identified a land deed for which the proper tax had not been paid for nearly a century.[118] One may speculate that litigants sometimes avoided presenting all their land deeds for fear of being exposed on such grounds—adding an additional incentive to keep a legal argument's focus on fengshui. Elite families certainly followed fortune: at the beginning of the Chinese Communist revolution, after many years of various tax and surcharge increases on land, David and Isabel Crook observed wealthy families hiding theoretically taxable lands through large gravesites.[119] Rising tax surcharges during and after the nineteenth century may have made large gravesites with good fengshui especially useful for those well-to-do families.

The identities of Sichuan's local tax heads add further intrigue. In at least two recorded instances from the nineteenth and early twentieth centuries, the tax head or township leader was none other than the resident geomancer for the local community. It is hard to say whether this arrangement was common, but the links between the two roles are hard to miss. Tax heads and geomancers were both familiar with the local secrets of landed property—a role that gives new meaning to the English phrase, "to know where the bodies are buried."[120]

Besides the moral, cosmological, or religious motives that drove the imperial state's protection of graves, tombs were instrumentally important to local administration because of an ironic feature of Qing land taxation: the names of long-deceased ancestors were often listed on rural tax registers, which were not regularly updated.[121] Nor were the titles of land plots regularly updated by property owners wishing to avoid paying the deed tax. The result of all of these arrangements was that ancestral graves, tomb inscriptions, genealogies, and ancestral tablets could be pragmatically important for identifying an area's living taxpayers.

Consider the following case. In 1847, members of the Maohu branch of the Deng lineage found themselves in debt and behind on their annual tax obligations. Devising a plan, the men adopted the surname Li and sought out the person designated by the state to collect tax revenue for the area to claim that the descendants of Deng Maohu, their ancestor whose name was written on the area's tax books, were all killed by a sudden malady. This fabrication would allow the tax head, the person responsible for collecting the land tax for their area, to remove their ancestor's name from the cadastral register, relieving the men of their liabilities.[122] The conspirators then destroyed their ancestral tablets, dismantled their lineage cemetery, removed the tomb inscription of their ancestor, and attempted to sell the cemetery's trees to timber merchants outside the county.

Their plan nearly worked, until a distant relation heard about the fengshui trees of a Deng grave hill being sold for sixteen thousand cash (about sixteen ounces of silver), a cost that, for comparison, was on the higher end of bride prices in the region.[123] This sale spurred him to present a lawsuit at the county court alleging grave destruction and fengshui sabotage. A clerk dispatched from the yamen confirmed the story, and after the men, now surnamed Li, were detained for punishment, the magistrate ruled that it was illegal for them to destroy their ancestral gravesite and sell the fengshui trees while ordering their tax obligations be reinstated.[124] What this extravagant plot reveals is that one way for people to escape land tax liabilities was to make the dead—meaning ancestors—disappear.

The actions of the Dengs may seem shocking, but there were understandable motives that may have informed their actions. The fact is, just as in our time and place, people in the Qing fell on hard times and got desperate, even if that meant sacrificing future fortune ("fengshui") for quick money. Consider also that people like the Dengs were likely aware of the links between tombs and taxes. During the resettling of Sichuan in the 1700s, some migrants to the province registered grave lands of other people under their own names so as to claim land or adopt tax identities with the state.[125] Other people even attempted to claim ancient graves as their own by insisting they were liable for the tax on the surrounding land through inheritance.[126] Was it feasible for officials to check the validity of such claims? People wagered that it was not.

In addition to matters related to moral authority, the Qing state had practical reasons to protect graves, not least because its governance depended on a tax-paying, sedentary agrarian population. Graves and farmland rooted people firmly in place, which is exactly where the imperial state wanted them to be.

Fengshui and the Three Judicial Offices

Since people occasionally killed in defense of their fengshui, or else severely damaged graves in attempting to obtain good fengshui, geomantic conflicts could make it all the way to Beijing. As demanded by Qing law, severe crimes required review by capital authorities, meaning that provincial authorities communicated the details of a crime in the form of a memorial to the emperor. In practice, these official communications were passed to the Three Judicial Offices in Beijing, which included the Board of Punishments, the Censorate, and the Court of Judicial Review. After reviewing the governor's recommendations, these three offices memorialized the emperor with a report of their findings. In most cases, the condemned were detained in holding cells pending review during the Autumn Assizes, when the emperor had the opportunity to review the list of prisoners awaiting capital punishment and grant pardons or reduce sentences. This section briefly introduces the Three Judicial Offices and then considers cases about fengshui sent for capital review.

The physical locations of the Three Judicial Offices in Beijing held special meaning in dynastic governance. In Yin-yang cosmology, east represents spring, which is the season of new life and growth. West represents autumn, or the season of life's ending. People in imperial times closely associated the western direction with offices handling crimes and punishments, thus symbolically and physically separating them from the other governmental organs. Accordingly, the central governmental offices associated with crime (the Board of Punishments, the Censorate, and the Court of Judicial Review) were all constructed adjacent to each other to the southwest of the Gate of Heavenly Peace (*Tian'an men*) on the western side of the Thousand Steps Corridor (*Qianbu lang*). Prior to the late Qing, the other five central boards (personnel, rites, revenue, war, and works) were located together southeast of the Gate of Heavenly Peace on the eastern side of the Thousand Steps Corridor.

By Qing times, this spatial cosmology already had a long history. Judge Bao (999–1062 CE), China's most famous jurist of the premodern era, became deified and associated with White Tiger astral bodies in later centuries, with White Tiger representing "west" and "autumn"—the season of the Autumn Assizes.[127] Beyond Beijing, this cosmology applied to government structures across the imperial bureaucracy: notwithstanding local variations in structural form, county and prefectural yamens were typically constructed with holding cells for prisoners, shrines to the prison deities, and shrines to Judge Bao on their western side.[128] The image from Kai County, Sichuan, (figure 1.8) offers

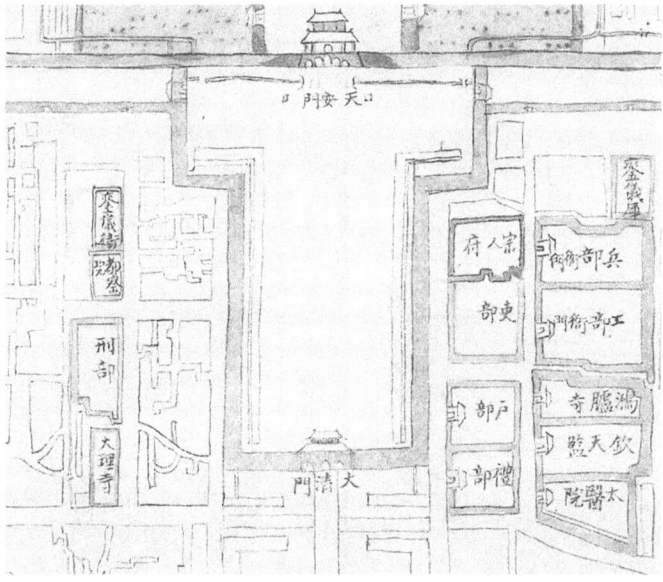

FIGURE 1.7. The Six Boards of the Imperial Government. This map depicts the key central administrative offices of the imperial government in Beijing. At the top center of the map is the Gate of Heavenly Peace (*Tian'anmen*); at the bottom center is the Great Qing Gate (*Da Qingmen*). Between the two runs the Thousand Steps Corridor, where Tian'anmen Square is located today. To the west are the Three Judicial Offices—including the Board of Punishments (*Xingbu*), the Censorate (*Duchayuan*), and the Court of Judicial Review (*Dalisi*)—as well as the Imperial Procession Guard (*Luanyi wei*). To the east are the Boards of Personnel (*Li*), Rites (*Li*), Revenue (*Hu*), War (*Bing*), and Works (*Gong*), as well as the Warehouse of the Imperial Chariot Procession (*Luanyi ku*), the Imperial Clan Court (*Zongren fu*), the Court of State Ceremonies (*Honglu si*), the Astronomical Bureau (*Qintianjian*), and the Imperial Academy of Medicine (*Taiyi yuan*). Reproduced map fragment of Li Mingzhi, *Beijing quantu* (d.u.; c. 1861–87).

one example of a western-situated women's prison, but many others exist as well.[129]

Cases destined for capital review began at the local level with a thorough investigation by a county magistrate. His initial report was passed up the bureaucracy and edited by prefects and circuit intendants until it reached the provincial government.[130] After a review at the provincial level by the judicial commissioner, governors composed a memorial to capital authorities and, when relevant, flagged fengshui as the source of the dispute in the opening line. One 1811 routine memorial composed by the governor of Hunan, Jing'an

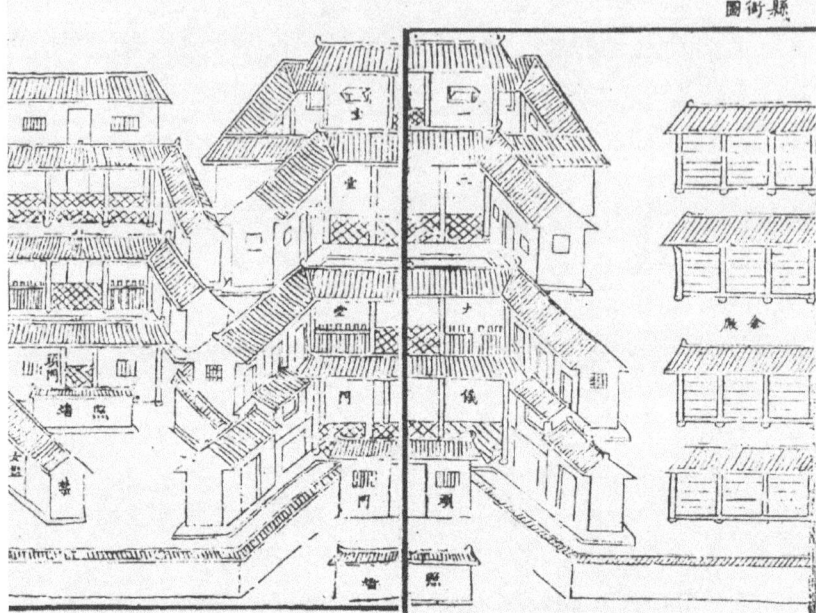

FIGURE 1.8. Map of Kai County Yamen with Women's Prison on the Western Side. This image depicts the layout of Kai County's government yamen, located in Sichuan Province. In accordance with imperial precedent, the holding cell for female prisoners (*nüjian*) was constructed on the western side of the complex, at the bottom left of the image. Map from *Xianfeng Kai xian zhi* (1853).

(d. 1822), read: "A memorial concerning the detainment and interrogation of Shaoyang County's Li Zhongsheng; because Li Zicai rented out land for a coal kiln and obstructed an earth vein, there ensued a fight where Li Zicai was killed. This case should be processed in accordance with the statutes, with strangulation after the Autumn Assizes."[131] In practical legal terms, these capital cases were not dissimilar from any other property dispute that ended in homicide.

A reason that capital reports might include information about fengshui is because such details established motive. Motive always mattered in cases involving serious punishments. In the wake of a grave's destruction in 1811, capital authorities rejected the punishment of strangulation recommended by Anhui's provincial officials because the defendant had not desired to destroy the grave for farming or to steal its contents, but rather, to find a burial place with good fengshui for his deceased wife. Since the grave was destroyed inadvertently, the Board of Punishments recommended a reduced sentence of military exile.[132]

On the flipside, the intentional act of harming fengshui as sabotage—a charge distinct from grave destruction—was recognized as a serious crime. In 1817, the governor of Zhejiang reported that a man named Pan had attempted to sabotage the Xu family's fengshui with the intention of summoning disease and hardship. Pan had nailed a peach tree branch on one of the Xu ancestral graves. The governor interpreted this act as commensurate with the article in the law code against "using spells and incantations to curse someone," and Pan was sentenced to a nonfatal punishment of eighty strokes of the heavy bamboo and two years of penal servitude.[133] From the two examples above, we can see that Qing law made a distinction between intentionally and unintentionally harming fengshui.

As previously mentioned, disputes over a grave or cemetery's fengshui were seldom reported to the palace bureaucracy unless they involved a homicide or an egregious instance of grave destruction or sabotage. Yet cases concerning fengshui were also reported to Beijing through other channels, including capital appeals by local communities alleging corruption among officials.[134] There was an upswing of capital appeals in the 1820s, the result of a general rise in violent conflict and perhaps also encouraged by the reformist atmosphere of the Jiaqing and Daoguang emperors.[135]

One such case originated in Yilong County, which neighbors Nanbu in Sichuan Province. After the appeal's submission, details about the matter were provided in a memorial composed by the governor-general of Sichuan, Dai Sanxi (1758–1830). The memorial is long and complex in its original form, but the key points can be distilled, as the case highlights ways residents unabashedly manipulated fengshui for their own purposes against the state's desire to protect the authentic fengshui of families.

The case began in 1824, when one affiliated family of the Cai lineage sought to sell trees around the ancestral cemetery to repair the county's City God Temple. A second affiliated family, represented by a man named Maoying, presented a lawsuit at the county court protesting the trees' sale, presumably citing fengshui. The magistrate discovered that the trees in contention were far enough away from the Cai cemetery and thus unconnected to fengshui, so he fined Maoying for concocting a false accusation. Maoying proceeded to appeal the ruling at the prefectural court, which rescinded the fine, forced the first family to split the proceeds from the trees, and ruled that the trees should never have been cut down.

Three years later, in 1827, when the Cais collectively agreed that the cemetery's trees had withered—and thus were unlikely to portend good fortune or showcase high status for the lineage—members of both Cai families decided to sell the trees for timber. Yet, because there was no consensus on how to divide the profits, Maoying again alleged that the first family was attempting to sell the fengshui trees illicitly. The county court investigated this plaint and found it false since no transaction had yet occurred. From here, the county court increasingly suspected Maoying was up to no good. Maoying skipped an official summons and went into hiding, rendering the county magistrate apoplectic at the continued machinations over the Cai lineage cemetery.

Later that year, Maoying's father fell ill while visiting a relative and passed away. Upon hearing this news, Maoying emerged from hiding and traveled to Beijing, where he presented a capital appeal to the Censorate, alleging that the first Cai family had killed his father, destroyed the Cai cemetery grove, and robbed his mother. He also claimed that county authorities had engaged in bribery and corruption over the City God Temple repairs. Qing authorities investigated these allegations and found them to be false. As a punishment for his string of false accusations and antics, the Sichuan governor-general recommended that Maoying be sentenced to military conscription and exile to a frontier area.

In his memorial, the governor-general also bitingly remarked that "members of the Cai lineage did not cherish and care for the cemetery and its trees, which are henceforth forbidden from being tampered with through illicit burials or further cutting."[136] The governor-general's suggested verdict aligned with imperial law, and the matter was referred to the Board of Punishments for a recommended outcome. Significantly, in emphasizing that the Cais did not properly care for their cemetery's fengshui, the governor-general delivered a message that the family's fate was self-induced. Plenty of textual evidence supported that impression. Esteemed geomantic manuals, such as Jiang Dahong's (c. 1620–1714) *Water Dragon Classic*, flagged military or penal exile to a frontier as resulting from poor fengshui and inadequate grave care.[137] As later chapters will make clear, the laws of the land often resonated with core concepts in imperial law.

Capital cases seldom focused on the cosmological details of fengshui itself. Rather, reports to Beijing tended to concentrate on the particulars of cases involving homicide, criminal theft, or corruption. Nonetheless, authorities at every level of the Qing government provided details about the earth veins and other aspects of the fengshui in contention because they saw such details as establishing the nature of criminal intent. Capital officials appear to have been aware of the immense pressure to obtain and maintain gravesites and the implications for the social order of doing otherwise.

To add local specificity to this observation, in provinces lacking large lineages with multiple ancestral halls and lengthy genealogies, such as Sichuan, the central physical site that bound a kinship group together was the communal ancestral cemetery and its surrounding trees. In such places, officials may have placed considerable importance on maintaining those sites. As seen in the case of the Sichuan Cais, even after a litany of blatantly false accusations from the dysfunctional lineage, the county court continued to accept plaints and to dutifully dispatch runners to inspect and write reports about their cemetery over the course of three years.

While one must keep the distinct geographical and social features of Sichuan in mind when considering cases from its archives, its residents were not uniquely obsessed with fengshui. There were perfectly understandable reasons for all people in Qing China to care about graves, to acquire grave land, and to fight for auspicious land. An episode from the most celebrated novel of the Qing era supports this conclusion. Toward the beginning of *A Dream of Red Mansions*, Qin Keqing warns the manager of the Jia family purse, Wang Xifeng, to acquire ritual fields and grave lands for the Jia family.[138] Grave lands, Qin argued, represented good, safe investments that would bring fortune to the Jia family, especially since such properties would never be confiscated by the state in the wake of legal wrongdoing. One can now understand, then, why Qin's strategy—not adopted by Wang Xifeng and the ill-fated Jias of the novel—was so compelling. Graves represented both cosmic fortune and material well-being to the people, who understood clearly that burial sites constituted powerful claims to land and space. This chapter has also emphasized a crucial flipside of that understanding: protecting graves was very much in the Qing state's interests.

More People, More Graves

The ways people discussed land in the Qing dynasty reveal a lot about how they understood their relationship to it. Did people view land as patrimony inherited from their ancestors? Or did people increasingly see it as a commodity that could be bought and sold with ease? The evidence from nineteenth-century Sichuan suggests that a plot of land could be patrimony in rhetoric and a commodity in practice and also hold cosmological significance, all at the same time. Families wanted multiple options for conceptualizing and presenting claims over land. An informed ambiguity by local players "in the know" was strategic and part of the game. Graves and ancestors could be faked or adopted for material gain. But graves and ancestors could also be real and thus require formal recognition and government protection. Earth veins and dead children could be the subjects of fraudulent accusations, but they could also be the heart of sincere claims. As we have seen, county magistrates sometimes concluded that children died or were at risk of dying because of damage to fengshui.

Qing markets in landed property were commercialized, with land constantly being bought and sold by private actors with relatively little government oversight. Within that context, it is noticeable how customary practices that placed checks on the land's alienability remained ubiquitous in local adjudication.[139] Consider grave fengshui, which tended to limit individual control and the free disposal or alienability of land around tombs. This feature of fengshui paralleled the impact of other well-known customary practices related to property: the conditional sale of land, including the right of permanent redemption and expectations of additional payments to account for the title's current market value, and the right of first refusal or first purchase by kin.[140] Although the Qing state had attempted to check excessive applications of those practices in the eighteenth century, their relevance did not disappear. In trying to keep the peace during the turbulent nineteenth century, local courts continued to uphold them.

The disorienting transformations that shaped Sichuan's land economy did not weaken religion or cosmology's hold over rural landscapes. They did the opposite, elevating powerful cultural forces that the Chinese state only suppressed when violently reforming and collectivizing land in the mid-twentieth century. Consider Michael Taussig's study of devil contracts signed by Colombian cultivators and Bolivian miners losing ownership over land during transitions to capitalist economies. Devil contracts were believed by farmers to increase wages, but the money earned could only be spent immediately on consumer goods. Wages could never be spent on land, which magically remained barren for any former peasant who made a pact with the devil. Taussig captures the importance of these contracts, writing, "Magical beliefs are revelatory and fascinating not because they are ill-conceived instruments of utility. . . . Magic takes language, symbols, and intelligibility to their outermost limits, to explore life and thereby to change its destination."[141] To be clear, I do not consider scholarly treatises on geomancy or the academic study of fengshui by Qing literati to constitute magic. Still, for the widespread social phenomenon of "fengshui obsession," Taussig's case is a helpful analogy.

Something profound caused people in Sichuan to become "obsessed" with fengshui at the turn of the nineteenth century.[142] Over the eighteenth and nineteenth centuries, officials serving in other provinces of southern China and even areas of northern China also diagnosed the same condition. As previously mentioned, more than one Qing official publicly lamented that half of all lawsuits in their jurisdictions concerned fengshui.[143] The social phenomenon's etiology in Sichuan may be appraised as follows. Around 1800, the province's population crossed twenty million, which is about the population of Sri Lanka today. At this time, the region ran out of new land to reclaim for cultivation, and the formal bureaucracy of the Qing state did not adjust in size or depth to manage the practical implications of those changes.[144] People inside and outside the government employed other strategies. This chapter showcased one of them: over the 1800s, both officials and litigants invoked fengshui to, in Taussig's terms, "change the destinations" of family fortunes when needed through the legal system. Unlike Taussig's case, officials did not seek to sanction a new relationship to land through invoking fengshui. Instead, they sought to maintain an ideal one aligned with the Qing state's interests.

To be sure, population growth is not the sole explanatory cause of the litigation seen throughout this chapter. Since fengshui had been invoked in law for centuries, geomantic disputes did not require the dramatic demographic conditions present in nineteenth-century China. A society can absorb demographic growth without causing environmental catastrophe or social unrest, and Thomas Malthus's theories about China's population have been thoroughly, and rightly, discredited.[145]

More immediate pressures in Sichuan influenced legal trends and outcomes, such as the shortage of cemetery space and grave land; the opening of marginal lands for farming; general resource scarcity; and rising fuel costs. In the historical record, Qing officials and scholars linked these factors to population pressures and even to property disputes over graves and fengshui. One early observation comes from the governor of the province, Hiyande (Xiande, d. 1740), who in 1727 noted the effects that migration had on Sichuan's legal system:

> In the past, the people of Sichuan were few and the fields were barren. When imperial rule was finally established in the area, people returned to their ancestral properties, but there never was a formal measuring of land, and thus many properties remained unrecorded. As the years went on, the province's population grew greatly. Now, there are cunning people who, with no [contractual] evidence for the boundaries of their land, repeatedly engage in litigation. Among the litigation records of Sichuan Province, for every ten lawsuits, seven or eight are about landed property, and without measuring the land, these cases cannot be properly adjudicated.[146]

The extensive measurements of land proposed by Hiyande never arrived, and the legal phenomena he described did not diminish in the following decades. Not long after Hiyande's memorial, the governor of Yunnan, Tulbingga (d. 1765) memorialized the emperor about the cost of rice, ascribing higher than normal prices to the growing population while noting that across the southern provinces, "many people have become obsessed by fengshui and are seeking massive spaces for their graves."[147] Tulbingga claimed that the popular desire to obtain good fengshui for their ancestors was taking valuable farmland out of the agricultural market.

Turning to the nineteenth century, Sichuan's local gazetteers fill in the picture with observations like "as the population grows ... the matters of living and dying grow, the number of those who attend to gravesites becomes many."[148] The 1813 and 1870 editions of a gazetteer from Pi County conveyed that people in the densely populated area had become "obsessed" to the extent that "cunning persons pretend that an ownerless ancient tomb is his ancestor's grave, or designate other people's graves as their own and invade or sell them. These actions cause endless litigation and are truly despicable."[149]

Yet another comment echoing those views came from Wang Dacai (b. 1829), an official serving in Guangxi Province during the 1860s. His description of litigation trends perfectly echoes the legal phenomena discussed throughout this chapter:

> When I had just began my tenure here, I examined the litigation files for the past few years, discovering that out of ten cases, at least three or four were about gravesites. When I carefully inquired into the origin of this trend, I found that the population had grown considerably so that lineage cemeteries were already full. Then, in the years after the Xianfeng era (1851–62), people were contesting fengshui greatly, having been deluded by popular geomancers. In every case of a cemetery dispute, someone would claim that the land had been reclaimed by their predecessors, or that they had purchased the land previously or that their ancestors had been buried here—but because of the fighting from wars over many years, there were no contracts left as evidence. There were illicit burials made at the extremities of fields and mountains, and even people who forged contracts and presumptuously offered tax payments for such lands.[150]

It is difficult to gauge whether the actual percentage of presented cases about graves or fengshui among all kinds of litigation was growing over time, either in Sichuan or elsewhere. I have not found statistical evidence of such growth in Nanbu's archival collection, which largely offers insight into the the nineteenth century, when disputes over highly prized burial lands may have already become more common relative to earlier periods. However, anxieties over fengshui, geomantic litigation, and related social phenomena were present to the extent that Qing officials and gazetteer compilers wrote explicitly about the subject.

The roles that fengshui played in adjudication over the nineteenth century parallel social trends identified by other historians of this era. Writing of the Lower Yangzi region, Anne Osborne observes that "long-term demographic, economic, and ecological trends" during the eighteenth and nineteenth centuries "altered the standards by which land was judged as valuable or wasteland," which in turn made "control of the landscape ... the subject of sometimes fierce competition."[151] As a result of these changes in the Lower Yangzi highlands, people relied more and more on community consent rather than officialdom to secure their claims to land. In turn, Qing officials increasingly relied on curated local knowledge to keep the peace.

When one steps back and looks at land adjudication in Sichuan, a picture resonant with Osborne's comes into view. As more lands were claimed without formal registration, as more contracts were left unstamped, as more families disputed lands without reliable genealogies, and as more people fought for limited amounts of trees, water, and mountain land using every conceivable trick and argument, Qing officials became more dependent on directly interpreting land without even attempting to adhere to the evidentiary standards of the law code.[152] By actively analyzing land during trials facing the public, officials could still investigate and solve cases despite the state's lack of information about rural property. There was no other realistic choice. Because courts were willing to engage contracts, inscriptions, genealogies, fengshui, and divination while taking the living and the dead alike as legal actors, the complex rural land tenure system remained broadly legible to officials responsible for adjudicating conflicts about it.

In legal practice, interpreting land meant governing the people who lived on it. Protecting a fengshui grove ensured a family would not later try to steal other people's trees out of desperation. It also dissuaded neighbors from conniving tricky schemes. Sending a clerk to map a temple forest demonstrated to community stakeholders that the government took the concerns seriously, even if the accused monks were found innocent of harming fengshui. Rejecting a false accusation about fengshui meant a powerful neighbor could not exploit an innocent farmer. Publicly condemning an untalented geomancer while explaining the precise reasons his skills were poor encouraged locals to up their game and seek out talented ones with reasonable literacy levels. Protecting an illicitly created grave meant a son would not believe his mother had become a hungry ghost. Having him purchase the fengshui around the site announced to the surrounding community that the land transfer was a done deal. Ensuring that ancestral properties were adequately protected encouraged residents to seek out the Qing state for aid in desperate circumstances rather than turning toward Sichuan's pervasive secret societies. Permitting families to repurchase a plot of land because it might have once been connected to fengshui gave kin a second chance to keep a family's line alive. Keeping kin together through one more winter meant another year without wandering recruits for the next rebellion.[153]

Across Chinese society and under Qing law, fengshui served simultaneously as a weapon of the weak, a privilege of the elite, and, most notably for our purposes, a powerful administrative vocabulary in a county magistrate's toolbox. Regardless of their personal beliefs, officials presiding over legal trials invoked it wisely, scolding the people for incorrect interpretations of land while offering ones they deemed correct.[154] In serving as the judge that could render verdicts for geomantic disputes, the imperial state affirmed its position as the highest authority of the lands under heaven.

2

Mapping Fengshui

READERS OF THE PREVIOUS CHAPTER may have noticed a commonality among the presented cases. Magistrates frequently asked clerks from the county's Department of Works to "carefully investigate the matter, draw a map, and attach an explanation." For most cases, judicial maps were discarded after trial, leaving only the clerk's dispatch order remaining in the archive. But the maps that survive in the archive reveal a remarkable cartographic trend. Of the 181 surviving litigation maps produced in Nanbu County during the last century of Qing rule, 42 percent depict houses and graves, many in dispute because of fengshui.[1] Another 21 percent represent the trees around tombs, houses, and temples, many in dispute because of fengshui. The county court mapped temples, graves, and trees more than twice as often during the course of adjudication as it mapped farmland (63 percent and 27 percent, respectively). The archive did not preserve mapped cases concerning fengshui at unnaturally high rates. Rather, the imperial state created these maps at high rates.[2]

Why did Nanbu's lawcourt often map landscapes that were in dispute because of fengshui throughout the nineteenth century? What practical benefits did the imperial state gain by mapping graves, houses, temples, and trees? Who else was composing maps and which landscapes did they choose to illustrate? Parallels between the actions of Qing officials and local elites emerge that offer significant implications for law and landed power. Through examining judicial maps of houses and graves, the county's atlas of local security units and rural militias, county gazetteers, and family genealogies, this chapter argues that fengshui informed much cartography in Qing Sichuan. The Qing state used mapping as an administrative technology to solve disputes that courts could not otherwise answer based on property records alone. But mapping did more than fill gaps in local records. It also produced new knowledge about the land by analyzing—and publicly announcing—how landscape components related to and influenced each other.

One assumption that has hindered the academic study of fengshui is that it was devoid of logic, substance, or coherence. The image of a geomancer spewing nonsense has remained center stage in the West since the earliest Orientalist depictions of the practice in the nineteenth century. But suppose fengshui completely lacked any logic or structure. In that case, courts could not have investigated, analyzed, and weighed fengshui with the frequency and rigor they did in Qing China. There would have been nothing to analyze. Courts would not have needed to spend valuable time and resources dispatching clerks to gather information about the locations of trees, streams, mountains, graves, and houses. But as seen in the case of termites invading a wooden coffin in muddy soil, there was geomantic substance for courts to analyze. People did not conjure technical knowledge about earthly principles out of thin air. It was written down and circulated in technical manuals based on real case studies composed through sustained observation at the hands of educated literati. And it was practiced by experts hired by the Qing state. It was discussed and acted upon in society. And it was negotiated and selectively recognized through courts of law.

The intertwined relationship between doubt and belief has become a major focus of scholarship on knowledge-making and natural studies in late imperial China. The theme is explored in Dagmar Schäfer's study of Song Yingxing (1587–1666), Carla Nappi's portrait of Li Shizhen (1518–93), and Francesca Bray's examination of the technologies that shaped pervasive ideas about gender.[3] If a consensus has emerged, it is one that urges decentering approaches to the history of science that affirm the primacy of Enlightenment-based histories buttressed by claims to absolute truth. Historians of late imperial China have demonstrated that knowledge about the natural world was not produced in a vacuum, but rather in an intellectual environment that did not take metaphysical concepts like *qi* and the five agents for granted and instead treated them as useful vehicles to explain and analyze material reality. One can extend this observation to the legal system, where judicial verdicts about land, property, and environments were also not rendered at random, but in courtrooms filled with rational people engaged in arguments colored by doubt, belief, logic, intelligence, persuasion, and deceit.

What was at stake in county courts was no less than the production and reception of practical knowledge about the natural world. Song Yingxing and Li Shizhen both engaged similar questions over their careers, albeit at a consistently elite level. That class difference is significant since historians have long recognized that elite and popular forms of knowledge diverged in imperial times. At the highest levels, classical studies listed fengshui alongside subjects like astronomy, music, and mathematics—fields that, in the words of Benjamin Elman, "represented a nuanced account of the natural phenomena as words in a text that needed to be decoded primarily through the analysis of language by connoisseurs."[4] The famed strategist Zhuge Liang (c. 181–234 CE) strongly agrees with that assessment in the celebrated late imperial novel *Romance of the Three Kingdoms*:

> To serve as a general and not be fluent in astrology, to be unfamiliar with earthly principles [fengshui], to not know the practice of divination, to be unaware of Yin-yang, to not examine diagrams of tactical troop formations, and to not understand strategic positioning—such a person is a mediocre talent![5]

Yet even with Zhuge Liang's fictional endorsement of cultivating geomantic knowledge, one is still faced with an obvious question: If many elites celebrated geomantic knowledge, what was the juridical problem for Qing courts? Since demand for ready-made geomantic remedies was immense, and much circulating knowledge about fengshui was popular, officials seriously doubted whether commoners could understand earthly principles to the extent that Zhuge Liang allegedly could. Put bluntly, Qing officials needed a way to verify the claims of the common people.

For their parts, litigants needed enough proper knowledge to make a persuasive case and get a lawsuit accepted for consideration. They were in luck, literally. Geomancers and litigation masters were right on hand to offer that knowledge.[6] By identifying changes ranging from the drastic to the subtle in rural landscapes, litigants could present coherent and compelling narratives to the courts.

Facing a constant trickle of claims resulting from such consultations, officials were compelled to employ cartography to appraise the merits of many allegations. On one level, through judicial mapping, fengshui became "useful, practical, or current knowledge" that supported local statecraft, which, above all, necessitated a performed monopolization of power to the people.[7] But on a deeper level, the government's mapping of graves, houses, and trees also involved the "skilled revelation of skilled concealment."[8] Through mapping, officials concealed from the people what the government lacked—factual information about property ownership drawn from land surveys—while revealing to the people what officials knew better than anyone else: knowledge shared by the educated elite, including an understanding of the law code and proper ritual, as well as a broad range of information that high levels of literacy allowed, including the basics of medicine and fengshui.

The sources at the center of this chapter are maps (*tu*). In imperial times, *tu* held a wide range of meanings. At its most basic level, *tu* could simply mean "image." Maps and paintings were likewise called *tu* without qualification.[9] Still, as shown by Schäfer, *tu* in conjunction with technological content served as blueprints or "templates for action."[10] This chapter profiles *tu* operating across this spectrum. Geomantic manuals offered technical *tu* to showcase examples of good and bad fengshui. Lineages composed geomantic *tu* that were included in genealogies to stake claims over broad stretches of land. Finally, courts commissioned judicial *tu* to appraise claims about graves, houses, fengshui, and other issues that involved land disputes.

Some maps profiled in this chapter are explicitly geomantic. These images, found in an ideal form in fengshui manuals and in practical deployment in family genealogies, displayed the technical intricacies of fengshui, usually through tracing earth veins or auspicious forms over a territory. Historians have seldom examined these images, but people produced many during the Qing.[11] Other maps did not portray earth veins or landscape forms, but rather depicted the placement or location of a site in relation to other elements of the land. Most judicial maps were not explicitly geomantic, but as seen in the previous chapter and made further explicit here, magistrates readily consulted these maps to answer questions about fengshui. Gazetteers could contain geomantic maps, but even when they did not, their images revealed places of significance (e.g., "famous sites," *mingsheng*) that often overlapped with places believed to have good fengshui.

The truth is, a clear distinction between geomantic and nongeomantic maps is hard to discern in some Qing cartography. Many maps not typically thought of as geomantic were created by using a geomantic compass. For example, Minister of War Yao Lide's (1718–83) preface to *Complete Conspectus of the Grand Canal in Shandong* (1776) conveyed that "we ordered maps of the Grand Canal drawn using the 'using squares to record scale' (*kaifang jili*) method, and for the orientations of water flow between the watergates, we employed a geomantic compass (*luopan*) to establish the orientations of *yin* and *yang*, so that when readers open to a page, they can immediately discern north, south, east, and west."[12] The mapped elements of *Complete Conspectus* do in fact label the geomantic directions for discrete parts of the Grand Canal, such as *qian-xun* (northwest-southeast).[13] The creators of those maps assumed that educated readers would recognize the directional significance of the trigrams, heavenly stems, and earthly branches. It is possible that a broad spectrum of Qing cartography paid attention to geomantic features, including river systems and water currents, as a matter of principle.[14]

The chapter begins with an overview of official mapping guidelines, which courts invoked to resolve disputes about fengshui. Then it examines maps of houses and graves created in court that resonate with the technical diagrams found in geomantic manuals. Then, turning to local genealogies, gazetteers, and atlases, it situates these judicial maps within the broader context of maps of county landscapes in the nineteenth century, noting how fengshui influenced these documents.

Regulations for Administrative Mapping

The Qing government commissioned maps of many kinds during its reign over the provinces of China. The Palace Collections of Beijing and Taipei contain maps concerning astronomy, general territory, rivers and lakes, river transport systems, military affairs, imperial tours, famed scenic spots, auspicious omens, tribute, salt affairs, temples and altars, the imperial tombs, and fengshui. While the grand geomantic maps of the imperial tombs stand in an elite category of their own, local governments also paid attention to fengshui. Over the 1700s and 1800s, imperial edicts and official handbooks encouraged officials to map graves and mountain lands to prevent or resolve disputes over their valuable terrains. Readers may recall that it was precisely during these centuries when judicial administrators in Sichuan and beyond issued notices that the residents of their provinces were "obsessed with fengshui." The results are abundantly evident in the maps of Nanbu's collection.

Besides the fact that imperial law strictly protected graves, a reason that Qing officials were compelled to map gravesites was due to the proliferation of private genealogical illustrations in late imperial society. Geomantic grave maps began appearing in family genealogies in Huizhou and other parts of south China following the emergence of the family lineage as the central organizational institution of rural society.[15] Accordingly, gravesite diagrams did not become a "standard component" of genealogies until around the sixteenth century.[16] By the nineteenth century, these privately produced maps could take on dramatic territorial scope, connecting many natural and built features of rural settlements through the depiction of dragon veins and auspicious landforms. For lineages, privately mapping fengshui served, among many things, to stake ancestral claims over territories that may have otherwise lacked the support of historical evidence. The map from nineteenth century Huizhou (figure 2.1), which portrays the auspicious landforms shrouding two south-oriented ancestral halls of the Bao lineage, provides an example of the genealogical images that proliferated in the mid-to-late Qing.

The Qing state placed considerable weight on mapping graves. In 1729, the Yongzheng Emperor ordered resident administrators to inspect any ancient tombs of historical importance within their jurisdictions by the close of every lunar year to ensure they were properly protected and maintained.[17] County reports of these inspections were sent to their provincial superiors, who were instructed to punish any official for failing to adequately protect local tombs.[18] This edict was enforced at the county level. Through the end of the dynasty, Nanbu's government annually received instructions from Baoning Prefecture to map the county's prominent tombs and draft a report on their upkeep.[19] In

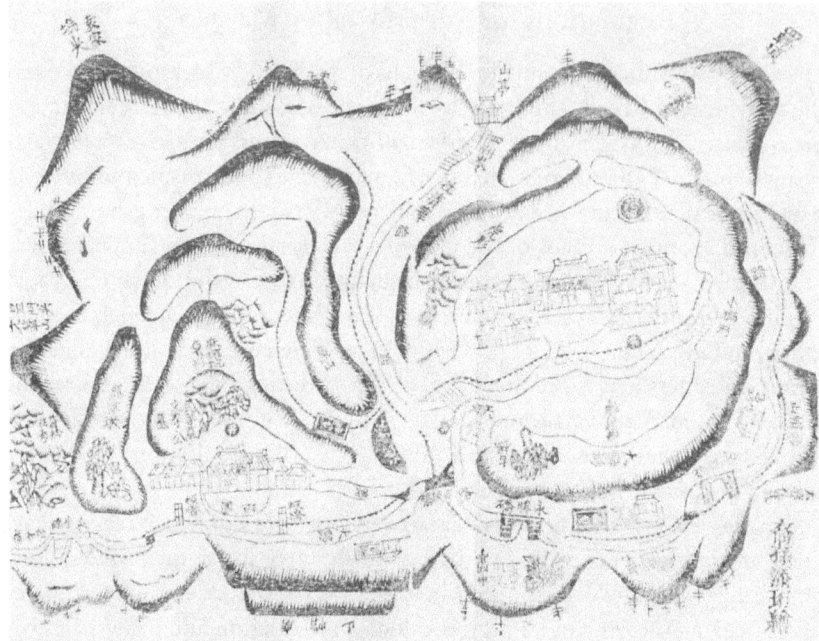

FIGURE 2.1. Genealogical Map of Bao Family Village (Huizhou). This nineteenth-century genealogical map depicts the sprawling terrains of the Bao Family Village in Huizhou. The mountains, bridges, well, ancient trees, and two ancestral halls are bound together in a landscape marked by shared kinship. Note also the identification of landforms characteristic of Form School fengshui, including lion (*shixing*), tiger (*huxing*), and elephant (*xiangxing*) landforms. The twin lion landforms, both located directly south of the main ancestral hall, may have invoked the pairs of stone lions that were constructed in front of prominent temples, governmental offices, and important tombs. Map from Yamashita, ed., *Chūgoku mokuhan fūsui chizushū* (2010).

addition to protecting the physical resting places of the dead, the act of mapping these tombs also, in theory, prevented opportunists from arbitrarily claiming them as their own. Of course, as seen in the previous chapter, government regulation did not prevent attempts at such claims.

In addition to the Yongzheng Emperor's edict, Qing handbooks for officials emphasized the need to examine fengshui in several contexts. As Huang Liuhong's (1633–1710) handbook for county magistrates, *A Complete Book Concerning Happiness and Benevolence*, relates, newly appointed officials were instructed to inspect a county's dragon vein prior to engaging in public works projects.[20] Another handbook, Tian Wenjing's (1662–1733) and Li Wei's (1688–1738) *Instructions for Magistrates Published by Imperial Order*, urged officials to "survey fengshui, clear up land boundaries, inspect the *baojia*, and examine carefully matters of agriculture and sericulture," with *baojia* referring to a community-based system of public security in imperial times.[21] The fact that two formative official handbooks of the Qing era recommended that magistrates inspect fengshui upon arriving in a county underscores the practicality and importance of such information for local administration.

There were also specific guidelines for judicial mapping. One handbook, identified by Pierre-Étienne Will as "perhaps the classic" standard magistrate handbook, Wang Huizu's (1731–1807) *Personal Views on Learning Government*, discusses the techniques employed to survey land during the adjudication of disputes.[22] Wang wrote, "As for the matter of surveying and measuring lands during litigation, there are four genres: one is called fengshui; one is called irrigation; one is called mountain plots; and one is called field boundaries."[23] Wang shared important information about these four techniques of land surveying. Cases concerning field boundaries and irrigation, he writes, "can be resolved by a glance." In other words, those cases were relatively easy to adjudicate since officials could, in theory, rely on tax registers or other written records to obtain key facts for a case.

But turning to cases about fengshui and mountain plots, Wang took a different tone, indicating that these cases feature "complicated insinuations and implications" and could be full of false information. Since fengshui cases were not easy to adjudicate, how could they be adjudicated in court? Wang recommended that officials "must first entreat the two parties of a dispute to map the site in order to identify precisely the mountain's identity and orientation, then once again carry out an investigation."[24] Wang Huizu's text was not the only handbook to recommend this method; an essay in Xu Dong's (1793–1865) *Book for Magistrates* offers similar advice for adjudicating grave conflicts.[25] Qing officials appeared to have agreed that one of the best ways to solve disputes about fengshui was to directly investigate lands in dispute through cartography.

A collection of the legal cases processed by Xu Shilin (1684–1741) showcases these principles in action. Over a quarter of Xu's collection of 102 cases, most either reported by county magistrates or appealed by local litigants during his service as the prefect of Anqing, concerns grave mountains and trees. When considering these cases, Prefect Xu consulted genealogical maps submitted by litigants, judicial maps created by county yamens, and commissioned maps of grave mountains in secret so that litigants could not pressure or bribe dispatched clerks and runners in favor of their personal interpretations of fengshui.²⁶ Xu scolded appellants greedy for the auspicious lands of other families, but he nonetheless explicitly recognized fengshui concerns in verdicts so as to keep community peace—doing so in one instance to protect fengshui even after he bluntly concluded that he did not think it had been disturbed.²⁷ Xu also occasionally criticized county magistrates and amended their verdicts when he felt they had perfunctorily analyzed grave lands in dispute.²⁸ Prefect Xu's judicial collection reveals that the act of mapping gravesites involved interpreting the environments around graves, on which the nuances of a case often depended.

The practice of mapping cases that hinged on questions of fengshui can be seen across Sichuan's judicial bureaucracy. The circuit intendant of eastern Sichuan, Wang Dingzhu, invoked the need for mapping in processing fengshui-related legal appeals during the 1820s.²⁹ In appealing county-level judgments over graves, litigations were in essence arguing that county magistrates had incorrectly interpreted fengshui. Here again, Circuit Intendant Wang took a measured approach. In considering the appeal of a man surnamed Hu concerning harmed fengshui, Wang ordered the Qijiang County court to oversee the mapping of the site in contention and respond with an affixed description.³⁰ But in another instance, Wang scolded an appellant by writing: "You have been requested to draw a map of the site and report back, but you continue to act disrespectfully with these lawsuits by presenting them with vague details about fengshui, brimming with trickery and presumptuousness."³¹ In this instance and others, Circuit Intendant Wang's administrative posture appears to have aligned with the guidelines found in official manuals.

The fact that courts commissioned maps of gravesites does not mean that they were flawless or objective pieces of evidence. In attempting to solve any legal case, the imperial state relied on the cooperation of local actors—gentry, Yin-yang officers, the heads of mutual surveillance units, and clerks and runners—who had their own incentives and biases. Litigants might lodge appeals over maps alleged to inaccurately portray the positions of houses, graves, and trees, and one suspects the clerk maps were occasionally less than accurate.³²

For their part, magistrates took the act of cartographic interpretation seriously. In a case from Taiwan's Dan-Xin Archive, which holds the administrative records from Danshui subprefecture and Xinzhu County dating from 1776 to 1895, an official inspecting a commissioned judicial map commented it was difficult to discern whether the grave was obstructed. He thus requested that a second map be drawn with a new analysis.³³ A magistrate of Ba County responded similarly to a clerk's report about the status of land boundaries for a case in 1852, here ordering the clerk to create a third map of the site in contention for the court:

As for whether Yu Dalin's grave land (*yindi*) along the wasteland was truly invaded and dug up by Yang Tou, this matter still has not been carefully investigated with a clear conclusion. Draft another report, and do not reply again perfunctorily. I return the [second] map to you.[34]

No magistrate wanted to walk into a public trial without a firm grasp of the environmental conditions of a gravesite in contention. As such, they made their clerks and runners provide a complete picture.[35]

Yet, to avoid overstating the potential for corrupt opportunism or gross inaccuracy in the creation of litigation maps, it should be noted that a majority of the 181 maps in Nanbu's archive were created by a limited number of people who were directly employed by the government. Annotated images were usually attributed to a named county clerk. In one instance, two clerks were sent to survey a dispute over property on different dates, with one tasked with mapping a gravesite and the other tasked with mapping trees.[36] Two separate reports were submitted to the court, which considered them side by side. A further example concerns a clerk from the Department of Works named Wang Qingxi, who was assigned to create at least sixteen litigation maps from 1886 through 1912, most of which depicted gravesites and trees.[37] Our knowledge of these mappers is limited due to the nature of the sources, but Wang's successful career spanning over two decades in the county government could have only been possible by maintaining a reasonably good reputation (see Appendix C). And although dozens of clerks staffed the county yamen in any given year, the fact that the majority of maps in Nanbu's nineteenth-century collection were composed by twenty-two named clerks at the Department of Works suggests that mapping, especially the mapping of gravesites, was a specialized field of practice.

Qing officials expressed concern about the potential dishonesty and trickery in legal allegations over fengshui. Nonetheless, they recognized fengshui disputes as a distinct genre of legal case and set out firm guidelines for appraising geomantic allegations at the county level. In practice, these guidelines were also followed at the prefectural, circuit, and provincial levels of the bureaucracy when disputes were disputed or appealed upward. The result of these regulations was that in some regions of China through the nineteenth century a considerable amount of judicial mapping was tied up in appraising cosmological-driven allegations about the dead, the living, community health, and land. Qing officials obtained the necessary information—the locations and orientations of houses, graves, temples, trees, dragon veins, and so on—to render verdicts that projected their knowledge, morality, and authority to the people they governed.

There was a flipside, however. The fact that the Qing state had to take fengshui into consideration speaks to the power of people to impose their own forms of knowledge on the imperial state. With good reason—and a range of intentions—people kept coming to court.

Mapping Family Residences

As previously discussed, residences for the living were viewed as the *yang* counterpart to the *yin* of graves. Family homes, termed *yangzhai*, could be mapped with annotations in the wake of, or even in preparation for, geomantic litigation. House mapping occurred with less frequency than the mapping of graves, both in court documents and in family genealogies, but several legal files from Northern Sichuan are worth considering.[38] In popular society, the term *yangzhai* generally referred to houses, although *yangzhai* fengshui manuals also contained information for the building of temples, schools, and governmental offices. After introducing the contents of these building guides, the section will trace their applications in the Qing legal system.

Because geomantic and building knowledge were intertwined in imperial China, establishing a clear categorization for such knowledge has proven a challenge since the early twentieth century. As Peter Carroll, drawing on the words of Republican intellectuals, has observed, "The disciplines of architecture and architectural history were entirely absent from traditional Chinese studies," with the term "architecture" (*jianzhu*) only entering the Chinese language via Japanese in the 1870s.[39] Nancy Steinhardt has similarly pointed out that there was no word in imperial China for "architect"; instead, there were terms for wood craftsmen and carpenters for whom literacy was not necessarily a requirement.[40] For historians writing in the wake of the New Culture Movement (c. 1915–25), such as Yue Jiazao (1867–1944), China's seeming "lack" of an architectural discipline seemed clear in light of newly translated foreign scientific and technological categories.[41] China did possess a longstanding and rich tradition of building and construction, but the emic frameworks of that tradition did not easily translate into twentieth-century imported categories of knowledge.

The large corpus of writing for constructing residences in late imperial China was filled with what today would be considered building information. This included encyclopedias for daily use, carpentry manuals, garden overviews, and fengshui manuals, the latter of which contained information on wood types, building plans, design features, and construction methods.[42] Beyond popular manuals and almanacs, imperial treatises on construction, such as the Song-era *Treatise on State Building Standards* (1103 CE), were influenced by geomantic texts and contain many references to fengshui.[43] In short, fengshui—particularly as expressed in geomantic manuals for the construction of dwellings—constituted an important component of architectural and building knowledge in late imperial times.

Fengshui was not synonymous with modern notions of architecture. More precisely, it guided the development of varied building traditions throughout the Han-populated regions of the empire by informing the positioning, construction, and orientation of structures. As shown by Ronald Knapp, every region of China possessed its own distinctive building styles: the garden complexes of Jiangnan, the symmetrical courtyard houses of Beijing, the delicately austere merchant complexes of Shanxi, and the fortress-like walled houses of Huizhou all speak to immediate environmental and demographic needs.[44] Shaanxi homes were built with roofs of a single sharp angle so that any timely rain would pool into the courtyard for the family's use; houses needed to be constructed in that way so water flowed directly to the family. Jiangnan families, blessed with ample water sources, often built homes next to river transport channels. The underlying commonality between these styles was the desire to maximize the good fortune and safety of the family who inhabited these structures. Throughout China, buildings were constructed by people who claimed to be guided by fengshui—doorways and side rooms were bestowed with carefully selected auspicious names or adorned with geomantic charms and amulets.[45] The market-driven publishers and writers of Qing-era fengshui manuals profiled a diversity of needs in their texts for the broadest possible readership.[46]

Rather than viewing fengshui as a fixed blueprint for residential construction, it is more fitting to take it as a language of power, one that that addressed many of the same issues that arose with the building and maintenance of graves. To be sure, the residents of houses in Sichuan did not possess contracts detailing the regulations or specifications for every fixture of a house's layout. But when problems arose, either between members of an extended family or among neighbors, recourse to fengshui was an expected, useful, and persuasive way to make a point or establish a claim. Precisely because the geomantic implications of a given physical structure might not be interpreted in the same way by all members of a family, or by its neighbors, fengshui was mobilized to voice concerns and resolve disputes in the interests of community harmony.

Two manuals for constructing houses are useful for understanding residential fengshui as it was used by people to select locations and design structures in late imperial times. The first manual, *Secret Essentials of Judging Houses*, published in Nanjing in 1595, is a compendium of descriptions and diagrams for the selecting of auspicious sites and for the building of a wide range of residences, including government offices, academies, temples, and houses. As a composite work, the text bears the names of several contributors, including

Xiong Weiyao, who composed the text's preface, and Yuan Shizhen (d. 1631), who went on to become a prominent official of the late Ming. The second manual, *Golden Mirror for Peaceful Living*, published in Hangzhou in 1780 by two Jiangnan geomancers for commercial distribution, focused specifically on building materials and the layout of houses.

These two texts provide glimpses into the development of geomantic concepts for houses over the Ming and Qing periods. Though filled with technical discussions of proper placement and positioning based on the eight trigrams and five agents, both texts had user-friendly applications, with ample case studies for building and construction drawn from lived experience. For instance, a passage from *Secret Essentials of Judging Houses* concerning construction related to the qualities associated with fire—one of the five agents—reads: "When the ridges of a house's roof are acutely pointed, its walls are long and sharp, its eaves are composed of kindling material, and its rafters are exposed and sharply slanted, then these features will summon a lawsuit, a deadly fire, a consumptive disease, or detention in prison."[47]

Here, the authors capture an observed natural phenomenon: when tall wooden structures with pointed roofs catch fire, the flames are difficult to extinguish.[48] The tall, pointed roofs of wooden structures were deemed susceptible to lightning or accidental fires. For this reason, such buildings became associated with other outcomes related to fire through the discourse of fengshui. Likewise, a house with overly long walls could make extinguishing fires difficult for people struggling to find an entrance into a burning compound. As people presented lawsuits contesting the construction of overly tall or pointed edifices—an experience that Western missionaries later came to know when constructing conspicuous church steeples in the late 1800s—another perceived outcome, litigation, was added to the list of inauspicious side-effects.[49] Finally, because the law code contained punishments for damage done by fires, whether accidental or started intentionally, a legal punishment—detention in prison—was added to the list.[50] Though some principles of fengshui may appear arbitrary today, many tenets possessed a readily apparent logic within the legal culture of late imperial China.

The fact that at least one of the contributors to *Secret Essentials of Judging Houses* became a prominent Ming official is worth considering further. Other fengshui-related manuals published during the Ming and Qing periods display prominent endorsements from officials serving in the government and prefaces written by credentialed degree-holding gentry. Ouyang Chun's *Two Works on Fengshui, On Forms and Material Forces, Organized by Categories*, introduced briefly in the previous chapter, furnished a preface by the director of Xintian County's academy in Hunan Province. Two well-regarded Qing manuals, *Five Secrets of Earthly Principles* and *Three Essentials of Residences*, were composed by Zhao Tingdong, an official stationed in Sichuan during the latter years of the Qianlong reign.[51] Elite authorship and prominent endorsements provided commercial publishers with confidence that a book's contents were not subversive and provided consumers with a guarantee of product quality. Nonetheless, over time, more manuals of varying quality and origin entered general circulation. This trend led the imperial state to produce its own fengshui manual, *Correct Doctrines of Fengshui*, which synthesized existing geomantic theories and practices in the mid-eighteenth century.

The Qing-era *Golden Mirror for Peaceful Living* shares general features with the Ming-era *Secret Essentials of Judging Houses*, but it also includes particularly detailed information on construction materials for buildings, courtyards, lightwells, and walls. The extended section on roads around a house and house trees underscores the importance ascribed a residence's built and natural environment.[52] Specific tree species appropriate for planting around a house included green pines and bamboo, which, by providing domestic shelter from strong winter squalls, invited wealth and health into the home.

Some trees were inauspicious for house environments. Dense tree cover that surrounded a house on four sides—the Qing equivalent of "a cabin in the woods"—was not advantageous.[53] Nor were trees that blocked the main entrance, were poorly located, or possessed certain physical features, such as lengthy vines (see figure 2.2). Dense foliage near a house's entrance was identified as an easy hiding spot for thieves. Likewise, if a tree with lengthy vines grew in front of a house, the resident family could expect legal proceedings that culminated in a judicial conviction of adultery or banditry, representing female and male crimes respectively. Houses with unkempt, lengthy vines were suggestive of low occupancy rates and thus taken as easy prey for wandering bandits. At the very least, neighbors might well complain about such a tree, provoking community quarrels that could culminate in an unfortunate lawsuit.

Structural problems of residences were called house illnesses (*zhaibing*), which rendered dwellings inauspicious for inhabitation. These house illnesses had the potential to make residents physically ill or else highly unfortunate. The most common symptoms of house illness included: litigation, sexual licentiousness, pregnancy issues, financial bankruptcy, failed harvests, and the sudden appearance of bandits. For example, a house at the intersection of busy roads was taken as inauspicious since the residence was overexposed to wind and noise, ensuring the onset of illnesses among inhabitants.[54] Changes to a residence's surroundings by the addition of poorly situated cooking areas, millstones, and barns could also induce house illnesses—and ultimately human illnesses.[55]

Women featured prominently in the sections on house illnesses, since a function of *yangzhai* involved concealing and protecting the wives, mothers, and daughters who dwelled inside, in effect reinforcing patriarchal ideals of

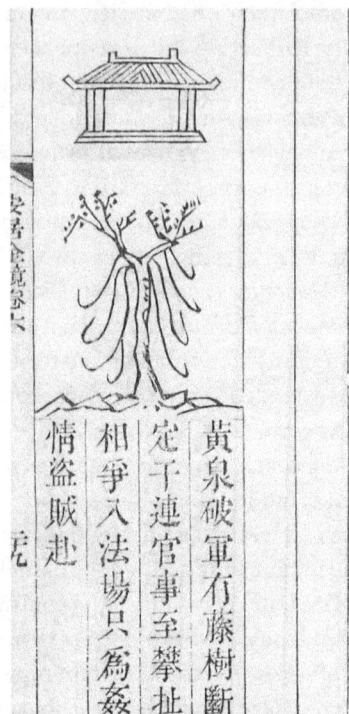

FIGURE 2.2. Ill-Placed Vine Tree Producing Litigation Involving Capital Punishment Geomantic Diagram. This diagram depicts an inauspicious vine tree in the Yellow Spring and Breaking Army Star direction. Fengshui manuals not only contained diagrams analyzing the physical structure of houses, they also lent attention to their surrounding environments, including trees and water sources. Diagram from Zhou Nan and Lü Lin, eds., *Anju jinjing* (1780). Original manual held at the Harvard-Yenching Library.

female domesticity.[56] As alluded to above, many house fengshui diagrams dealt explicitly with women's health, including not only the ability to conceive and bear children, but also the well-being of the mother in the wake of childbirth. One inauspicious diagram was labeled Breast-Feeding Style House, which exhibited a protruding structural form said to induce a wife's death in childbirth.[57] Another diagram (figure 2.3) was labeled "Scoundrel [Summoning Lawsuit] House Form," which, by obstructing the house's front gate by the addition of a detached room or small dwelling, was said to induce a miscarriage and hurt women's health, in addition to summoning expensive litigation at a magistrate's court.[58]

House layouts were not only linked to physical well-being, they were also closely associated with legal affairs. When litigation was mentioned in a

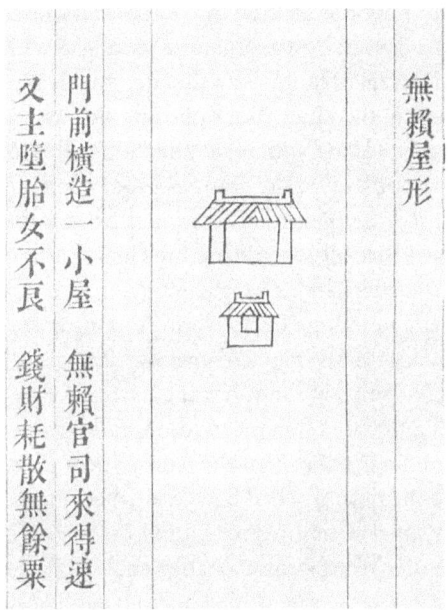

FIGURE 2.3. "Scoundrel [Summoning Lawsuit] House Form" Geomantic Diagram. This diagram offers an example of an inauspicious house. Its caption reads, "If one builds a small dwelling in front of a residence's main gate, a lawsuit from a scoundrel will arrive soon. Additionally, the structure will cause miscarriages, and women of the house will not be healthy; money and wealth will depart, and there will be no surplus of grain." Note that the four-line caption rhymes at the last character of the first, second, and fourth lines (*wu*, *su*, *su*), as do many other geomantic descriptions of the manual. This feature suggests a broader oral tradition of geomantic formula that people recited. Diagram from Zhou Nan and Lü Lin, eds., *Anju jinjing* (1780). Original manual held at the Harvard-Yenching Library.

geomantic manual, it was always in the context of an inauspicious outcome, linked to bankruptcy and legal punishments. There were no auspicious lawsuits in Qing China. The fact that the residential manual contains dozens of references to formal litigation leads to the conclusion that the text was not only a work of fengshui, but also of practical legal knowledge. The geomancers who used this text were providing legal counsel to families. In fact, there are legal cases in which geomancers were offered as witnesses.

One such case involved a dispute among members from two branches of the Wang lineage over the health of their residence. In this instance, one branch of the joint family accused its relatives of installing a cooking area and ox barn to the south of the residence's commonly managed central hall. This side claimed that these actions severed the central hall's geomantic vein, which was attested to in the lawsuit by a geomancer serving as witness. Their lawsuit alleged that the harmed geomantic vein threated the lives of all the people of the house, both men and women. The opposing side asserted that the structures were built on private land with no connection to the communally owned home. A contract was offered to the court as proof of private ownership.

The court accepted the plaint and commissioned a clerk from the local Department of Works to analyze the structure's fengshui and map the house.[59] The department dispatched clerk Zhang Zhongyou, who was a regular mapper for the court.[60] Zhang found the ox barn and cooking area constructed to the southwest of the house. In his map and analysis of the residence, which bore resemblance to the "Scoundrel [Summoning Lawsuit] House Form" diagram discussed above, Zhang stated that these additions had indeed obstructed the central hall, concluding, in effect, that the family's geomancer was correct and that the fengshui of the property had been disturbed. However, the clerk also noted that the contested structures were built on private land, not territory common to the broader Wang family. Zhang's analysis thus carefully captured the validity of both parties' contentions: it appeared that the fengshui of the house was disturbed, but the disturbance had been created on land that truly was privately owned. With the map and analysis submitted to the court, it was up to the magistrate to render a verdict.

The magistrate, upon analyzing the map and appended description, ruled that "Wang Guoshun et al. last year created a room on the level ground in front of the house. This room was lower than the central hall, so how can be it said to pose geomantic harm to the people in the house?"[61] In the magistrate's view, fengshui had not been disturbed. But just in case, and with an eye to family harmony, the magistrate still ordered the cooking area and ox barn moved to a new location at the conclusion of the trial. In arriving at this verdict, the magistrate had based his ruling on his own knowledge of fengshui. Although the clerk had drawn the house in a way that appeared to align with the "Scoundrel [Summoning Lawsuit] House Form" diagram, which flagged the addition of a room in front of a house's front gate as inauspicious, the magistrate interpreted the site through a different diagram. This diagram was titled High Front House (see figure 2.5), which was described in *Golden Mirror for Peaceful Living* in the following terms:

> This house is called a High Front House. Its front is high, and its rear is low. Its master's land will be sold to other surnames, and the house will produce orphans and widows; the house will hurt children, summon disease and fire, and every year will be marked by mourning. For suitable living, the front must be lower, and the back must be higher . . .[62]

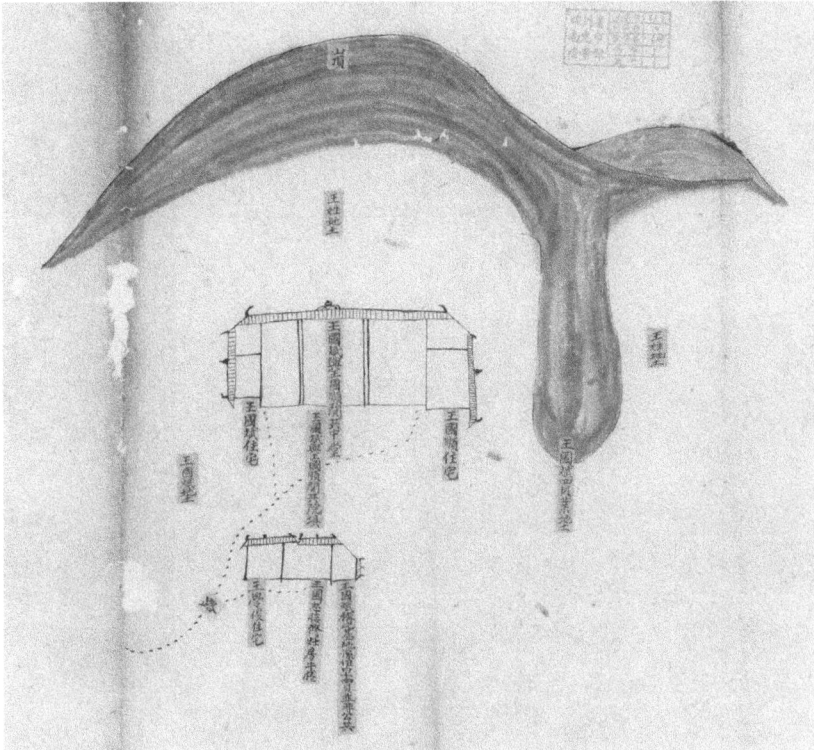

FIGURE 2.4. Commissioned Court Map of House in Dispute. This map was composed by a clerk from the Nanbu yamen, Zhang Zhongyou. The top center of the map, like many judicial maps found in the archive, depicts a mountain, beneath which was land held under the Wang surname. The largest structure depicted on the map was the Central Hall, which was the common property of the two branches of the Wang lineage involved in this case. Surrounding the central hall were two attached residences—one to its right and one to its left—each owned by one branch of the joint family. Below the Central Hall were the structural additions that had inspired the conflict. Map from Nanbu County Qing Archive.

By placing emphasis on this principle, the magistrate was able to contend that fengshui had not been harmed.

The significance of the rear of a house, termed the Main Hall or Central Hall (*zhongtang*), derived from its frequent overlap with the domestic family shrine, where the ancestral tablets were stored and the portraits of the ancestors were hung. In Sichuan, the coffin was laid here without the body—symbolic of the process through which the recently deceased became an

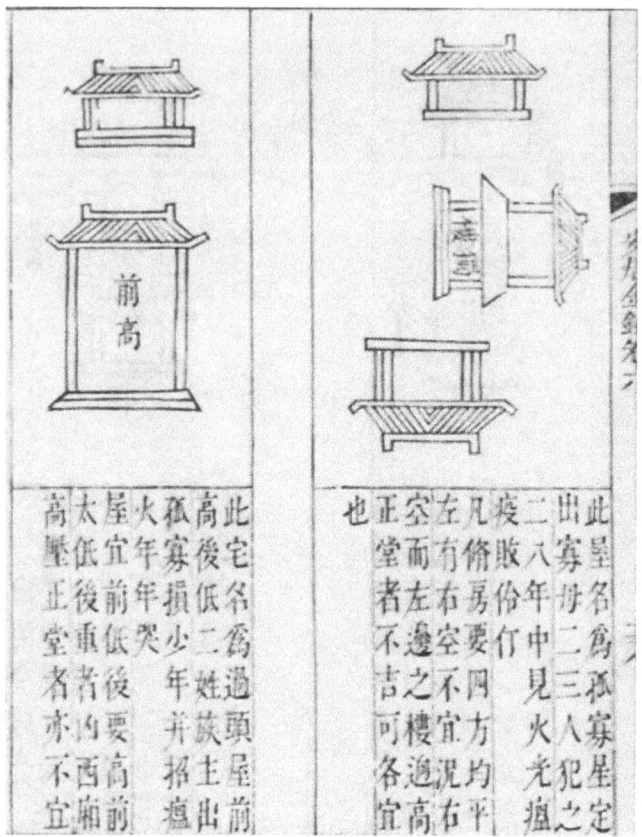

FIGURE 2.5. "High Front" House Geomantic Diagram. The "High Front" Diagram and description are found on the left side of the image. Diagram from Zhou Nan and Lü Lin, eds., *Anju jinjing* (1780). Original manual held at the Harvard-Yenching Library.

ancestor and indicative of the close relationship between "houses of the dead" and "houses of the living."[63] There were practical reasons informing this principle as well, with the elevated rear of the house carrying waste drainage from the living quarters out through the house's lower front. For these reasons, it was widely held that the rear of buildings had to be higher than the front—a practice that the Jesuit missionary Matteo Ricci (1552–1610) noticed in his journey across China.[64] The principle of lower elevation in front of structures and higher elevation in the rear can still be seen today in many old buildings, from the Palace Museum of the former Imperial City to the rural courtyard homes of Sichuan's ancient towns.

The Wang family case reveals one important way officials adjudicated disputes over residential properties. In this case, a property dispute witnessed the summoning of a geomancer to provide witness testimony. This geomancer formulated a lawsuit that attested to the infringement of well-recognized fengshui principles that could easily be found in Qing-era house fengshui manuals. The clerk from the county's Department of Works discovered that the house's fengshui had indeed been disturbed, confirming the geomancer's claim. Likewise, the judicial map created by the clerk depicted the Wang's courtyard in a manner broadly similar to house diagrams found in geomantic manuals. However, the map created by the clerk also revealed a critical detail of elevation that the magistrate, drawing on his own knowledge of fengshui—specifically, the principles espoused in the High Front diagram—was able to cite in concluding that fengshui had not been harmed. But in a surprising detail one should not overlook, the magistrate still ordered the allegedly inauspicious kitchen moved to avoid the potential of future conflict in the months and years ahead. Even though it was officially dismissed on a factual basis, the lawsuit alleging that fengshui had been disturbed was ultimately effective.

Readers might wonder whether county government staff consulted *A Golden Mirror for Peaceful Living* to compose litigation reports. The truth is, they did not need to. The exact same or very similar house diagrams were found in other Qing-era geomantic manuals, such as the 1882 reprint of a Ming-era text, *Ten Books on Residences*.[65] Useful information traveled widely through new privately published manuals and reprintings of older texts. It is unclear where or how the clerk dispatched by the yamen to map the Wang house cultivated his knowledge of fengshui, though we may speculate that the state's Yin-yang officer employed at the Yin-yang school near the county yamen offered some training or advice (see chapter 3). At the very least, the clerk, Zhang Zhongyou, knew enough about property disputes and geomantic principles to create a detailed map of the house that analyzed both the cosmological and contractual dimensions of the dispute. The same clerk went on to create at least five other maps for Nanbu's court, three of which concerned tree and grave disputes.[66]

Most geomantic lawsuits over houses concerned the fengshui trees felled to create their foundations or expand their structures, but there was a wide range of lawsuits about home layouts and environments.[67] Some of those were mapped in the style of the Wang house above. Like any other type of lawsuit, some fengshui plaints were accepted for consideration and some were rejected. One lawsuit that involved a manure pit created near a house was accepted and adjudicated, while another involving alleged injury to the vein of the Zhang ancestral hall from nearby construction was rejected.[68]

In reviewing cases to accept, the court looked for nuances in the narratives of lawsuits and paid attention to the timings of inauspicious discoveries. The lawsuit about the Zhang ancestral hall was rejected because the court was skeptical about why family members did not protest the construction when it began.[69] The case about manure was accepted because it involved a family in shock at the discovery of a new fertilizer pit next to their house. Since their plaint successfully conveyed outrage over this new development, the court got involved quickly to quiet neighborly tensions. And even in the rejected case of the Zhang ancestral hall, one can still see that the principle of residential dwellings possessing their own veins was broadly accepted by law courts.

What did it mean for houses to channel geomantic veins? Legal cases suggest that the veins of houses were understood as disturbed when a previously auspicious residence became afflicted by a house illness. One map, composed by the Liu family and presented at a county court, captures this transformation in action. The case involved a property dispute with a person named Ma, who acquired the residence in the wake of land reform during the establishment of the Northern Sichuan Soviet Area (1932–35). Following the departure of the communists, the Lius accused Ma of disturbing the ancestral shrine in their courtyard house, which had been constructed during the Daoguang era (1821–51).

Although not stated explicitly in the case file, the Lius likely levied the charges to get Ma out of the house. Many property deeds had been destroyed after the arrival of the Red Army, meaning emphasis on historical settlement, ancestry, and fengshui may have been the best option available for the Lius. The Lius' inability to simply seize back their house after the communist departure speaks to the possibility that land reform had been messy and even locally accepted by some parties. This case's periodization is admittedly outside our scope of inquiry, and national legal standards had changed by the 1930s. Yet the house map in this case was richly drawn in the presence of local elites to serve as evidence that Ma had disturbed the Lius' residence and ancestral hall.[70] This map provides the clearest and most detailed illustration of a house in the wake of a fengshui dispute, so I introduce it here.

The annotated illustration of the Liu courtyard reflected diagrams found in house fengshui manuals. It depicted—as did the clerk's map from the previous case—the residence on a north-south axis. Of note was the "Lotus Pond," situated in the front of the house. This pond reflected a layout found in manuals that emphasized the need for flowing water near the residence, with the presence of blossoming lotus flowers thought to signify a healthy source of water. Semicircular lotus ponds were not only featured in homes, but also in grander forms at the front of academies and Confucian temples, where they were termed *panchi* (*pan* ponds). Passing the qualifying exams for the provincial examinations was to "enter the *pan*" (*rupan*), which literally involved walking over an arched stone bridge in the front portion of a prefectural Confucian temple.[71]

Other elements of the Liu house blended geomantic principles with the local architectural styles of Northern Sichuan. The carved latticework on windows and beams in Northern Sichuan houses like the Lius' often depicted scenes from popular operas. For the opulent residences of elites, who were often salt merchants in Sichuan, some carved scenes depicted operas performed at the opening of salt mines, recalling a property owner's successful brine drills. One ancient house surviving in Nanbu County, which resembles the image in figure 2.6, possessed wooden doorways that showcased auspicious characters and phrases. Carved bats, pomegranates, pines, and white cranes spelled out the characters for "gold and jade fill the hall" (*jinyu mantang*).[72] The latticework in the figure 2.6 is not visibly detailed, but it was surely present in the physical house.

Although the artfully crafted wooden carvings of a house held significance beyond fengshui for residents, their presence announced that there was no doubt the space was auspicious. It was not always possible to radically alter a house's form in pursuit of better fengshui, thus auspicious creatures like bats or cranes carefully carved into a geomantically questionable wall offered a promising compromise for property owners. Over time, practices such as these became standard fare in Sichuan and beyond, as elites left no stone unturned in pursuit of peaceful living—and genteel messaging.

Even roofs held significance. Roof tilers added stone-carved ornaments to the top of elite residences that often signified the three asterisms for "blessings" (including good fortune in scholarly and family life; *fu*), "prosperity" (including financial security; *lu*), and "longevity" (including good health; *shou*) to people standing outside of the house. Similar roof ornaments to the one in figure 2.6 may still be seen today in the old houses of Langzhong's ancient town, located adjacent to Nanbu County.

A final point of consideration concerns the lightwell (*tianjing*). Fengshui manuals such as *Golden Mirror for Peaceful Living* contained guidelines for a lightwell's ideal depth and dimensions, but actual constructed sizes were heavily influenced by region.[73] Starting from the North China Plain and moving south, the exposed surface areas of house courtyards diminished significantly.[74] Courtyard space in Beijing, Shanxi, and Shaanxi was comparatively expansive. By contrast, courtyards in south China, where land was more expensive, were condensed and took the term "lightwell." Functionally, northerners needed the rain, and families took as much of it as possible when it fell; southerners by contrast opted for the ideal amount of light in their multistory residential structures.

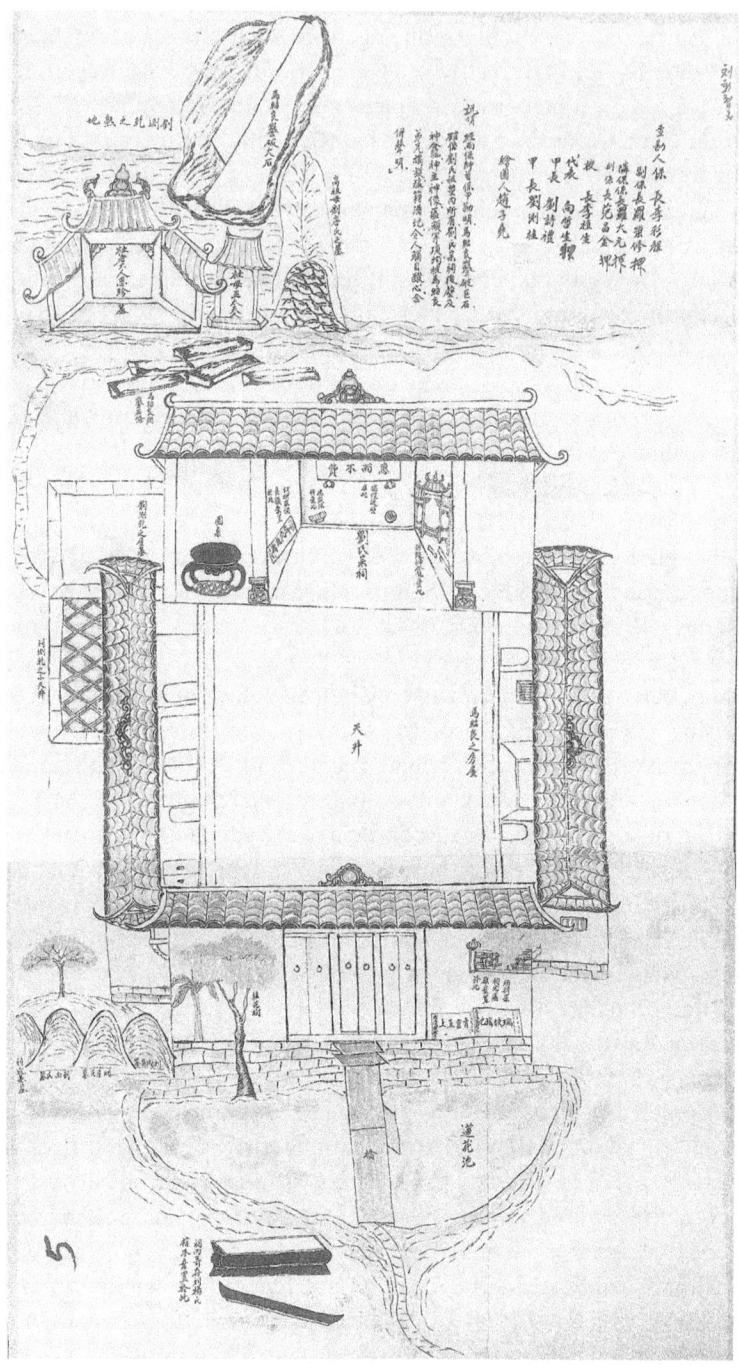

FIGURE 2.6. Mapped Courtyard Residence of the Liu Family. This undated image was created in the 1930s, not long after the dissolution of the Northern Sichuan Soviet (1932–35). The list of names and accompanying seals at

Baoning Prefecture can be considered quintessentially "middle-ground" by virtue of its location in a southwestern province, but one that shared its boundaries with northern provinces such as Gansu and Shaanxi. The local result was a distinct blend: courtyards here were called lightwells following the southern style, but still occupied relatively large spaces on floor plans. Sichuan's courtyards, known for being smaller than the north's but larger than the south's, structurally reflected the distinct climatic conditions of the province. A courtyard needed to be large enough to allow the sun to shine through during Sichuan's overcast winter months, but small enough to provide shelter from the sun during the hot summers.

The Lius had a long list of complaints against Ma: moving inscribed boards from their proper locations, removing a burial casket previously stored in the house's rear shrine, and quarrying stone from behind the residence. To make these points clear, the map showing the stone allegedly quarried by Ma depicted it in enlarged dimensions next to the Lius' ancestral graves, and the adjacent tree was encircled to identify its geomantic significance. The charge boiled down to the idea that Ma had illicitly acquired the residence and subsequently disturbed its fengshui, which the true owners of the compound would never dare to do. The mobilization of residential fengshui by elite families like the Lius reflected broader patterns of rural class tensions, which also applied to gravesites.

Mapping Family Graves

Most of Nanbu's litigation map collection consists of images created in the wake of disputes about graves. This section explores maps used by the government to settle lawsuits over graves, examining the ways officials interpreted these maps. The number of grave maps commissioned by Qing law courts was

FIGURE 2.6. (*continued*)

the top right includes the witnesses present upon the map's creation and submission to the local court. The map's illustrator, Zhao Ziyao, was not a member of the Liu lineage. The annotation to the left of the names describes the crime scene, mentioning the large stone that Ma had cut near the Liu family graves, as well as damage done to the Liu ancestral hall during the Land Reform campaign. Important details include the coffin and inscribed boards, previously stored in the house's Central Hall, that Ma had allegedly tossed in the front of the house. The Central Hall's altar, shrine, and ancestral tablets had all been tossed on the ground or else discarded. The room that Ma occupied was annotated on the right side of the courtyard's lightwell. Map from Langzhong Municipal Archive.

impressive and worth careful consideration. Officials spent valuable time and effort on these disputes because, as indicated previously, graves mattered, both to the fortunes of the people and the stability of the state. An examination of judicial maps from local archives, genealogical maps, and imperial maps stored in the palace collections collectively suggests gravesites were among the most abundantly illustrated landscapes in Qing China.

I have selected a few cases that involved intensive interpretation by magistrates. These maps related to the graves of powerful local notables in the nineteenth century, and their analysis allows for an understanding of the ways tombs of the elite were appraised and regarded. It should be emphasized that most of the judicial maps of commoners' graves were not explicitly geomantic and that in many instances officials simply judged whether a grave was obstructed by looking at a clerk's map. For instance, in one case concerning a dispute over trees resulting from "selling the land but keeping the graves," a magistrate ruled that, "upon inspecting the submitted map, the Pu family graves at present have a boundary stone marking the border of the cemetery; the Liang family is not permitted to invade the territory to cut trees within this stone, while the land outside of the stone shall remain the Liangs' property."[75] This verdict reveals a commonplace use of judicial maps by magistrates in the resolution of geomantic litigation. By contrast, the tombs of local elites who were powerful in rural society sometimes called for more technical analysis by county magistrates.

Consider an example from a published collection of legal cases, Fan Zengxiang's (1846–1931) *Judgments of Fanshan*. This record of cases during Fan's long tenure as a magistrate in Shaanxi Province from the 1890s allows us to observe similarities between fengshui's legal invocations in Nanbu County with those of another province located to the north. Magistrate Fan, a native of Hubei, was suspicious of popular geomancers who urged lawsuits over graves. On one plaint, he exclaimed, "Tian Jianghai does not understand fengshui! He is cheating you for an honorarium payment; immediately cut off ties with him and do not give him money or else you will regret it."[76] This comment reveals that Magistrate Fan was confident in his own understandings of fengshui. He certainly could provide ample advice from the bench.

Commented on by Magistrate Fan over several plaints, one dispute involved two neighboring families, the Gaos and the Zhangs. The Gaos owned a purportedly auspicious burial spot near the Zhang family property. The Gaos had not yet buried a body in that spot, but had spent a considerable amount of money to acquire it as a future investment. The site was called a *shengkuang*, or grave built before one's death. The "fake" grave dispute of the Zhong lineage from the previous chapter concerned one of these structures. In this instance, Zhang wished to construct a wall on his own property next to his family residence, which was located to the west of the Gaos' land. As sometimes happened in similar cases, the Gaos claimed this action would hurt the fengshui of their auspicious land, thus ruining their investment in the grave spot. A map of the site was submitted to Magistrate Fan, who analyzed it carefully.

Magistrate Fan ultimately ruled that the Zhangs could construct the wall. To begin with, he expressed incredulity that the Gaos had not buried a person in the spot yet had the audacity to commence litigation. As there was no real grave in question, there was nothing for the state to protect. But since both sides persisted in presenting lawsuits, Magistrate Fan engaged more deeply, drawing attention to the orientation selected by the geomancer hired by the Gaos for the future grave. Upon analyzing the map, which was not included in Fan's published collection, Magistrate Fan responded: "Your future grave spot is located on the *xun* mountain facing the *qian* direction, how can you say that its earth vein comes from the west?"[77] This comment offers quite a bit of technical information, which is understandable with the aid of fengshui manuals.

Xun and *qian* are the names of two trigrams named by the *Classic of Changes*. Together with two other trigrams, the ten heavenly stems, and twelve earthly branches, they composed the twenty-four directions of the Chinese geomantic compass.[78] Graves were positioned with two geomantic elements in mind: a root "mountain" where the grave was located and an orientation based on the flow of *qi* toward and from the site along earth veins. These two elements were described with two Chinese characters—in this case, *xun* and *qian*— each marking the heavenly stem, earthly branch, or trigram that represented the specific compass direction.

Most grave-oriented fengshui manuals published during the Qing contain a compass diagram, with the southern direction pointing toward the top, and the north toward the bottom. There were different kinds of compasses used by Qing times, but many fengshui manuals included a basic diagram (see figure 2.7). Fengshui practitioners did employ physical compasses to analyze landscapes, but ideal models could be reproduced in print. One printed compass is found in a manual published in 1692 by the Zhejiang native Kong Wenxing (1620–1705), a Buddhist monk who was also recognized as one of the greatest geomancers of the early Qing.[79] Kong's manual will be discussed in the next chapter for a closer analysis of its contents, which were closely related to successful examination outcomes.

The Gaos had contended that the proposed wall, located to the west of their land, would hurt the fengshui of their acquired spot. The logical issue that Magistrate Fan uncovered was that the spot was located on a southeastern mountain oriented to the northwest. Bear in mind that in fengshui, the slightest bit of difference in orientation matters significantly, since there are many potential combinations for a grave's geomantic positioning. That is, "north"

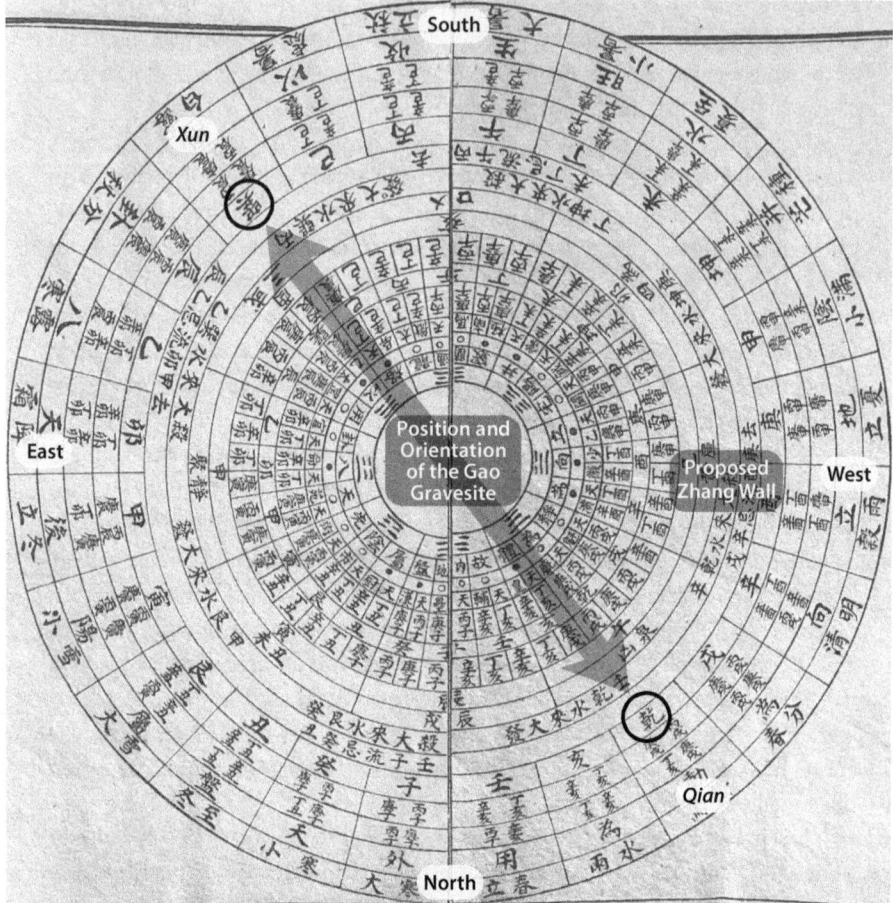

FIGURE 2.7. Compass Annotated for *Xun* Facing *Qian*. This is an annotated compass for a gravesite oriented on a *xun* mountain (southeast) facing *qian* (northwest). While it is possible some officials consulted a geomantic compass when investigating fengshui, surviving judicial comments suggest that officials were familiar key directional-pairings and could discuss them impromptu. Diagram from Kong Wenxing, *Huitu dili zhizhi yuanzhen* (1692/1875–1912).

and "northwest" are not the same thing, otherwise people would not have spent money paying for expensive geomantic consulting to obtain ideal directions. In rejecting the Gaos' petition, Magistrate Fan was contending that in theory a grave on the *mao* (direct east) mountain facing the *you* (direct west) direction would have a valid claim against the Zhangs' proposed wall, but *xun* facing *qian* did not. Magistrate Fan emphasized in his final comment that his logic derived from geomantic cartography: "Speaking of the map, Gao Mingde's future gravesite is on the *xun* mountain facing the *qian* direction; as the Zhangs' house is over ten steps to the west, the direction of the site's dragon vein is not affected, so why do you keep presenting lawsuits?"[80]

By laying out Magistrate Fan's logic, repeated over several lawsuits, one can see that his apparent dismissal of some claims about fengshui could be easily misinterpreted. When Magistrate Fan remarked he "did not believe in fengshui," he was not speaking in a Western context of "all-or-nothing" identitarian belief, but rather about the logic of specific technical claims.[81] Magistrate Fan consistently demonstrated over the course of legal proceedings that his knowledge of fengshui was fair-minded, and he was well-acquainted with geomantic principles. Profit-seeking and poor knowledge among commoners and even some gentry made educated officials like Fan skeptical about popular practitioners and frivolous lawsuits. But Magistrate Fan never dismissed the importance of a well-situated burial site for a parent or spouse.

In discussing geomantically related lawsuits, officials like Magistrate Fan generally preferred to invoke nouns such as principle (*li*) or earthly principles (*dili*) rather than the term fengshui to emphasize the learned, empirical content of their judgments. In one comment to the Gaos, Magistrate Fan confessed that he thought the site selected by the Gaos' geomancer was inauspicious even without the proposed wall and should never have been purchased for burial in the first place. In another instance, Magistrate Fan forbade a farmer from tilling barren land because the action would potentially "obstruct" (*ai*) a grave's earth vein.[82] Fan's published collection plainly reveals that fengshui was a language of power used to discuss and analyze landscapes in the law. Because of its manifest relevance to rural communities, officials selectively engaged that language to protect the people and project the state's power through the courts. Officials also did so because their own knowledge of fengshui was almost invariably greater than that of the audience they faced in court.

One geomantic map created at a yamen during a trial demonstrates a similar logic, suggesting that Magistrate Fan's methods of mapping and analyzing gravesites were known to the wider legal system. And because this case occurred in a settlement that was also mapped in the state's atlas of the county, it is possible to compare two cartographic images—one geomantic and the other nongeomantic—side by side. The case involved two branches of the Li lineage, which was a relatively prestigious surname in Qing Nanbu. Seven ancient tombs associated with eminent persons from the Ming named Li were said to have survived in the county, more than those of any other surname from that dynasty.[83] These tombs were recognized and protected by the local government, and some Qing-era kinship groups bearing the surname Li had taken advantage of their existence to construct family narratives that placed their forebears in the county prior to the great migrations into Sichuan during the eighteenth century.[84] It is unclear whether the Lis involved in this legal case participated in that project, but it is safe to say there were a number of powerful Lis in nineteenth-century Nanbu.

In 1887, a man named Li Zaitang presented a plaint against his cousin Li Jiran over unpaid debts.[85] Though seemingly a simple monetary dispute, the court quickly ascertained that the decline of familial relations had been precipitated by Zaitang's insistence that a millstone needed to be moved away from the area adjacent to his father's ancestral grave. Millstones were large essential tools for the grinding of grain and were expensive to acquire. Because families often pooled money to create a millstone for common use, disputes over their placement were not rare.[86] Because Zaitang, a wealthy member of the Li lineage, presumably had put up most of the funds for its acquisition, he felt that the millstone belonged to his immediate family. His move of the millstone onto his own private property sought to restrict access to it from poorer lineage branches.

To understand the fengshui of the Li homestead, one must appreciate something important about the geography of the place they called home. In place of long-standing nuclear villages of the kind found in other parts of China, Sichuan saw its residents live in satellite "dispersed farmhouses or farmhouse clusters" around a central site, which was usually a temple.[87] These temples hosted markets that met on a fixed schedule, thus attracting people from near and far for trade. Because of its distinct history of largescale inmigration during the Qing, Sichuan was home to thousands of these rural market towns, like the Lis' homestead at Pig's Trough Pass.

Pig's Trough Pass was located a few days' journey (around 110 kilometers) overland from the county seat of Nanbu. From the map and description of the market town provided in a county atlas (see figure 2.8), it is clear that by the mid-nineteenth century three lineages dominated: the Huangs, the Mas, and the Lis. Of the three, the Lis were ascendant. As a small, remote settlement in Nanbu, the entire area sported only one security unit and a single militia. From the late nineteenth century through the early stages of the communist revolution, people surnamed Li controlled both.[88]

Drawing on the map of the settlement in the county's atlas, someone from the county court, presumably the magistrate, a legal secretary, or a clerk, employed a geomantic compass to compose a map (figure 2.9) of Pig's Trough Pass during the adjudication of the Lis' dispute. Though its exact place of composition and authorship is unstated, the map was cited by the magistrate in his verdict. The mapper began by drawing the dragon vein of the Li lineage, which extended through all the properties owned by Li lineage agnates, with temples

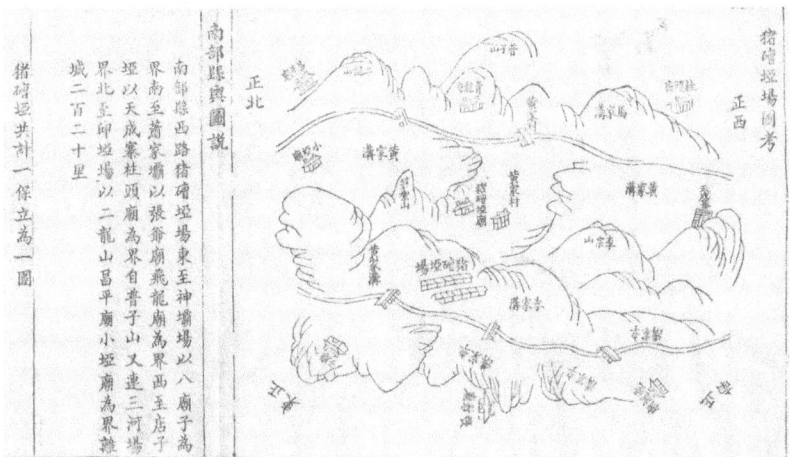

FIGURE 2.8. Atlas Map of Pig's Trough Pass. Pig's Trough Pass was a remote market town accessed by the north road out of Nanbu's county seat. The Li lineage dominated the area south of the central market and its namesake temple. The lower river and its leftmost bridge, as well as the boundary-marking religious structures of the "Eighth" Temple, Zhang Fei Temple, and Flying Dragon Temple were all depicted in the court's map of the site along and within the Li lineage's dragon vein (See Figure 2.9). Map from *Nanbu xian yutu shuo* (1853/1869/1896).

patronized by the lineages forming the contours of the vein. Reflecting the overlapping geographies of consequence mentioned above, the Li lineage's dragon vein precisely overlapped with the southern and eastern boundaries of the market town recorded in the county atlas.

Zaitang's parents were not buried in the common ancestral cemetery on the Li ancestral mountain but rather on a separate, individual mountain—a spot that evoked the power of Zaitang's family and his differentiated fortune from other lineage branches. The three elevated semicircles in the center of the map represented the graves. The tomb farthest left was Zaitang's mother's grave, next to which on the right and center was his father's grave. Farthest right was an unlabeled gravesite that was not central to this geomantic dispute. Zaitang's description of his ancestral graves revealed to the magistrate his elevated status and wealth in the family. Not only were these graves surrounded by a grove of lush Qinggang Oaks, but a river ran across the land, bending just before Zaitang's ancestral graves. This natural feature allowed the water to pool in front of his father's grave, securing fortune for Zaitang's family. The mapper also ascertained the location of the millstone, which was located close to a

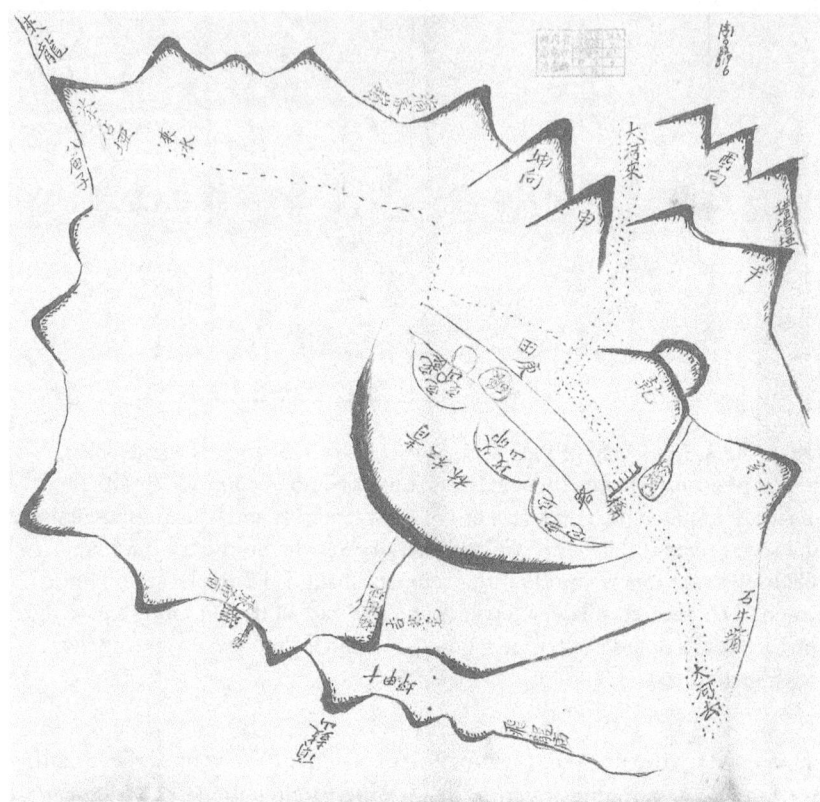

FIGURE 2.9. Judicial Map of Li Dragon Vein. This annotated map of the Li lineage's dragon vein covers the bottom third of the market town depicted in Figure 2.8. The map's top right corner marks the central market of Pig's Trough Pass. The top left corner of the map contains the words "Arriving Dragon" (*lailong*). The undulating line, representing the dragon vein, continued (rightward) in two directions. Below, the vein passed through the three temples seen in the lower edge of Figure 2.8. Above, the vein traveled through the key mountains used to orient the graves. These peaks were labeled with the *kun*, *shen*, *you*, *xin*, *qian*, *hai*, and *ren* directions (see Figure 2.10). The graves are seen in the center of the map, represented by three continuous semicircles. Water, depicted on the map by a stream of dots, flowed toward the graves, traveled under the bridge, and then continued flowing through the bottom right side of the image. The circle with the character *nian* located at the center of the map marked the location of the millstone. It was labeled as lying in the *geng* direction. Map from Nanbu County Qing Archive.

large cypress tree—a significant detail that would later matter for the magistrate's verdict.

Zaitang told the mapper that his father's grave lay on a *mao* mountain facing the *you* direction. Confirming the direction with a compass, the mapper identified the distant mountain located in the *you* direction. This was the central peak anchoring the father's grave. Zaitang's mother's grave was divined on the *gen* mountain facing the *kun* direction. These directions were also appended to the map. With these two compass directions fixed, the mapper turned to the surrounding mountains, identifying peaks locating at *qian* and *xin* directions on the compass. An ideal, diagrammed compass that explicates the mapper's analysis can be seen in figure 2.10.

Then came the hard part. The millstone was situated in front of, but slightly between, the graves of Zaitang's parents. With the father's grave facing *you* and the mother's grave facing *kun*, only two possible compass directions remained between them: *shen* and *geng*. The compass identified a distant mountain slightly left of the millstone as *shen*. Suddenly, something remarkable happened. The mapper ascribed the only remaining possible direction on the compass, *geng*, to the millstone itself. The *geng* direction was a heavenly stem envisioned as directly next to *you* but distinct from *you*. The compass's needle could not lie.

In front of eight representatives of the Li lineage, the mapper, who was presumably the county magistrate or one of his clerks, proved through annotated cartography and careful geomantic calculation that the millstone did not hinder either of Zaitang's parents' graves. With Zaitang's fortunes firmly secured, the millstone could remain in place for use by all lineage members. To make the point clear to Zaitang, the magistrate in his final ruling focused on the unmarked cypress tree shown on the map next to Zaitang's residence and reproached him for not previously realizing its geomantic significance:

> The court rules ... Li Jiran and Li Zailin (another relative) must return home and clearly and earnestly settle each person's accounts with Li Zaitang with the lineage as a witness; the return of the funds should be done in a set time limit within the year. Regarding Li Zaitang's seizing of the common millstone, it should remain for the common use in its previous location. By the entrance of Li Zaitang's house is a large cypress tree that is forbidden to be disturbed— it may not be cut down. I order Li Zaitang to consider the relationships within the lineage. It is not necessary to take the ordering of accounts that seriously. Both sides pledge to follow this ruling, return to their homes, settle the family finances, and avoid harming peace and harmony.[89]

According to the magistrate, that cypress—unlike the millstone—was influential for his fengshui and thus needed careful protection. The county court in Sichuan, just like Magistrate Fan's court in Shaanxi Province, drew on

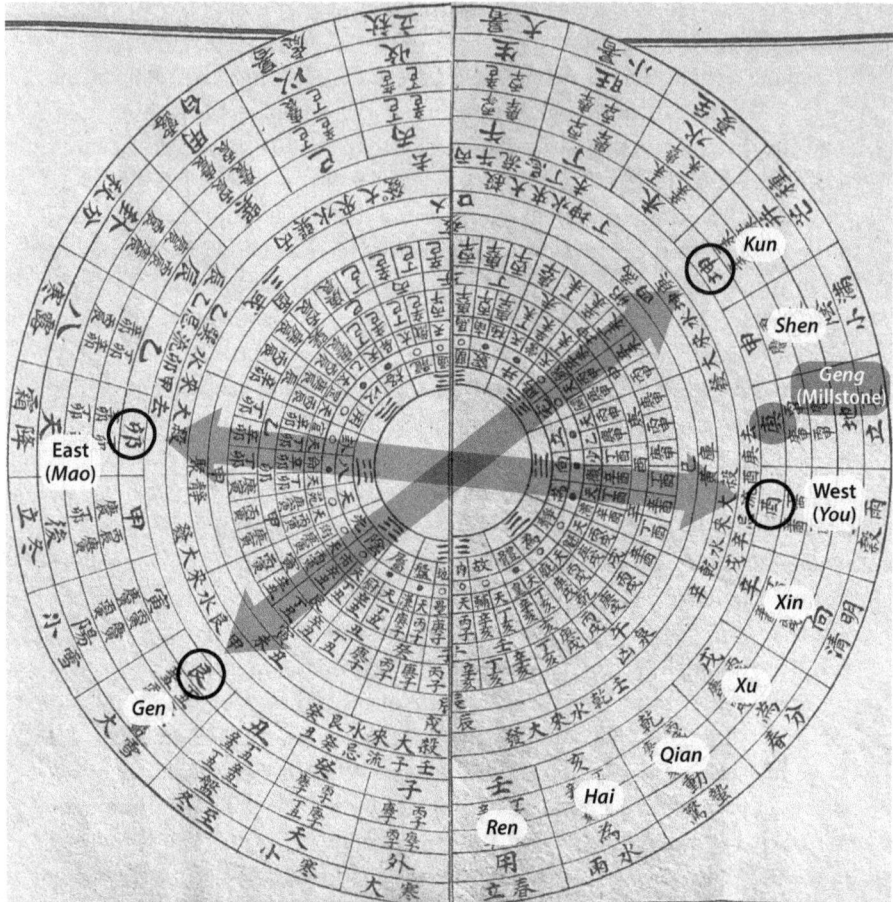

FIGURE 2.10. Compass Annotated for Mao Facing You with Geng Highlighted. On the compass seen above, the two arrows represent the grave orientations of Li Zaitang's parents: *mao* facing *you*, and *gen* facing *kun*. The creator of the map in Figure 2.9 labeled the distant mountains with the *kun, shen, you, xin, qian, hai,* and *ren* directions. By ascribing the *geng* direction to the millstone, the mapper demonstrated that it did not hurt the fengshui of the existing graves. Diagram from Kong Wenxing, *Huitu dili zhizhi yuanzhen* (1692/1875–1912).

cartography to ultimately discover that a litigant had misinterpreted the geomantic significance of a grave. There are even further resonances. Recall the capital appeal of the Cai lineage discussed in chapter 1. There, as in this case, Qing officials scolded people for failing to cherish the trees around key dwellings.

These two cases involving judicial maps that were created to appraise grave fengshui reveal stark realities about rural life in the nineteenth century. Wealthy elites unabashedly invoked fengshui to control the environments around prominent gravesites and even around lands deemed auspicious without graves. In the first case from Shaanxi, a member of the elite attempted to block the construction of a neighbor's wall by referencing a future burial spot. In the second case from Sichuan, a wealthy member of the Li lineage attempted to move a millstone for his personal benefit, essentially claiming the stone for his own family against other branches of the lineage. One imagines that this phenomenon occurred more often than the legal record suggests and that elites got away with it if courts did not intervene.

In both the Shaanxi and Sichuan cases, the county magistrates knew fengshui well, and they knew they could win arguments and enforce the state's agenda with that knowledge. Through analyzing fengshui, officials could disarm bullies by proving that their claims were wrong. Hence, courts mapped grave lands in dispute, carefully analyzed them, and attempted to redirect the cosmic forces at play along rural landscapes toward ends the state deemed ethical.

To be sure, expedience at times led Sichuan's courts to lean toward the interest of wealthy local families. A Daoguang-era (1821–51) case from Jingyan County, located about three hundred kilometers southwest of Nanbu, showcases a completely different outcome from the Li trial. There, the Hu lineage claimed that a newly dredged waterway and watermill, which were built by a person not local to the county, hurt the fengshui of their ancestral graves. The court investigated the matter and ultimately gave the Hus control over the mill.[90] The outcome may have been related to the fact that the builder of the mill was not a local person, or perhaps the court truly did find the Hus' fengshui disturbed. Maybe it was both. But this case appears in a local gazetteer, which suggests that not only was the verdict widely known, but also that the Hus were quite influential in Jingyan County.[91] The Hus made sure that the literati compilers included that verdict in the gazetteer's contents—perhaps to remind incoming magistrates about the legal precedents that mattered to them.

Considering different genres of historical material together allows for a broader picture of property dynamics in nineteenth-century Sichuan. Archival case files containing state documents naturally project an image of magistrates who were fully in control of trial proceedings and completely free of corruption. Yet, from judicial compendia of higher-ranking officials like Prefect Xu Shilin, who amended the verdicts of local magistrates, to information in local gazetteers, to accounts in elite writings on statecraft, and to popular novels and plays, one gets the distinct impression that a county yamen's projection of moral purity may have not always been present on the ground, especially in the decades of weakening state power during the 1800s.[92] Magistrates could sometimes challenge elites and reject their expansive claims on the basis of fengshui, but it is clear that they could just as easily recognize and uphold them.

Lest one gets the impression that officials understood fengshui as a method to expediently justify a verdict they already had in mind, it is important to affirm that there are many cases where fengshui was sincerely analyzed by Qing officials and served as the means through which they arrived at conclusions. This practice was not limited to county magistrates or resident administrators. Cases preserved in the published judicial comments of Li Rong (1819–90), a native of Sichuan's Baoning Prefecture who served as lieutenant governor of Hunan, record the official's extensive engagements with legal disputes involving fengshui. By Li Rong's own admission, he was talented at fengshui, and readily provided advice from the bench.

In one judicial comment on an appeals case, Li Rong carefully explained that the site of a newly created grave for the mother of a man named Li Changgeng was extremely inauspicious. After extrapolating on the landforms, orientations, and *qi* flows around the site through examining a judicial map, the judge commented, "Whether Li Changgeng moves the newly created grave of his mother, in three years it will be clear whether this site is auspicious or not; if my words are verified, thus living up to my reputation for being good at fengshui, it will be too late for you to believe in them."[93] To understand this judicial comment, one may recall the verdict rendered by a Nanbu magistrate in the 1879 over termites in poorly drained soil: according to well-established laws of the land, a grave's geomantic impact on living descendants was thought to be revealed to a family within three years. More important for our purposes is the fact that Li Rong—like other Qing officials—directly practiced fengshui for resolving a case.

Several patterns emerge from Sichuan's archives and published judicial compendia. Lands in dispute because of fengshui were regularly mapped for consideration by courts across the Qing bureaucracy and across China's eighteen provinces. Some magistrates who claimed to not believe in fengshui possessed an uncanny ability to break down the elements of a landscape and carefully explain their significance to families in court. At times, officials cited fengshui to justify a desired outcome, and other times they actively decided cases based on their understandings of fengshui. Even in cases of the former, it is still notable that magistrates felt compelled to frame their verdicts with reference to fengshui rather than to Qing law. When fengshui did influence an outcome, it was up to officials to determine which information—directions, orientations, tree cover, and so—was compelling and believable. The methods Qing courts employed to interpret these maps of land lay bare the ways people interpreted the world around them and the questions that mattered to their lives.

Mapping Family Genealogies

When a magistrate arrived in a county to assume office, he would want to read a bit about the place he was assigned to serve. Magistrates serving in Jiangnan had plenty of materials to consult, including local gazetteers and the genealogical prefaces of well-to-do families. Competition between elite families in southeastern China meant that the quality of these sources was higher than in other regions, and extravagant claims could only be pushed so far before other elite families called foul.

Prior to the 1840s, none of those textual sources existed in Nanbu. The county had seen the publication of a gazetteer during the Ming, but it had been lost during the wars of conquest in the seventeenth century. Several magistrates had attempted to organize committees to compile a local gazetteer, but none had borne fruit. For the first two centuries of Qing rule, the county government had little information about the area's hinterland, which had been completely resettled during the 1700s. In the decades after the White Lotus Rebellion (1796–1804), however, local authorities redoubled their efforts to shore up the county's mutual surveillance and security units, and in the process of gathering strategic information, produced some written and cartographic information about the territories they governed.

Starting in the 1840s, there was a flurry of publishing activity. The first surviving gazetteer of the county was published in 1849, followed shortly thereafter in 1853 by an atlas of the county's market towns, temples, surveillance units, and rural militias officially commissioned just after the outbreak of the Taiping Civil War (1850–64). Not long after, in 1858, the discovery of a long-lost genealogy of the Chen family was announced. This genealogy was the most notable one published in the county during the Qing.

The Chens had burst onto the scene with a remarkable tale. Surviving several dynastic transitions, the Chens alleged to have been resident in Nanbu since the Song dynasty (960–1279), with the fengshui of tombs mapped and annotated in their genealogy to prove it. Their favorable fengshui was joined in print with many names of virtuous women associated with the family. A few Chens sat on the gazetteer publication committee and made sure key information related to their eminent ancestors, ancient graves, and virtuous women was all incorporated into its contents.[94] The government came to accept the Chen claims as valid and issued orders for the protection of the tombs allegedly dating to the Song dynasty.

In late imperial China, there was no word exactly corresponding to "geography" in the modern sense. One counterpart was *dili* (earthly principles), which had two principal meanings. The first evoked the administrative names of counties and prefectures, or the names of mountains, rivers, and other prominent features of an area.[95] The second was the study of the land's cosmic powers, i.e., fengshui. The closer we look at geographic knowledge of Nanbu during the Qing, the more these two aspects of *dili* become blurred, even inseparable. This section examines the relationship between the three sources published in the county between 1849 and 1858—a gazetteer, an atlas, and a genealogy—to reveal that the geomantic claims of prominent lineages shaped semiofficial and official records of the county.

Since Nanbu witnessed a strong trend in ancestral hall construction during the 1800s, with dozens built across the county in that century, we can assume that some genealogies that were once created have not survived.[96] The genealogies that do survive from Sichuan typically contain geomantic maps of ancestral territories. The Lis of Cangxi, a county neighboring Nanbu, justified the mapping of their lineage's dragon vein by emphasizing that their map would prevent "people with different surnames" from invading their territory.[97] The felling of trees, quarrying of stones, and creation of new burial sites were all policed by the Lis within the territories demarcated by their dragon vein. The publication of a genealogy was an opportunity for a lineage to stake such claims in public and in print. For lineages with landholdings that spanned several townships and whose illustrious surnames burnished dozens or hundreds of property deeds, these genealogical records conveyed the big picture of ancestral settlement.

The Chens' ancestral settlement included thirty sites where they claimed gravesites dating from the Song, Ming, and Qing dynasties. The Chens' geomantic diagramming of land portrayed these tombs with visual depictions of auspicious landforms and dragon veins, in the same way that the Lis had described their holdings for the court's map of Pig's Trough Pass. Such maps conveyed vital information by construing elements of a landscape—dimensions, orientations, and placements—to showcase a hierarchy of importance within a web of environmental relationships. Prominent features on these maps included old tombs, new graves, courtyard houses, ancestral halls, prominent trees, bridges, millstones, stone forts, and flowing water. Visually adjoined mountains, temples, and trees explained the success of the lineage over time, providing cartographic evidence for the Chens' harnessing of the natural landscape to sustain the physical structures attesting to their family's story.

One geomantic map in the genealogy depicted two ancient trees with labels dating them to the Tang (618–907) and Song periods. These trees served as physical evidence to claim that this was the resting place of Chen Yaozi (970–1034), a first-ranked palace degree-holder (*zhuangyuan*) from the Song dynasty, whom the lineage claimed as an ancestor.[98] The same location for the trees was also offered in Nanbu's county gazetteer.[99] It was essential for the Qing-era Chens to identify at least one ancient tree as dating from the late Tang dynasty to prove the grave was Chen Yaozi's, since the genealogy's Qing-era compilers assumed that his family would have demanded a spot with good fengshui—marked by an already ancient tree—by the time of his burial during the Song dynasty.

As previously mentioned, ancient trees were strong signifiers of social status in Sichuan, and the genealogy implored descendants to plant trees on any barren hillsides linked by a Chen dragon vein as investments for future fortune.[100] Other genealogies of elite Sichuanese lineages forbid descendants from cutting these trees down "when building manors and mansions," with the implication that allowing the trees to grow to old age was a powerful symbol of elite status.[101] Much like the lineages examined by Michael Szonyi, Sichuan's families cited a search for "good fengshui" as a narrative device to explain their migration to the province during the Ming or early Qing dynasties.[102] The case of the Chens suggests that at least some lineages fostered fengshui as tangible evidence for those historical narratives. When property disputes arose within the territory of a mapped dragon vein, the Chens could invoke fengshui to present a claim and try their luck in court. Legal negotiations over landed property articulated in such terms are documented in the genealogy.[103]

The Chen geomantic maps also emphasized the size and dimensions of certain landscape elements as well as the relationships between sites present on the land. In one rendering (figure 2.11), titled "A Map of the *Yinzhai* and *Yangzhai* of the Chen Lineage in Building Prosperity Market," the Chens traced ancestral properties along adjoined undulating lines, representing veins and auspicious landforms. In a manner broadly resonant with the Huizhou geomantic map (figure 2.1), houses, graves, and ancestral halls were enveloped within the embracing protection of flowing, connected landforms. Showcasing the Chen lineage's influence in the town, the central market itself was depicted in modest dimensions in the upper right of the image, next to the town's namesake temple, Building Abundance. By contrast, the Chen Family Corner and ancestral hall were generously depicted, with approximately the same dimensions as the town's central market.[104]

The temple to Erlang—a patron deity of the province and a favorite device of the Chens for displaying their ancient Sichuanese heritage—appeared at the center of the map, with additional notations for the graves, houses, and the branch's ancestral hall. Also included was Protecting the People Nunnery, which the Chens financially supported during the White Lotus Rebellion to supplicate for the surrender of the rebels—a deed also recorded in the county's gazetteer.[105] It was not the intention of the Chens to diminish the importance of the town's central market or its namesake temple; above all,

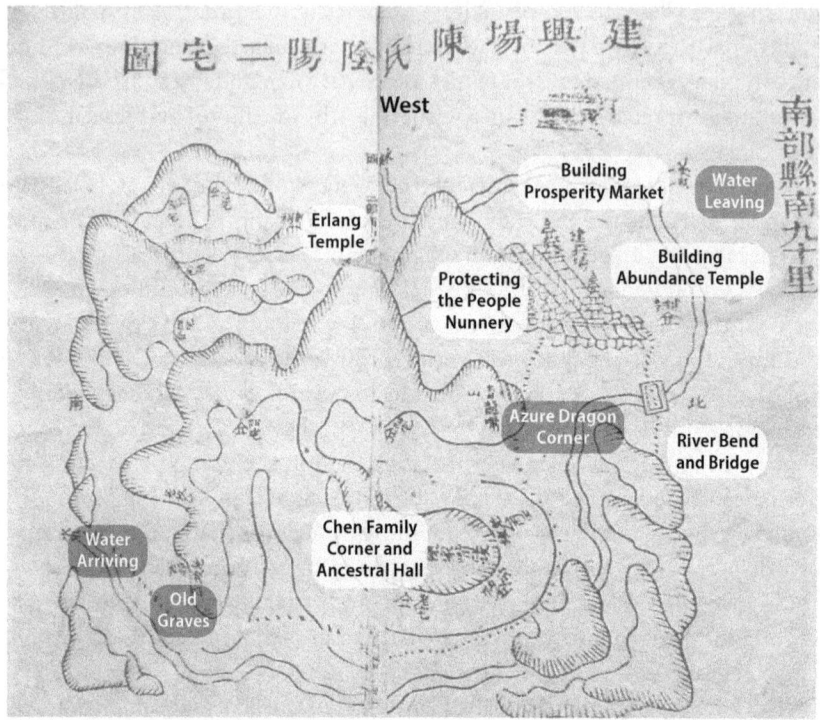

FIGURE 2.11. Chen Genealogical Map of Building Prosperity Market Town. This map showcases the Chen lineage's fengshui in Building Prosperity Market Town. The top of the map marks the western direction, while the bottom marks the east. The Chen Family Corner and Ancestral Hall are centered at the bottom of the map, just below the Azure Dragon Corner, while graves and houses dot across the undulating vein. Erlang Temple is centered above the Azure Dragon Corner and Chen Family Corner. A water source, Tall Building River, was drawn as "arriving" (directional flow) in the lower left side of the map. The river bends at the lower right side before continuing under a bridge and across Building Prosperity Market, which is shown in the upper right corner. Map from Chen Tingxian, *Sichuan Nanbu xian Chenshi zupu* (1858).

they convey that their lineage had helped these sites, which were important for all town residents, to prosper.

As in almost all maps of geomantic landscapes, the Chen genealogical maps emphasized the sites' water sources and their directional flow, specifically, the direction by which water arrived and departed from the central site. At Building Prosperity Market Town, the local water source was Tall Tower River, which was pictured in the Chen genealogy without straight lines as required by the tenets of fengshui. Every bend in the river was thought to pool good fortune for a family. Where rivers were not in evidence, irrigation ditches were mapped in the genealogy to fulfill the requirements of fengshui. Although classical geomantic texts recommended naturally flowing water sources—streams and rivers—near significant sites such as tombs, the abovementioned cartographic details reveal how geomantic theory was adjusted through the centuries, as families sought to meet auspicious requirements for settlements in landscapes transformed by centuries of intensive agriculture.

A final detail for understanding how these geomantic illustrations were intended to be read lies in their orientations. Many of the lineage's geomantic maps were drawn on a west-east axis, with the west at the top and the east at the bottom. The logic of this orientation was likely related to the fact that gravesites were often oriented on such an axis, represented respectively by the White Tiger and Azure Dragon. House geomancy, by contrast, was divined on a south-north axis with most major buildings situated in the north facing south—a principle related to the maximum capture of natural sunlight since China is in the northern hemisphere. This principle reflected a "standard" fengshui model, but there were variations. The geomantic map of the Bao ancestral halls from Huizhou (figure 2.1) was oriented on a north-south axis, and other renderings placed south at the top. Geomantic maps fit no single model: a diversity of styles, and an image's orientation and scale, depended on the features that an illustrator wished to highlight.

The image of figure 2.11 can be compared with a map of Building Prosperity Market Town from the county atlas (figure 2.12). The two images were created around the same time. The state's cartographic rendering of Building Prosperity Market Town was not explicitly geomantic, and many places other than the Chen Family Corner were depicted, including settlements named for the Lis, Hus, Zhaos, Yangs, Dus, Huangs, and Wus. As in the case of Pig's Trough Pass market town, the atlas identified temples as the recognized boundaries of Building Prosperity and its security units. Most intriguing is the fact that the county atlas reproduced many of the sites that were included in the genealogy as encompassed within the Chen dragon vein, including the Chen Family Corner, Erlang Temple, Protecting the People Nunnery, and local water sources.

What did county authorities choose to include in the atlas—created during an effort to strengthen and map rural security units—and why? Obviously, the atlas emphasized prominent temples, mountains, bridges, shrines, and trees, which were the same features well-to-do lineages tended to represent in their genealogies. But the government had a particular interest in strategic defense. This interest is seen in a chapter on mutual surveillance units copied into the collection, *Compilation of Writings on the Statecraft of Our August Dynasty* (1826). The chapter contains a passage from the early Qing official Huang

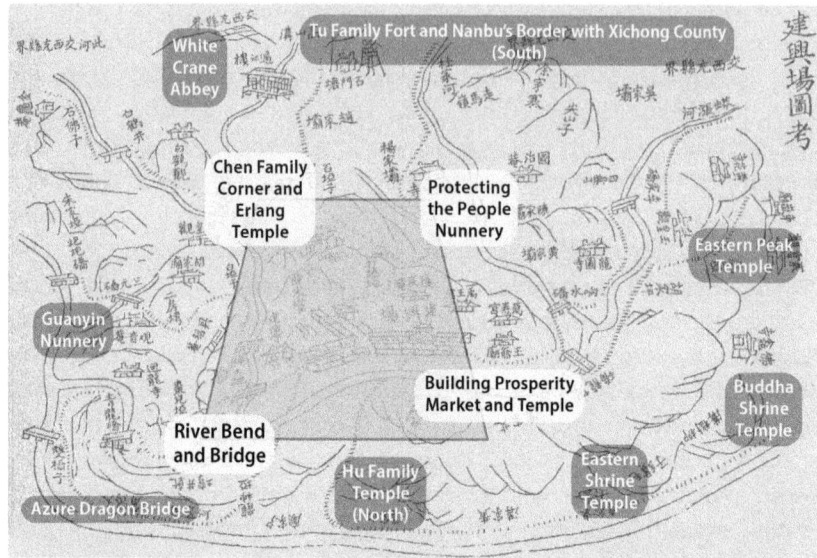

FIGURE 2.12. Atlas Map of Building Prosperity Market. In contrast to Pig's Trough Pass (Figure 2.8), Building Prosperity Market Town was large and densely populated. The town was divided into eight mutual surveillance units (*bao*), each forming their own militia (*tuan*). The Chens occupied central lands in the town. The area covered in the geomantic map seen in Figure 2.11 occupies the left-of-center portion of this map. The atlas depicts the Chen Family Corner, Erlang Temple, Protecting the People Nunnery, and even a local bridge in roughly the same locations as seen in the geomantic map. Note that the top of this map in the atlas represents south, and the right side of the map is west. Map from *Nanbu xian yutu shuo* (1853/1869/1896).

Liuhong that recommended specifically surveying fengshui for the construction of defensive walls and moats.[106] In that essay, Huang was referring to natural features of the landscape—specifically, strategically elevated points.

With Huang Liuhong's administrative recommendations in mind, the contents of Nanbu's atlas make sense. As local control became more difficult for officials to maintain, managing rural surveillance units involved mapping both natural and human landscapes, in no small part because elite families with the funds and resources to organize security units and militias were also the ones who claimed good fengshui. Accordingly, as the government sought to shore up control through security and surveillance units, it needed more information about the region's geography.[107] That information largely came from the participation of elite families and local gentry who already exerted significant control over rural landscapes.

But in an important way, the Chens also needed the Qing state. The Nanbu Chens had no relation to the famous Song statesmen Chen Yaozuo (963–1044), Chen Yaosou (961–1027), or Chen Yaozi. None of the tombs they claimed dated to the Song dynasty, and a close reading of the historical record suggests that the Song-era Chens were never buried in Sichuan. Imperial sources such as *Records of the Unity of the Great Ming* provided contradictory locations for the Chen burial sites, offering both Henan Province near the old Song capital of Kaifeng and Northern Sichuan in different chapters.[108] As Cai Dongzhou has demonstrated, the Henan location is the historical site of their resting places considering the strong epigraphic burial record there.[109] By contrast, Sichuan's local sources were never consistent with where, exactly, or how many Chen tombs existed. Nonetheless, ambiguity and contradiction in voluminous imperial histories allowed an enterprising lineage to exploit the possibility of a Sichuan burial site.

Following the resettlement of Sichuan in the decades after the Ming-Qing transition, a family named Chen took the Song dynasty brothers as their ancestors and identified ancient graves in the county as their own. Notwithstanding the imperial histories that the Qing-era genealogy liberally cited, most of the genealogy's earliest transcribed local sources dated to the early nineteenth century, when lineage leaders got involved in establishing ancestral halls and shaping the family narrative.[110] As seen in the previous chapter, commoners sometimes claimed ancient graves as their own property and were challenged in court for doing so; here, people at the elite end of the socioeconomic spectrum engaged in the same practice on a grander scale and with a much greater likelihood of success.

When the Nanbu county gazetteer was compiled in the mid-nineteenth century, these ancestral claims came to pervade its contents and served as the foundation for the genealogy, parts of which may have predated the gazetteer's publication.[111] The mountains listed in the gazetteer's section on territory profiled many auspicious landforms and peaks that were considered to have geomantic or religious significance, including the alleged burial sites of the Chens and other elite families.[112] Several prominent bridges in the county were named in reference to the Chens, and three separate genealogical prefaces, allegedly dating to the Song, were included in the prose section of the gazetteer.[113] The same three prefaces also appear in the genealogy. The Chens drew on local myths that were already present in Sichuan and elevated the county's folklore to new heights.

The Chen's family influence also extended into public records of gender and female virtue. Cosmologically, fengshui and marriage matchmaking were closely linked via a person's eight characters (*bazi*) of birth time, which were thought to determine a person's fate. As it turns out, those with good fengshui in Qing society tended to obtain good marriages. No less than seventeen daughters and wives surnamed Chen were recognized as exemplary women for their chastity, virtue, and loyalty in Nanbu's gazetteer—the second highest number for any surname in the county.[114]

Connections among female virtue, women's health, and fengshui were well recognized. Geomantic manuals and medical treatises alike emphasized these links.[115] To give just one example, an undated fengshui manual from Sichuan, *Preface to the Origins of Earthly Principles*, described one inauspicious grave configuration by writing that "the men of the family will become bandits and women will become prostitutes; the family's fate will involve military exile, strangulation, or decapitation, and after conviction there will be no imperial pardon [during the Autumn Assizes]."[116] The genealogy and gazetteer demonstrate that elite lineages like the Chens owned the greatest number of ancient tombs, enjoyed some of the best fengshui, claimed connections to the county's oldest trees, and reared women who were recognized for their virtuous conduct at eyebrow-raising rates. In the nineteenth century, they also managed to get it all down in print.

Over the 1800s, the county government came to adopt the Chen genealogical claims in its annual reports on ancient tombs to Baoning Prefecture, with the Chen ancient tombs receiving orders of protection.[117] In these governmental reports, there was never factual consistency on where exactly the tombs were located or even how many there were, but the Baoning prefect seldom pointed out the discrepancies in submitted files. Only once—in 1875—did a prefect reply that the county had not supplied the tomb maps required by law.[118] In truth, there were no grand Song-era mausolea in Nanbu to illustrate; what existed was a patchwork of recently created Chen tombs and some generously illustrated fengshui maps. The power of the Chen genealogy derived not from its accurate telling of history, but from records of graves and fengshui that turned ancestral myths into geographical facts on the ground.

Admittedly, the Chens were not a typical lineage in Sichuan, where most kinship groups did not compile or publish such extensive genealogies. But examining the contents of the county atlas, one suspects the Chens were not alone in successfully transforming lineage landscapes into county geography. The Lis of Pig's Trough Pass Market Town, the Chens of Building Prosperity Market Town, and the Xianyus of Numinous Treasure Mountain Market Town—the latter of whom claimed gravesites allegedly dating to the Tang dynasty that were also listed in the gazetteer—all participated in the revamping of the county's security and surveillance units in the decades after the White Lotus Rebellion, but they also got something in return. In ways large or small, the elite's claims to land as marked by graves, trees, shrines, and temples were recognized by the Qing government, which lent power to their ancestral claims.

Mapping Temples and Shrines

As previously discussed, because of Sichuan's distinct history of migratory resettlement during the Qing, Buddhist temples and Daoist shrines patronized by networks of resident families came to serve important roles in rural organization. These roles were reflected in Nanbu's atlas, which mapped many temples that served as marketing centers and administrative boundary markers for mutual surveillance units and local militias. Temple inscriptions from across Sichuan highlighted the geomantic and religious significance of their locations as well as their importance for boundary and jurisdiction markers.[119] The spatial layouts of temples crept into administrative geographies because no other structures were of comparable size and importance to serve such functions for the state.

The trouble was, the county government had no comprehensive record of temples. Nanbu's gazetteer offered the names and locations of around fifty prominent temples and nunneries, but, in truth, there were hundreds of temples across the county. The county atlas provided much more detail than the gazetteer, but even it was not comprehensive. Neither of those sources accounted much for the changes of rural landscapes over time, for instance, with the building of new religious structures or the cutting down of ancient trees. These changes mattered—for religious reasons, obviously, but also for security units, militia organization, and a range of other services with which the state had to concern itself.

As readers might anticipate, the geomantic properties of these temples could be manipulated in law by interested parties, just as graves were. Because large Buddhist temples were patronized by groups of families in Sichuan, this manipulation required a substantial degree of coordination between allied kinship groups, but it happened. An 1857 incident captures one such case. That year, residents around Gaoguan Temple traveled to the yamen of Baoning Prefecture with a dramatic claim.[120] Around the Buddhist temple grew a large grove of cypress trees that was connected to the fengshui of five resident lineages. The five lineages had been longstanding patrons of the temple, and some had ancestors buried around the monastic grounds. They contended that the resident Zhaos, one of the five lineages, wished to cut down sixty-five of the cypresses and sell them to timber merchants.

A Nanbu magistrate had ruled that the temple's fengshui needed to be protected three years earlier, but the Zhaos claimed that this verdict did not apply since, in their view, the temple and surrounding community were in Langzhong County, outside of Nanbu's jurisdiction.[121] The Zhaos had already taken the matter up with Langzhong County's government, which had signaled its willingness to let the Zhaos cut the trees down with the understanding that Gaoguan Temple was in Langzhong. The four remaining families resolved to take the issue up with the prefect of Baoning, who oversaw governance of both Langzhong and Nanbu counties.

The arguments of the Zhaos and the other four surnames clarify how locals understood the Qing government's relationship to fengshui. Both the Zhaos and the other families agreed on one principle: an order of fengshui protection or permission to cut fengshui trees down could only be granted by the imperial state. The question was who had the authority to rule on the matter, especially when two neighboring county governments had offered contradictory resolutions. There was no official record of the community around Gaoguan Temple in the gazetteers of either county. The temple had not been illustrated or mentioned in Nanbu's atlas, which had only been created four years earlier, when the temple surely had already existed. Now that the government had to answer a question about the temple's fengshui, it first had to ascertain where Gaoguan Temple was located.

The prefect dispatched a clerk to analyze the site and determine its county affiliation.[122] Upon surveying the temple and its surrounding grove of cypress trees and gravesites, the clerk reported that the temple was in Nanbu County but located less than one Chinese mile (around one-third of an English mile) from the Langzhong border. While the clerk's report does not reveal exactly how he reached this conclusion, we do know that tax records were not part of his calculus, since many families had registered a bit of land with both counties to obtain access to two law courts in the event of disputes—like the one at hand.[123] Upon receiving the clerk's report, the Baoning prefect transferred the case back to Nanbu, and a new order for the protection of fengshui was issued by the magistrate.[124]

Nearly a century later, in 1942, land surveyors in Sichuan stumbled upon the rural community of Gaoguan Temple.[125] They discovered that the locals—all named Zhao, by then the most powerful lineage in the area—claimed to live in Nanbu County and pointed to their namesake temple as evidence. The surveyors found this claim perplexing, since the rural settlements surrounding the temple were all located in Langzhong, well over a kilometer from the county's border with Nanbu. What happened next is anyone's guess. The surveyors may have discovered the notice of fengshui protection from the Nanbu magistrate and filled in the gaps accordingly. In any case, the surveyors set out to transfer the settlement to Langzhong, where it remains to this day. But for nearly a hundred years (1857–1942), an interpretation of a temple's fengshui by a county clerk had enabled a community to protect a temple grove and claim that they lived in Nanbu even though they inhabited an enclave surrounded by the territory of the neighboring county.

Gaoguan Temple offers a striking example of the imperial government's increased reliance on local knowledge for administration in the nineteenth century. It also complements previous legal cases that saw maps created by state affiliates for county-level trials, as well as the official recognition of the fictional Song-era Chen tombs. Many officially commissioned maps were not intended for long-term instrumental use or to sustain an expanding body of geographic information. The atlas of the county's hinterlands somewhat approached those goals, but even its maps were incomplete and scarcely updated in the decades after its printing in the 1850s. The primary goal in the county's commissioning of maps was to address immediate problems (e.g., militia organization), resolve immediate conflicts (e.g., legal disputes), or perfunctorily fulfill an annual regulatory requirement (e.g., mapping significant tombs). The uneasy tension present in local governance was that most maps were not intended for long-term reference, even as the judgments drawn from them could have long-term consequences.[126]

Mapping Qing Law

Over the twentieth century, some historians posited a decline in cosmological mapping in China following the arrival of the Jesuits, who brought Western cartographic techniques to the imperial courts of the Ming and Qing dynasties.[127] Historians later challenged this view, arguing that the limited appropriation of specific Western mapping techniques in late imperial times did nothing to diminish the production of cosmologically oriented maps.[128] When considered together, technical geomantic diagrams, genealogical illustrations of territory, gazetteer maps of auspicious sites and astronomical constellations, judicial maps, and maps from the imperial palace represent an enormous body of cartographic artifacts. Historians of science may well ask: what forces motivated people across social classes to map, record, and analyze the natural world during the Qing? Fengshui was not the only factor, but it was a significant one and remained so through the last century of the dynasty, both inside and outside the legal system.

Why was that the case? Fengshui did not remain a significant theme in Qing cartography out of deference to ancient tradition. The picture that emerges from the mid-1800s is hardly one of an unchanging "folk culture." While geomantic mapping could and sometimes did focus on representing idealized landscapes, there were many grounded, practical incentives for private actors and the Qing government to map sites associated with fengshui.

The significant number of genealogical maps produced during the dynasty, with many published during the nineteenth century, leads to the conclusion that elite families were not innocently motivated to illustrate beautiful landscapes for their private enjoyment. There was considerable money, influence, and power on the line in the private production of genealogical maps of ancestral graves and ritual fields. Wealth was invested, hidden, and mystified within the sprawling landscapes of actual or imagined ancestors. These maps served as evidence at trials in regions where genealogies were common, despite the law code's officially banning their consideration. Even in Sichuan, where comprehensive genealogies were less common than in Jiangnan, the geomantic maps published in the nineteenth century had considerable political influence due to a lack of alternative written records.

As geomantic maps of territories proliferated in private genealogies, the Qing state needed to act to maintain order and project its unchallengeable authority. Official handbooks produced during the eighteenth century offered a pragmatic solution: the court-commissioned mapping of landscapes in dispute because of fengshui. These recommendations were implemented at the county level across Sichuan, where courts consistently mapped and analyzed grave, house, and temple environments to the very end of the dynasty. The Qing demonstrated its moral obligation to the people through judicial mapping, showing them that the government did not tolerate illicit tampering with gravesites, temples, and ancestral halls. By mapping lands in dispute, officials could draw on their knowledge to analyze rural landscapes, neutralize charlatans, unmask fraudsters, and assuage worried litigants with informed, persuasive knowledge that their children were safe and their fortunes were secure. Even some seemingly mundane grave and house infringements were painstakingly mapped and thoroughly investigated by the court, leading one to conclude that nineteenth-century authorities did not find those cases to be trivial.

A good number of related cases drew upon information contained in the diagrams and descriptions of fengshui manuals. These sources circulated widely as guides for general consumption, with many published by commercial presses or written in manuscript form.[129] Though one could say more about this topic, the decentralizing of elite fengshui knowledge over the Ming and Qing dynasties through the proliferation of commercially published manuals may have paralleled trends in pharmacological and medical knowledge, which was also driven by private investigation over that time. Anxiety over diagnostic theory and skill spanned the spectrum of recognized knowledge. Following centuries of private studies in medical pharmacology, the Qing court under the Qianlong Emperor issued the *Golden Mirror of Medical Orthodoxy* in 1739, which, among many things, "became a venue of legitimization of preexisting regional trends of medical practice."[130]

The existence of a plethora of private geomantic manuals likewise appears to have galvanized the state to issue the *Imperially Commissioned and Sanctioned Book of Harmonizing the Times and Distinguishing the Directions* that same year (1739) and compile its own fengshui manual, *Correct Doctrines of Fengshui*, which synthesized the Form, Compass, and other regional schools of geomantic practice into a single guide in 1740. These works were all compiled in the first decade of the Qianlong reign, which, in the words of Marta Hanson, was marked by an "obsession with defining orthodoxy (*zheng*) in all realms of Chinese knowledge as a tool of Manchu control over both Chinese culture and the Chinese [people]."[131] The existence of these official texts underscored the government's role as the final authority on related knowledge.

The degree to which *Imperially Endorsed Treatise on Harmonizing Times and Distinguishing Directions* influenced the everyday practice of Qing law is debatable, but privately published manuals had considerable effect on the ways people envisioned legal practice. House and grave fengshui manuals consistently cited formal litigation as a kind of misfortune arising from poor residential fengshui and offered practical advice on how to avoid litigation. Yet, in practice, people mobilized those exact principles to formulate lawsuits in court. For their part, geomancers drew on these manuals to provide legal advice, and as we saw from the house lawsuit of the Wang family, some directly helped draft the cause-and-effect logics of submitted lawsuits.

Geomancers and other ritual specialists may have helped spread knowledge about the Qing law code by linking geomantic analysis and cosmic repercussions with imperial law, including punishments such as exile, strangulation, and beheading. Even the imperial pardons of the Autumn Assizes may have entered the popular imagination through fengshui manuals, which warned readers that a pardon might not come a person's way if his or her fengshui was in poor condition. By engaging the laws of the land during family rituals of funerals, burials, marriages, and house construction, people in practice learned a bit about the laws of the dynasty under which they lived.

3

Examining Fortune

THE STORY GOES that a scholar named Ren Wenyi of Northern Sichuan received a provincial degree in 1736. He then set his sights on the metropolitan examinations. In the autumn of 1744, some residents of Suining County set out to construct a dam to help irrigate the countryside adjoining Ren's property. Scholar Ren fiercely opposed the project and presented a lawsuit in court, arguing that the dam would disturb his ancestral graves. The county magistrate dutifully set out to inspect the site. After analyzing the landscape, the magistrate told Ren that the dam's construction would improve his fengshui by directing water flow around the graves in a more auspicious direction. The initially skeptical scholar accepted the magistrate's logic and withdrew his legal complaint. After all, the magistrate had passed the examinations at their higher levels, indicating that he was a man who possessed many talents and understood the ways of fortune.

Suining residents completed the dam in 1745, which was a metropolitan examination year in Beijing. Ren Wenyi took the exam. Following a seventy-five-year drought of awarded *jinshi*—the highest and final degree of the imperial examinations—in the county, Ren passed and was assigned to serve as a magistrate in Guizhou Province. When the extraordinary news of his examination success reached Sichuan, people gathered by the dam to catch a glimpse of the famed Ren graves. The compilers of the county's gazetteer asked, with evidence like this, "how could one not believe in the theories of fengshui?"[1]

The civil examination system in Qing times consisted of three primary levels. Possession of a degree at any level brought an individual "gentry" status. In addition to preliminary county exams, held twice every three years, there were prefectural, provincial, and metropolitan exams, each held triennially.[2] Preparation for these exams, each a prerequisite for advancing to the next level, was extraordinarily taxing, time-consuming, and stressful. The testing process was brutal, requiring candidates to spend up to three days locked in dark examination hall cells. In the end, the vast majority of them failed. As Benjamin Elman has observed, recourse to religion, including direct appeals to local gods, and the employment of fortune-tellers and fengshui experts, were common ways for examination candidates to cope with the immense psychological pressures brought about by the prohibitively tight examination market.[3] These expedients may have eased the minds of many scholars, but the use of fengshui transcended self-therapy. Fengshui, after all, could tangibly improve one's prospects.

To a significant degree, the Qing state acknowledged that good fengshui was related to examination success. Individual officials disagreed about how great a role fortune played, with some downplaying fengshui's impact on success and others emphasizing it as essential. Yet the Astronomical Bureau–compiled *Correct Doctrines of Fengshui* included historical case studies from across the empire demonstrating good fengshui leading to positive examination results for aspirants to officialdom.[4] Local governments examined, licensed, and hired Yin-yang officers to interpret the constellations of heaven, select the sites of prominent buildings, and monitor the health of earth veins—all in the interest of cultivating a community's fengshui and emphasizing the state's control and expertise in these domains. Likewise, many officials in Sichuan actively tried to improve the fortunes of residents. Magistrates were also generally receptive to considering reasonable gentry requests concerning zoning, infrastructure, and development related to fengshui and the examinations. Keeping a county's scholars involved with local governance and on amicable terms with the state was part of a magistrate's job.

The spectrum of outcomes for examination hopefuls in imperial China had two readily identifiable extremes. At one end of the spectrum stood scholar-officials like Zeng Guofan, who received his *jinshi* degree in 1838 at twenty-seven and went on to a brilliant career. At the other end, some candidates became so disgruntled by their failures that they joined heterodox groups and launched insurrections against the state, like the anti-Tang rebel Huang Chao (835–884) or the Taiping leader Hong Xiuquan (1814–64). Between the two ends of this spectrum was a huge number of scholars (see below) who neither obtained posts, either by examination or purchase, nor started rebellions. Individuals in this vast middle ground often considered themselves reasonably talented but unlucky. With a growing population of men desiring official posts, success in the Qing exam market, much like good land on the property market, was becoming a very scarce resource.

By the nineteenth century, most people believed that some combination of talent and fortune determined examination success. Fortune stemmed from the accumulated merit of a living person's ancestors, petitions to deities of the examinations such as Wenchang, and the fengshui of one's ancestral graves, house, and native village or county.[5] Good fengshui could be collective (extending to an entire county or beyond) or individual (a house or grave), as could poor fengshui.[6] For those who did manage to pass the upper-level exams and become officials, or those who purchased an office, highlighting one's talent without mentioning a bit of help from ancestors or the role of good fortune was considered poor form in polite society. A subtle acknowledgment of one's talent, merit, and fortune was the mark of a proper Qing gentleman. The key was to show others these virtues gracefully—and not to tell them directly.

Fengshui mattered to all sectors of society because, unlike ancestral merit, one could immediately improve it with the correct advice, knowledge, and techniques regardless of one's family pedigree—at least in principle. Tangible "evidence" for efficacious applications of fengshui suffused Qing society. People saw this evidence daily: in the grand facades of elite residences, along the waterways of rural villages, in the academies their sons attended, and across densely forested temple grounds. Government buildings were carefully laid out and monumentally constructed, projecting the authority of officials and the dynasty they served. As seen in the opening story of this chapter, those who passed the exams at their higher levels invariably tended to live in the most excellent houses and claim the best gravesites. When news of a homegrown success arrived from Beijing—an exciting event in rural Sichuan—some locals went out of their way to pass by the scholar's ancestral homestead. It was always a good idea to check if any natural or built features stood out. Curiosity was contagious. After a surprising success, people might identify a new structural addition in the county seat that could explain the sudden fortune. Sometimes these acts were written down, producing new knowledge about the land and its relationship to people.

With this experiential evidence came doubt: no one—not officials, not gentry, and not commoners—believed in every theory of fengshui. Everyone knew that some tales of fengshui, as narrated by some people, applied to some settings, and ascribed to some events, were nonsense.[7] The same applied to medical techniques and practitioners during the Qing.[8] People were careful with their money and skeptical of popular experts. Yet most folks thought there was something very real about fengshui, especially when a person of proven knowledge, like a degree-burnished magistrate, interpreted a physical site or a situation from his seat of authority, projecting sincere concern for the people. Fengshui surely legitimized regional, community, and family inequality, but it did so in a way that allowed a broad and diverse array of people to participate meaningfully in imperial culture and aspire to its key symbols of institutional power—even if they never had a realistic chance of passing the examinations.[9] Accordingly, although some officials might scold the people for being fortune-obsessed, quite a few overtly acknowledged the role that fortune played alongside talent. The reality was that almost no one could become a Zeng Guofan, and the Qing state did not want an exam failure to become a Hong Xiuquan.[10]

Because much prior research has focused on Jiangnan, where the number of examination successes was higher than anywhere else, the dynamics surrounding the examinations in poor peripheries have yet to be fully explored. Lacking legions of local luminaries or esteemed ancestors to point to (with some fictitious exceptions) for encouraging scholars, one of the only governing strategies left for officials serving along the western frontiers was offering the protection and fostering of auspicious fortune. To be clear, Sichuan had its share of famous scholars, like Su Shi (1037–1101) and Yang Shen (1488–1559); the issue was that outside of Chengdu—as in Nanbu County, for instance— these luminaries tended to have lived in the Song or Ming dynasties.[11] Magistrates arriving from wealthy, well-connected backgrounds could not publicly conclude that repeated examination failures proved that the scholars of Sichuan's peripheries and hinterlands lacked talent. Many officials posted to these remoter parts of the country found it more advantageous to suggest politely that the scholars in the area were just temporarily unlucky and that one or two would pass in the future if the proper environmental conditions were met. These officials then acted, repairing academies, moving schools, or constructing pagodas in auspicious locations to help examination prospects. Magistrates serving in Sichuan attempted such projects into the final decade of the Qing, until the imperial state abolished the examinations in 1905.

In this social context, Sichuan's scholars presented many petitions to the court about activities and sites deemed to harm examination fortune, with repeated poor results directly offered as legal evidence. This evidence was more persuasive in court than one might anticipate. Over the 1800s, gentry targets included certain types of businesses, temples, mosques, bridges, and commercial activities like tree-felling and mining. Magistrates accepted some petitions and rejected others. Often, resolutions resulted in mediation and compromise. The petitions show that fengshui was a vehicle for members of the local elite to remonstrate with their government or protest some activity in terms that the state did not typically deem subversive. Officials needed the local public to see them considering those pleas, even though they did not accept every request or complaint.

Some non-Han people also participated in these negotiations. Fei Huang has shown that in the southwestern frontier province of Yunnan, which adjoined Sichuan to the south, Qing officials and Han literati cited fengshui as a civilizing discourse in their ongoing effort to assimilate non-Han peoples.[12] Similar dynamics were also present in Sichuan. Yet there, non-Han peoples sometimes invoked fengshui to protect their properties from theft or exploitation by Han Chinese. Members of the Muslim and Yi (Lolo) communities received substantial guarantees from the Qing state when they presented fengshui claims in court. The findings here align with historians now arguing that forces beyond Confucianism played substantial roles in extending "Chinese" culture into frontier spaces.[13] The physical acts of holding the civil examinations, constructing academies, and building schools in border regions were important. However, discourses of property, space, and power that surrounded the exams, like fengshui, did quite a bit of the heavy lifting on the ground. Fengshui offered skin in a game that, by Qing times, a diverse range of people were forced to play.

The following sections first consider the history of the imperial examinations in Baoning Prefecture and the role of a Sufi shrine in fostering local fortune. The analysis then turns to officials' efforts to improve fengshui through construction projects in Sichuan and the state's hiring of Yin-yang officers. The latter sections look at legal petitions about fengshui and the examinations from Nanbu, Ba, and other counties of Qing Sichuan.

The Gentry of Northern Sichuan

As indicated briefly above, members of the gentry (*shenshi* or *xiangshen*) were a group of people characterized by their standing in relation to the examination system. Legally speaking, the term *shenshi* applied to anyone who held a scholarly degree, whether gained by passing one or another of the state civil service examinations, or by purchase. These individuals and their immediate families enjoyed special legal, economic, and other privileges that were not extended to commoners (*shumin, liangmin*, etc.). Although distinct from scholar-officials (*shidaifu*, or, more generally *guan*, or officials), who had already secured positions in government, members of the gentry class—particularly those in possession of higher degrees—enjoyed a social status similar to that of officials. Most importantly, as a matter of social fact though not law, they provided the pool from which the vast majority of officials would be drawn.

At the bottom of the examination system, a series of qualifying tests weeded out most potential candidates for the higher levels.[14] As Isabella Bird (1831–1904) learned during her visit to Nanbu County in the late 1890s, fewer than one in ten children and teenagers who spent time studying at the county's central school even competed in these entry-level examinations.[15] Those who did pass those tests—*tongsheng* (qualified candidates)—took the prefectural exams, the first of the three major levels of examined achievement. Those who succeeded became known as *shengyuan* (budding officers) or *xiucai* (cultivated talents). The next step was the triennial provincial level exam. Successful candidates at this level were known as *juren* (promoted gentleman) and were technically eligible to serve as county magistrates or as the directors of county schools.[16] At the top of the examination structure were a series of metropolitan exams, also held triennially, which culminated in a palace exam (*dianshi* or *tingshi*) and eventual status as a *jinshi* (presented scholar).

By 1850, some four and a half million people across the empire were actively participating in the examination system, which awarded only 300 to 350 *jinshi* degrees every three years.[17] The wealthy Jiangnan provinces of Jiangsu and Zhejiang dominated, respectively producing 2,920 and 2,808 *jinshi* degree holders over the course of the dynasty. Sichuan produced less than a third of those numbers, with only 763 *jinshi* degree holders, which was one of the lowest aggregate numbers of the empire.[18] Within Sichuan, even modestly well-off Ba County only produced twenty-one *jinshi* degree holders, which for context can be compared to the 379 hailing from Renhe County of Zhejiang.[19] Sichuan's performance can be partially explained by its low population in the first century of the Qing and its degree quota, which was fixed at around sixty-three degrees per provincial exam prior to the 1820s.[20] More degrees were regularly awarded at the turn of the twentieth century, by which time the province's quota had been raised to 103 per cycle. Nonetheless, Sichuan never approached Jiangnan's rate of success even though its population was more than double that of Zhejiang's by the close of the dynasty.

Although some regions produced fewer civil officials than others, nearly all prefectures and counties where the exams were administered had a group of men who could identify as gentry. Rather than conceiving of Northern Sichuan as lacking a gentry base, it is more accurate to say that the area had a dearth of elite degree-holders.[21] In 1906, a Nanbu gazetteer counted 3,251 men as qualifying for some kind of gentry rank out of a total population of around six-hundred thousand people by that time.[22] If gazetteer numbers are accurate, on average less than ten living men of Nanbu County held the provincial degree at any given year in the 1800s, with the number slightly rising during the Taiping Civil War. The rest of the local gentry held lower degrees.[23] By contrast, an average county in Zhejiang might have five or six times as many provincial degree holders at any given time during the dynasty.[24] Despite their small number in Nanbu, these men were highly influential and served roles in education, public works, local defense, tax collection, medicine, fengshui, and petitioners in law courts. Sichuan's gentry often had ties to the salt industry, blurring the boundaries of merchant and scholar across the province. The heads of surveillance units and local community leaders in Sichuan also took on roles that in Jiangnan were usually reserved for members of the gentry.[25]

Northern Sichuan had not always been a region of poor examination outcomes. In the early Qing, there were a sizeable number of local degree-holders produced in Baoning Prefecture, where Langzhong, Nanbu, Cangxi, Guangyuan, Zhaohua, Tongjiang, and Nanjiang counties as well as Jian and Ba departments (not to be confused with Ba County) were located. Seventeen of Langzhong's twenty-two Qing-era *jinshi* degrees were granted before 1737, while three of Nanbu's five were granted before 1743. Two counties of Baoning Prefecture, Guangyuan and Zhaohua, produced no *jinshi* degree holders during the entire dynasty. There is no consensus on provincial degree statistics, since Sichuan's local gazetteers tend to inflate numbers by adopting successes from neighboring counties as their own.[26] County border and affiliation fudging—as seen in the last chapter—were not rare. Regardless of the exact number, examination performance in Northern Sichuan was only encouraging during the first few decades of the Qing and was poor after the 1740s, even though the population continued to grow dramatically through the end of the dynasty. Nanbu's two *jinshi* degree holders from the final 170 years of Qing rule were awarded during the Taiping Civil War, when large parts of Jiangnan were temporarily destroyed. Overall, Northern Sichuan's examination outcomes during the second half of the dynasty appear to have been slightly worse than the regions of western Shandong associated with the Boxer Uprising.[27]

The reasons for Northern Sichuan's early examination success and later decline were interwoven with the Qing's westward conquest in the 1600s. Baoning was the first major town in Sichuan captured by the Manchus, and it became the temporary capital of the province in the early Qing, serving as a strategically important pivot point lying on the communication routes that linked Beijing, Xi'an, and Chengdu. With the rest of Sichuan in ruins, Baoning hosted Sichuan's provincial examinations for almost twenty years (1646–65), before Chengdu was rebuilt and restored to its former administrative position as the capital. During this era, the scholars of Northern Sichuan had a small golden age. Thirty-three men of Nanbu passed the provincial exams by 1666, while Langzhong managed about twice as many. Neighboring counties like Tongjiang also saw some rare successes in the early Qing when Baoning hosted the examinations. As seen below, the region's promising examination performance in the early Qing became locally attributed in part to a most unlikely source.

Fortune and a Sufi Shrine

In the wake of destruction resulting from the Ming-Qing transition, Muslim adherents of the Qādirīyah Sufi network arrived in Sichuan from Central Asia with desperately needed merchant capital, helping Baoning rise from the ashes.²⁸ The Sufi network's leader was a charismatic teacher named Khoja Abd Allāh, a twenty-ninth generation descendant of the Prophet Muhammad. Although not many details are known about the Khoja's life, he was drawn to Baoning in 1684 upon the invitation of a Muslim military commander stationed in Northern Sichuan's main garrison. According to Chinese-language sources, the Khoja was struck by the austere beauty of Northern Sichuan's valleys, rivers, and mountains and sincerely enjoyed residing in the riverside town of Baoning. The Khoja lived in a local Buddhist temple for four years, sharing his knowledge of astronomy, medicine, and fengshui with local scholars, who were impressed by the breadth of his erudition. Readers should bear in mind that, Chinese writings of this sort assumed that the Khoja, as an educated person, would naturally be well read in astronomy, medicine, and fengshui—hallmarks of a respectable literatus.²⁹

Baoning had a Muslim community composed of a few dozen families during the Khoja's lifetime and in the centuries thereafter. Yet most local gazetteers composed by local Han Chinese did not reflect extensive knowledge or even curiosity about Islam, only observing that Muslims butchered lambs and cows, often joined the local army regiments, and traded with merchants from Gansu.³⁰ But a lack of interest in Islamic theology did not seem to matter when it came to real estate acquisition. When the Khoja died in 1689, Baoning's gentry invited his disciples to bury him on one of the most sacred mountains of the area, Coiled Dragon Mountain. Coiled Dragon Mountain for centuries had a strong Daoist presence and was thought to align with the dragon vein of the examination-proctoring walled town. Though few records survive from the time of the shrine's creation, in a stone inscription dated 1747, the Muslim community made a striking claim about the shrine's impact on the fengshui of the region. The inscription read:

> Just a few years after the Khoja's burial, the local culture and literacy rose and the number of people who performed well in the civil service exams grew. The people and myriad things in Baoning all started to see great improvement. If our teacher had not had the brilliance by which he viewed and understood the abstruseness of the patterns of heaven and identified suitable veins of the earth, how could this place have produced such talent and success?"³¹

The target audience of this commemorative stele was Baoning's gentry, the great majority of whom were not Muslim. The Muslim community had made significant contributions to the area during a period when it was reeling from decades of war—and it made sure the gentry of Baoning remembered it, in terms of fengshui. Such a bold, public claim about the shrine's geomantic impact on the entire area could only have been made if locals found it plausible.

Local scholars did find the claim plausible. For the remainder of the Qing, the mausoleum of Khoja Abd Allāh played an outsized role in the political, religious, and cultural life of Baoning and Northern Sichuan. Magistrates and prefects visited the shrine to pray for rain during droughts. Sichuan's officials, including a governor-general, Ding Baozhen (1820–86), bestowed gifts of calligraphy on the shrine during examination seasons, celebrating the Islamic shrine's contributions to the moral cultivation, good fortune, and general improvement of the area. Baoning's literati composed poems for the shrine comparing the Khoja to a great Daoist teacher.[32] Commemorative steles, which were carved almost any time a group of scholars rebuilt or repaired a temple, indicate the shrine's reception and importance. A greater number of dedicatory inscriptions and calligraphic boards from Qing officials were provided to the Islamic shrine than to any other religious structure in Baoning through the eighteenth and nineteenth centuries.[33] This considerable state and elite patronage occurred even though Muslims constituted a small minority of Baoning's overall population.

In the image in figure 3.1, from the county's gazetteer, the "Muslim Temple," is shown next to the town's preeminent Confucian academy, Brocade Screen, as well as Baoning's Star Gazing Platform and the Ming-era Daoist Hall of the Perfected Warrior. Today, the heavily forested shrine takes up over three square acres in Langzhong Municipality, where it is part of Coiled Dragon Mountain National Park, a nature reserve created by the government in 2013. No one in the Qing legal record ever contested the Muslim community's claim to the valuable and important mountain. For the Muslims living around Khoja Abd Allāh's shrine, their contributions to rebuilding the town, their continued involvement in local trade, and their invocation of fengshui collectively allowed them to maintain land on a significant mountain and obtain recognition for that property acquisition by local scholars and Qing officials alike.

There are, of course, limits to what the Baoning shrine can reveal about the history of Sino-Muslim peoples in relation to fengshui. The Sinophone Muslim populations of the Qing Empire were diverse and cannot be lumped together into a single group. In fact, some Muslim literati emphasized that their communities did not practice fengshui as it related to funerals and burials.[34] Such writings underscored the idea that Islamic burials were modest, simple, and quick, thus assuring wary officials that the religion was not a source of difficult property litigation. Islamic law does indeed require fast and simple burials, and the Muslim literati's emphasis on this specific point may imply that some members of their congregations were attracted to lavish burial plots.

Likewise, the case of the Baoning shrine does not detract from the tensions on the ground between Muslims and non-Muslim Han Chinese in Sichuan

FIGURE 3.1. Gazetteer Image of Khoja Abd Allāh's Shrine on Coiled Dragon Mountain. This gazetteer image depicts the area immediately northwest of the prefectural seat. Starting from the bottom right of the image (closest to the walled town) and moving up, a Wenchang Palace is located on the right and several Daoist shrines are on the left along with the Altar of Agriculture. Continuing onward (up and left), one passes Brocade Screen Academy on the right and the Hall of the Perfected Warrior and Palace of the Eastern Peak on the left. Finally, upon reaching the highest elevation, one encounters the Star Gazing Platform on the right and the "Muslim Temple" (Khoja Abd Allāh's shrine) on the top left. The dense tree cover that surrounds the Islamic shrine is still present today. Map from *Langzhong xian zhi* (1851).

and elsewhere. Han Chinese mobilized fengshui against mosques and Islamic shrines as a rationale for ordering their closure. For example, during the 1850s, after the local gentry of Ninghe, Zhili Province, complained that a newly built mosque hurt the county's fengshui, the presiding official ordered the structure demolished and forbid it from being rebuilt.[35] Dru Gladney cites similar

FIGURE 3.2. Photograph of Entrance to Khoja Abd Allāh's Shrine. This photograph captures the front gate of the Khoja's shrine in Langzhong. Behind the gate lies the path leading up to the central mausoleum of the Khoja and his disciplines. The path is shaded on both sides by the large, dense forest, which covers the shrine. Occupying over thirteen thousand square meters, the shrine on Coiled Dragon Mountain is adjacent today to the larger Sichuan Langzhong National Forest Park.

tensions between Muslims and Han over fengshui from his fieldwork in northwest China.[36] The trend of opposing the creation or existence of mosques likely accelerated in the nineteenth century, when ethnic and religious conflicts, some involving large Muslim populations, broke out across the empire.[37]

One defensive strategy used by Muslims involved purchasing older temples, schools, and even government structures to convert into mosques—thereby defanging an argument that an area's fengshui was changed by a mosque's creation.[38] It was not unknown for Muslim communities to advertise their mosques' positive geomantic positions, and a great many more included the auspicious timing of a mosque's construction or repair on dedicatory stone inscriptions.[39] The magistrate of Nanchong, next to Nanbu, presided over the erection of a dedicatory inscription at the main mosque on "an auspicious day during the first week of autumn's ninth month in the first year [*renxu* in the sixty-year cycle] of the reign of the Tongzhi Emperor [1862] of the Great Qing."[40] Islamic community leaders understood the importance of picking dates for construction that would also not overlap with inauspicious days for building or repairing structures (*xiuzao*) in the imperial calendar.[41] Similar defensive invocations of fengshui by other non-Han groups were leveraged in Sichuan's courts during the nineteenth century.[42]

The success of the Qādirīyah Sufi network and Baoning's Muslim community in rooting themselves in the social, economic, and religious life of Northern Sichuan had many supporting factors, of which fengshui was one. Fengshui tied an important Islamic shrine to the examination system and thus to the non-Muslim gentry of Baoning. Undoubtedly, the Shrine of Khoja Abd Allāh had a distinctive history that saw many contingent factors converge around its construction and development in the early-to-mid Qing. Nonetheless, its history offers answers to some broad questions: how to obtain valuable property, how to sustain ownership over time, and how to survive in the examination-obsessed society of the Qing dynasty. "Ancient tomb," "auspicious time," and "dragon vein," were not just boilerplate on a land deed or stone inscription. They were narrated components of the Qing legal order, protecting those who invoked them. The Khoja's shrine is testament to their power across time.

Constructing Fortune in the Early-to-Mid 1800s

After the White Lotus Rebellion (1796–1804), a series of memorials sent to the Jiaqing Emperor described desperate social conditions in Northern Sichuan. Expensive grain resulting from disrupted agricultural production led the governor-general, Changming (d. 1817), to recommend opening state granaries to try and stabilize prices, although he warned that even such actions might not solve the issue of hunger.[43] Although Sichuan had exported surplus rice to other provinces during the 1700s, the foundations of that largesse had eroded by the turn of the nineteenth century.[44] On the persistent problem of banditry, another report described the vast territory of Baoning Prefecture with similar pessimism. Poverty was rampant, heterodox sects propagated alluring promises of fortune from gods and spirits, and very few scholars—who were described as wearing "shabby dress and unhemmed gowns"—were passing the civil examinations.[45] The subtext was clear. Baoning needed an elite that was loyal to the state to help the government administer the region and address the area's social problems. Government messaging mattered, and the future needed to appear better than the present, lest the people turn to illicit teachings that promised a better life.

Officials posted to Baoning Prefecture after the White Lotus Rebellion went straight to work. Assuming the post of circuit intendant of Northern Sichuan in the early 1800s, the Hunan native Li Xuejin (1776–1838) formulated a plan of reconstruction. Li donated his "nourishing honesty" silver to repair key structures around Baoning and encourage local notables to contribute to the effort.[46] The city wall was reinforced to prevent seasonal flooding exacerbated by siltation along the Jialing River. The burial shrine of the Three Kingdoms hero, Zhang Fei, was rebuilt, as was a pavilion associated with the eminent Song dynasty official Sima Guang (1019–86) and the town's central tower. The examination hall was refurbished, Brocade Screen Academy was restored, and a new academy was constructed to meet the needs of new students from rural market towns. In an inscription erected next to the prefectural town's examination hall, Circuit Intendant Li was explicit about his intentions: "Everything we have done has been to foster fengshui, with the intention of cultivating talented people."[47] The elite that Circuit Intendant Li Xuejin needed would arrive ideally through the examinations.

During the Qing, successful officials were celebrated in local sources for their contributions to the prefectures and counties in which they served.[48] Improving fengshui was a common trope in the governing achievements (*zhengji*) sections of Sichuan gazetteers. When located within the broader social and historical context of the era, these annals shed light on the governing strategies of resident administrators during the 1800s, when Northern Sichuan recorded an ever-greater number of droughts, floods, and inauspicious phenomena.[49] As a result of such strategies, a number of physical structures in rural, county, and prefectural towns—including schools, pagodas, and shrines—came to be identified as having geomantic significance related to examination success.

By the time that the 1870 edition of Nanbu's gazetteer was published, over 120 men had already served as magistrate since the beginning of the Qing, but only fourteen were singled out as having governing achievements. Of those singled out, eight magistrates had focused on education. Magistrate Du, who assumed office in 1824, set up twenty-four charitable schools for poverty alleviation by levying a surcharge tax on wealthy households. Official efforts to expand access to education for poorer students were possible but required sustained monetary extraction from elite families, which was a challenge.[50] The harsh reality was that provincial and palace graduates, which the county desired, were not likely to emerge from charitable schools. A gentry petition from 1896 noted that the county's charitable schools had been out of commission for a long time.[51]

The remaining seven educational achievements involved projects associated with the civil examinations. Some of these projects invoked fengshui in the process of construction while others were designed to foster fengshui. This difference requires explication. Constructing an academy was not an act explicitly intended to foster fengshui, even if geomancers were consulted during the building process. Building an academy or school was first and foremost intended to provide education, with any additional geomantic benefits deriving from a carefully chosen location a welcome extra.

By contrast, the creation of structures bearing the title of pavilion (*ge*), storied building (*lou*), or tower (*ta*; also pagoda) often endeavored to bring benefit to a locality by improving fengshui. People understood these towers—which could be constructed to Wenchang or simply called fengshui pagodas (*fengshui ta* or *fengshui lou*)—as especially helpful for fostering examination success.[52] The use of stone towers as geomantic interventions to improve civil examination prospects started in the Song dynasty (960–1279).[53] By Qing times, nearly every county and prefectural seat had at least one and often more of these structures. These towers bore the physical shape of pagodas found in Buddhist temples without necessarily having a direct link to Buddhism or a monastic institution. In fact, some literati appear to have constructed towers that were physically indistinguishable from pagodas in order to compete with Buddhist domination of a cityscape.

With this difference in mind, consider the following table containing the magisterial achievements related to the examinations, beginning with the construction of a Wenchang palace on Mount Aofeng.[54] Most of the achievements directly or indirectly invoked fengshui.[55]

What do these fragmentary and gentry-centric local records reveal? Clearly there were limits to what resident administrators could do, given the shortness of their tenures and the vastness of the territories and human populations they administered. Magistrates were unfamiliar with the county, were probably not thrilled to be there, and had to manage with limited governing budgets. But for those who expressed an interest in the county, improving fengshui with a building refurbishment, a structural reorientation, or a stone pagoda was an achievable and popular project during their short tenures in office. Most importantly, by improving fengshui, magistrates gave support to the idea that local gentry could pass the high-level civil service examinations, thus recognizing the dignity and earnest efforts of scholars living in an impoverished periphery of the empire.

The question of financing is worth further consideration. Constructing new academies on prime real estate was expensive, as was the cost of hiring regular teaching staff. Nanbu's academy, the most important educational structure in the county, faced routine funding issues, which was why promises to improve fengshui were so central to its subsequent upkeep campaigns. Constructing stone pagodas to Wenchang, another astral deity associated with examination success, or moving Kui Star pavilions, were acts that required no additional teaching staff, but only a startup expense of a few hundred taels. Some residents thought it a more efficient use of funds to improve collective examination prospects by building pagodas and pavilions than to attempt to expand literacy through expensive charitable schools in inauspicious circumstances,

TABLE 3.1. Magisterial Achievements Related to the Examinations in Nanbu's 1870 Gazetteer

Official's Name (Native Place; Degree)	Date	Achievement
Wang Xun (Guizhou; *jinshi*)	1755	Constructed Wenchang Palace on Mount Aofeng
Li Yuanfen (Hubei; Provincial)	1778	Constructed Aofeng Academy
Li Wende (Fengtian; Provincial)	1815	Moved the Kui Star Pavilion—with Kui Star denoting an astral deity associated with learning and scholarship—to a place near the walled town's east gate known in fengshui as a water exit (*shuikou*), a geomantic term for the site where water departs a settlement.[i]
Wei Gengyang (Shanxi; Provincial)	1827	Selected wood from ancient cypresses to repair and reorient the Confucian Temple.
Li Yundong (Jiangsu; Provincial)	1832	Continued repairing the Confucian Temple and composed a dedicatory inscription that read, "Some say the school's decline and the earth veins' lack of vitality are the roots of the county's poor examination performance in recent years."
Wang Zhongxuan (Hanjun; *jinshi*)	1841	Constructed Qing-era Nanbu's First Examination Hall
Huang Qiyuan (Fujian; *jinshi*)	1863	Constructed a Fengshui Pagoda on Mount Aofeng

[i] For a discussion of water exits, see Clément, Clément, and Shin, *Architecture du paysage en Asie orientale*, 130–34.

which seldom yielded examination successes. Qing-era fengshui manuals endorsed the former approach, and one 1844 manual published in Fuzhou, Fujian, set its sights on no less than the national market:

> Wherever, in the capital, the administrative seats of provinces, prefectures, counties, or rural villages, scholars are not benefitting from environmental conditions, those who are not appearing on the registers of examination success can use the directions of *jia* (east leaning north), *xun* (direct

southeast), *bing* (south leaning east), and *ding* (south leaning west) to select auspicious land and establish a literary peak ("a sharp peak in the shape of a writing brush"; a fengshui pagoda). If the pagoda is higher than the surrounding mountains, successful examination candidates will emerge.[56]

Whether from county magistrates or other geomantic experts, the scholars of Northern Sichuan—who lived some 1,900 kilometers away from Fuzhou—got the message.[57]

How can one be sure they got the message? Proof is found in the pagodas. Pagoda construction in Qing Sichuan did not correlate with county- or prefectural-level examination passing rates. More fengshui pagodas—structures bearing the title *ta* (tower)—were constructed in Northern Sichuan during the dynasty than in any other region of the province. The prefectures of Baoning and Tongchuan, both affected by the White Lotus Rebellion, each saw twenty or more fengshui pagodas built, while Chongqing saw around seventeen and wealthier Chengdu Prefecture saw twelve.[58] Caution must be applied to such comparisons, since Baoning Prefecture covered a larger geographical area than Chengdu Prefecture. Nonetheless, Baoning and Tongchuan still outproduced Chongqing over roughly commensurate land areas. One suspects that Qing officials arriving in these areas after the White Lotus Rebellion felt anxious about improving fortune—and doing so quickly.[59]

Westerners in the nineteenth century also picked up on this pagoda-building trend. During his travels up the Yangzi, Archibald John Little (1838–1908) observed that pagodas designed to foster fortune appeared to be constructed with far greater speed and numbers in central and western China than in Jiangnan:

> Poor as it is, however, its inhabitants have come to the conclusion that its poverty is due—not to its isolated situation among barren mountains, nor yet to the prohibition of the authorities of modern appliances for improving its coal output, but to a defect of its "Feng-shui." Great sacrifices have therefore been made to remedy this defect, and evidences of these are seen in six completed stories of a new pagoda, which is being built on a conspicuous point of hard white rock on the left bank, a mile or more below the town. In eastern China, where we see the pagodas mostly neglected and in ruin, we little imagine that new pagodas still continue to be built in other parts of the empire. Nearly every town on the river has a pagoda ... which is supposed to prevent the wealth of the town being swept past it by the rapid current for the benefit of the cities below.[60]

Little's observation does seem to suggest that pagoda construction was a strategy of Qing statecraft in peripheral areas, where officials arriving from wealthy coastal regions set to work improving fengshui.

FIGURE 3.3. Photograph of Baoning Fengshui Pagoda. The White Tower (*baita*), as Langzhong's premier fengshui pagoda is called, was constructed in the Ming dynasty. However, the tower remained much beloved through the Qing, hence its appearance here. Photograph by Wei Yuangang.

Both officials and scholars were invested in the success of a fengshui pagoda's construction and reception. Sometimes, magistrates proposed building pagodas or pavilions, or else chose their geomantic orientations.[61] As we will see below, gentry also petitioned magistrates for constructing a pagoda. Once constructed, if a pagoda did not deliver good fortune or if it became associated with misfortune, it could be torn down or altered. The year 1889, when the gentry of Mianzhou alleged that a pagoda constructed in 1870 was putting the area's fengshui at risk, three stories were removed from the structure to decrease its height.[62]

Regardless of who conceived the idea, magistrates got involved in pagoda construction. They were often pressured to do so by local gentry, who understood the value of an official's endorsement of a project that required a successful fundraising drive. In one instance, a provincial degree-holder petitioned the county government to permit fundraising for a tower's construction under the banner of improving fengshui.[63] Officials also became involved due to the litigation that ensued over the large sums of money donated for the construction or repair of a tower.[64]

In either case, officials needed to tread carefully, since this governing strategy, like other construction projects, came with ample opportunities for

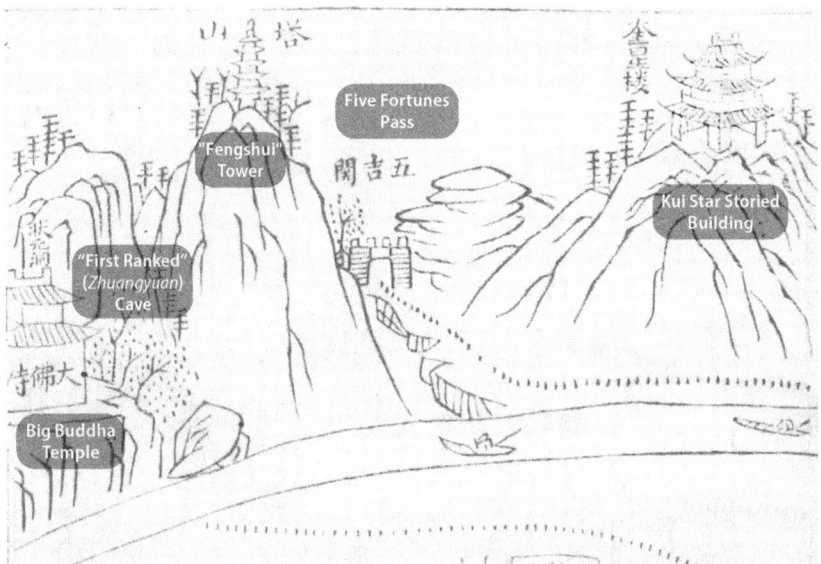

FIGURE 3.4. Gazetteer Image of Baoning Fengshui Pagoda and Kui Star Storied Building. This gazetteer image depicts the south bank of the Jialing River, directly across from the county and prefectural seat of Langzhong (Baoning). The fengshui pagoda of Figure 3.3 is seen on the mountain in the center of the frame, under the title of Tower Mountain (*tashan*). On the far left is [Great] Buddha Temple. On the far right is a Kui Star Storied Building (*Kuixing lou*). Map from *Langzhong xian zhi* (1851).

corruption. One case preserved in *The Conspectus of Penal Cases* tells of a scholar who, with the blessing of a prefect, raised funds across rural areas with a campaign to improve local fengshui by repairing a Kui Star pavilion, constructing a bridge, and dredging a canal.[65] But he pocketed most of the money, and the repairs were shoddy. The incident proved to be a scandal for both the prefect and the scholar, since residents had believed they were contributing to a "local public matter" (*difang gongshi*) by improving the area's fengshui for the examinations. With money flowing in for these public projects, officials needed to keep their eyes peeled for possible corruption.

Besides new construction, other options for improving local fortune involved identifying existing Buddhist, Daoist, and Islamic shrines, old woodlands, city wall gates, or even walled fortifications as linked to examination performance through geomantic connections.[66] In 1859, Zhejiang-native Magistrate Xie ordered a forest of ancient cypresses protected to secure the fortunes of Nanjiang County.[67] His order was inscribed onto a stele erected at the examination hall, refurbished through a fundraising campaign that allowed timber to be purchased from further afield. References to city wall gates and fengshui were also common. Peng County's gates were closed and moved over the eighteenth century and into the early nineteenth century to foster examination fortune and summon rainfall.[68] Figure 3.10 contains a visual rendering of one of Hengzhou's city wall gates permanently closed; many other visual examples of this practice exist.

By Qing times, the perceived links between pagodas and examination performance became so accentuated that they influenced people's understanding of famous historical landscapes. Magistrate Chen, hailing from Zhejiang, presided over the repair of a rural temple in Northern Sichuan's Santai County in 1811 and suggested that the ancient Buddhist sites of medieval Chang'an, such as the famous Great Goose Pagoda, had been constructed to improve fengshui for scholars taking the imperial examinations, which began in the Tang dynasty (618–907):

> When I was young, I went from Hangzhou to Beijing seeking employment as an official, I passed the North China Plain and crossed the Taihang Mountains, where I visited all the famous places of Guanzhong in Shaanxi. There, even when I saw places where the landscape had deteriorated, there were always Great Buddhas towering, standing tall above the clouds. These places had been originally constructed with the blessing-producing theories of the Buddha in mind, but those talented at earthly principles [fengshui] say that these structures were a pretext for improving deficiencies in the landscape.[69]

Magistrate Chen was wrong, but he reflects a common view held by the Qing elite of the practical functions of pagodas.

William Rowe has shown that many high Qing officials railed against Buddhist and Daoist influences that had found their way into Confucian ritual.[70] Perhaps fengshui offered another way to exert government control over Buddhist, Daoist, or even Islamic sites and to tie them into the state's civilizing projects and legitimizing institutions. Perhaps in self-defense, Buddhist monks sometimes embraced geomantic understandings of their temples. Doing so provided additional opportunities to solicit donations. Daoist priests also brandished their geomantic skills when needed.[71] Moreover, as seen in chapter 1 and as we will see more in chapter 4, residents employed strategies of identifying temples with fengshui to exert legal influence over monastic estates in court.

When scholars offered to invest in repairing structures associated with the exams, some magistrates objected that they had become too obsessed with fengshui and encouraged them to focus on the moral dimensions of repairing buildings, such as cultivating Confucian virtue. On one occasion, after the gentry of Baoning presented the county magistrate with a long list of requests for the Confucian temple school, including a reoriented screen wall, a new hall for changing into scholarly robes, and a Ling (Wenqu) Star Gate—the latter referring to another astral deity associated with examination talent.[72] In response, the magistrate urged them to focus on studying the classics rather than on fengshui.[73] Nonetheless, the magistrate still agreed to support most of the structural additions through a gentry-led fundraising campaign, since he deemed the improvements necessary for other reasons.[74] The magistrate's comment may have reflected a more general perception among some officials that local scholars had become so cynical about the examinations that they believe success depended completely on luck.

The civil examinations were a pillar of legitimacy of Manchu rule over China. The Qing examinations were also different from those that came before it.[75] Passing rates were historically low during the dynasty's second half, when "the odds for success in all stages of the selection process was one in six thousand (.01 percent)."[76] The increase in competition resulting from a growing population changed the metrics of official evaluation by placing higher requirements on memorization by the late 1700s.[77] The same involutionary dynamics also influenced the profile of *jinshi* degree holders, rendering them older than their Ming counterparts, having failed more times on average previously.[78] As Lawrence Zhang shows, many families with money skipped the system and purchased official posts.[79]

Small wonder, then, that frustrated scholars—especially those in outlying regions—tried everything possible to improve their examination chances. The examples above by no means exhaust the full list of efforts to improve fortune in Northern Sichuan during the 1800s, and one suspects many more unsuccessful attempts were never recorded.[80] In the elite academies of Jiangnan, fortune, merit, and talent may have been well balanced in discussions about the examinations. Along the western frontier of Qing Sichuan, however, a different governing strategy was necessary. With two key ingredients in short supply, arriving officials encouraged local scholars to, quite literally, try their luck.

Yin-yang Officers

One way the government regulated fengshui practices and managed examination anxiety was by hiring professional specialists in geomancy. This administrative practice dates to the Mongol Yuan Dynasty (1271–1368), when the government set up Yin-yang schools to regulate popular diviners. According to the *Collected Statutes of the Great Qing*, every county and prefectural yamen was required to hire one officer of the Yin-yang school to propagate and monitor correct practices related to astrology and geomancy across the empire.[81] Yin-yang officers at the county level did not hold a designated rank in the bureaucracy, while officers at the prefectural level received the lower ninth rank (*cong jiupin*; 9B)—the lowest rank in the official hierarchy.[82]

The term Yin-yang officer may perplex scholars of Chinese religion. In parts of north and northwest China, geomancers and Daoists often fell under the same category as Yin-yang masters (*yinyangsheng*), which referred to a range of ritual specialists who practiced Daoist arts and fengshui.[83] While the distinctions between ritual practitioners were often blurred on the ground, popular Yin-yang masters and Yin-yang officers differed in that the latter had an official license and had specific occupational duties. One case file spelled out expectations in the following terms: "The establishment of the [school for the] study of Yin-yang is limited to the investigation of matters relating to meteorological analyses, fengshui, divination, and astrology."[84]

According to imperial regulations, county and prefectural yamens had to evaluate and hire competent, literate individuals to be licensed as Yin-yang officers, thus differentiating them from the scores of popular local diviners of questionable erudition. In the Mongol Yuan (1271–1368), examinations for aspirants to the posts tested the fundamentals of geomancy and astrology. These officers then supervised practitioners belonging to state-classified Yin-yang households (*yin-yang hu*).[85] The Ming inherited these Yin-yang households, from whom Yin-yang officers were drawn, and established schools for the study of Yin-yang at the county and prefectural levels.[86]

By the Qing, the formal system of Yin-yang household registration had disappeared, but the governmentally appointed posts of Yin-yang officers remained, as did many physical Yin-yang schools (see figure 3.5). Consider the mid-nineteenth century experience of a candidate named Chen Yourong, who was recommended to the Ba County yamen upon the death of the previous Yin-yang officer. According to the archival record, Chen had to demonstrate his literacy, his knowledge of Yin-yang, and his lack of a criminal record to obtain the position:

> The yamen has examined the Yin-yang master Chen Yourong. From his youth he has studied sincerely the techniques of Yin-yang and is fluent in them. He is an appropriate candidate to fill this position, so the clerks should compile his age, appearance, place of birth, as well as any evaluative comments into a file.[87]

After a candidate like Chen was approved, the magistrate or prefect would then issue a license with his seal, certifying that the man could be hired without concern that he was a charlatan. According to the *Collected Statutes of the Great Qing*, Yin-yang officers were "forbidden from deluding the people with

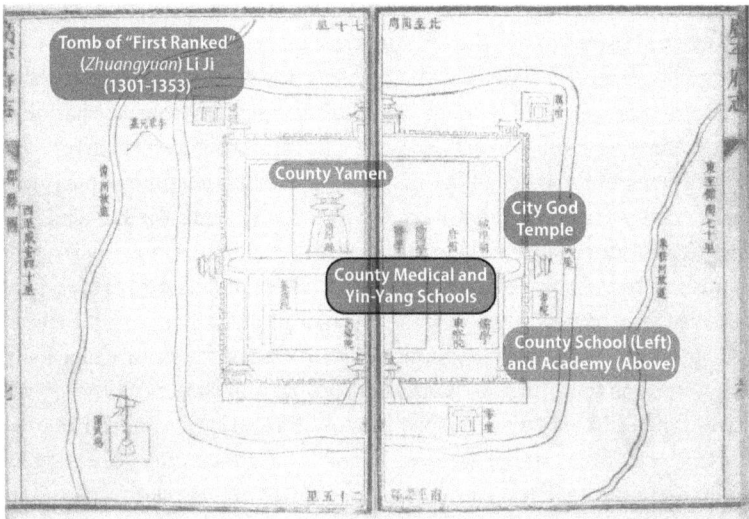

FIGURE 3.5. Yin-yang and Medical Schools of Guangping (Zhili Province). Image References to Yin-yang and medical schools and their attendant officers are common in Qing gazetteers. Sometimes, these schools were visually represented in gazetteer maps. This image from Guangping, Zhili Province, is one such case. For a sampling of visually depicted Yin-yang Schools in local gazetteers from the Qing dynasty, see the front materials of the following sources: *Kangxi Haifeng xian zhi* (1670; Guangdong), *Kangxi Shaoxing fu zhi* (1683; Zhejiang), *Kangxi Sui zhou zhi* (1690; Henan), *Qianlong Yicheng xian zhi* (1771; Shanxi), *Qianlong Taiping xian zhi* (1775; Anhui), and *Qianlong Nanchang fu zhi* (1789; Jiangxi). Map from *Kangxi Guangping fu zhi* (1677).

fantastical theories" (*jin qi huanwang huo min*) while offering competent advice for selecting locations, orientations, and times.[88] By hiring these experts, the government sought to limit potential disputes and litigation over fengshui with clear standards and expectations.

Yin-yang officers served much longer than magistrates, so they could provide valuable information about a county's ins and outs. The 1812 edition of An County's gazetteer offers the names of six individuals who assumed office as Yin-yang officers in the years 1730, 1752, 1765, 1789, 1796, and 1802.[89] The list likely does not bear the name of every individual who assumed the post of Yin-yang officer, but it is safe to say that the position could be held for long stretches of time. The aforementioned Chen Yourong, for instance, served as the county's Yin-yang officer for seventeen years.[90] In some counties, the post may have been bequeathed from father to son. In Dingyuan County for instance, two Yin-yang officers surnamed Wang contributed to the writing and editing of *A Thorough Explanation of the Geomantic Compass* (*Luojing toujie*), which was published in the 1820s under the family name.[91]

After being hired, what did Yin-yang officers do? As mentioned above, the officers handled tasks related to meteorological analyses, fengshui, divination, and astrology. A brief examination of their astrological duties before turning to fengshui provides a sense of the range of their official responsibilities. One requirement of Yin-yang officers was to attend ritual ceremonies for "rescuing the sun" and "rescuing the moon" during solar and lunar eclipses. These high-profile rituals counteracted the negative impacts of the eclipses, which were understood as inauspicious events. Whereas solar eclipses (*yang* force-sun) portended ill for a ruling emperor, lunar (*yin* force-moon) eclipses indicated problems in the empire's legal order, namely a warning that criminals had evaded proper punishment.[92] The Qianlong Emperor was well aware of the respective political and legal significance of solar and lunar rescue rituals and placed considerable weight on them, deeming them necessary to prevent restless commoners from interpreting celestial phenomena themselves—to the potential detriment of the Manchu dynasty.[93]

County-level ceremonies overseen by magistrates, Yin-yang officers, and other functionaries were impressive spectacles, demonstrating the Qing state's power and authority to the people. Drawing on the Jesuit-introduced "New Western Method" for calculating eclipses, the Astronomical Bureau in Beijing provided exact times for the beginning and end of eclipses to the emperor in advance.[94] Unlike the Ming (1368–1644), which only calculated eclipses for the capital region, the Qing Astronomical Bureau provided the specific eclipse times relevant to each province.[95] Naturally, this process included Sichuan, which received the calculated times for impending eclipses in Chengdu Prefecture, the provincial capital.[96]

For example, in the thirteenth year of the Guangxu reign (1887–88), after Beijing's Astronomical Bureau notified Sichuan (via the Board of Rites) of an impending lunar eclipse, the provincial government instructed prefects to organize their staff for lunar rescue rituals, offering the day and time that the ceremony needed to begin and end.[97] On the fifth day of the eleventh month, the Baoning Prefect alerted Nanbu County of a lunar eclipse occurring the following month (twelfth month, seventeenth day), which in the Gregorian calendar fell on January 29, 1888.[98] The Nanbu magistrate dispatched a summons to the Yin-yang officer along with the heads of the Daoist and Buddhist associations, who were tasked with bringing musicians, incense burners, candles, and a felt covering for the ritual table, among other accoutrements. The lively ceremony would begin at the county yamen with the lighting of the candles at exactly twenty-two minutes after four o'clock in the morning, at the first sight (*chukui*) of the eclipse. The official summons emphasized that prompt attendance was required.[99]

FIGURE 3.6. Western Depiction of a Qing Official and Ritual Specialists Presiding over a Solar Rescue Ritual in a Government Yamen. This image, published in Justus Doolittle's *Social Life of the Chinese* in 1865, depicts a Qing official presiding over a solar eclipse rescue ritual. The inspiration of the image may have been a rescue ritual observed by the author in Fuzhou, Fujian Province. The official is accompanied by two ritual specialists, identified as "priests of the Taoist sect." Note that the candles, incense, and table, and musical instruments closely match the description of the ritual preparations in Nanbu County. Image from Justus Doolittle, *Social Life of the Chinese* (1865).

Another site of Yin-yang influence is found in the astrological chapter that began most Qing-era local gazetteers. Yin-yang officers typically did not compile these chapters, but were responsible for interpreting the knowledge they contained. Astrological chapters enumerated classical texts on astrology and geography in support of a particular correlation between specific stars and territory, thus drawing on sources of imperial authority to craft relevant narratives for counties and prefectures. From antiquity, twenty-eight astrological constellations (*xiu*; lodges) were correlated with the territories of twelve ancient feudal states to identify portents relevant to their areas. By the late imperial era, every county and prefecture were assigned constellations in accordance with this field allocation system (*fenye*; lit. allotted countryside).[100]

There had been a number of scholarly critiques of *fenye* cosmology over the centuries, but the central government continued to support the system through the mid-Qing with the inclusion of astral correlations in official geographies.[101] Marking a break with imperial tradition, the Qianlong Emperor openly criticized the *fenye* system in the 1770s when he declared that the empire's territories extended too far beyond the correlative fields of antiquity.[102] With Xinjiang's inclusion in the system, the Muslim northwest, along with Shaanxi, Gansu, and much of Sichuan now shared the same pair of lodges: Well (*jing*; approximately Cancer) and Ghost (*gui*; approximately Gemini). With two out of twenty-eight lodges correlated with nearly half the empire's territories, Qianlong deemed the system unworthy of belief.

Yet, *fenye* retained some degree of importance in local governance, not least because Yin-yang officers trained in astrology remained employed in county and prefectural yamens. Gazetteer writers in Sichuan did not simply include astrological chapters in the nineteenth century—they updated them. When the new edition of Chengdu's gazetteer was published in 1873, for example, its first chapter covered the history of Sichuan in astrological terms, with particular attention paid to the traumatic wars of the seventeenth century under the rebel Zhang Xianzhong (1606–47), who destroyed much of the province in the years prior to the Qing conquest. Citing the works of the celebrated Sichuanese scholars Fei Mi (1625–1701) and Peng Zunsi (d.u.; 1740 *jinshi*), the compilers enumerated textual sources to explain the political chaos of the late Ming period in relation to astrological phenomena. One cited episode addressed controversies in the Kangxi era surrounding the precise positions of two lodges. Here, the affirmatively cited source was none other than the Western Calendric System (*Xiyang lifa*) produced by the Jesuit priest and court astronomer Johann Adam Schall von Bell (1591–1666).[103]

Astrology was never a static field during the Qing, especially at the provincial, prefectural, and county levels. Huiyi Wu captures the underappreciated Jesuit contribution to Qing-era literati records and gazetteers in observing, "Ironically, the Jesuits were both dismissive of the *fenye* system as contrary to the non-Sinocentric cosmography they tried to popularize, and adamant that their work for the Imperial Astronomical Bureau should not serve Chinese 'superstition.' However, it was precisely thanks to the resilience of *fenye* as a cosmological framework that the Jesuits and the notion of 'the West' were perpetuated in local gazetteers as contributions to Qing imperial astronomy."[104]

Imperial institutions and practices, whether in the form of eclipse predictions or *fenye* correlations, were updated—at times with considerable interest in the latest sources and knowledge available.[105] Consider, for instance, a fengshui manual compiled by two Yin-yang officers, surnamed Wang, from Dingyuan County, Sichuan (1823 preface), titled *A Thorough Explanation of the Geomantic Compass* (*Luojing toujie*). The manual drew on a Kangxi-era (1662–1723) text that considered updating values on the geomantic compass based on the Jesuit-introduced Western Calendric System.[106] These changes would thereby adjust

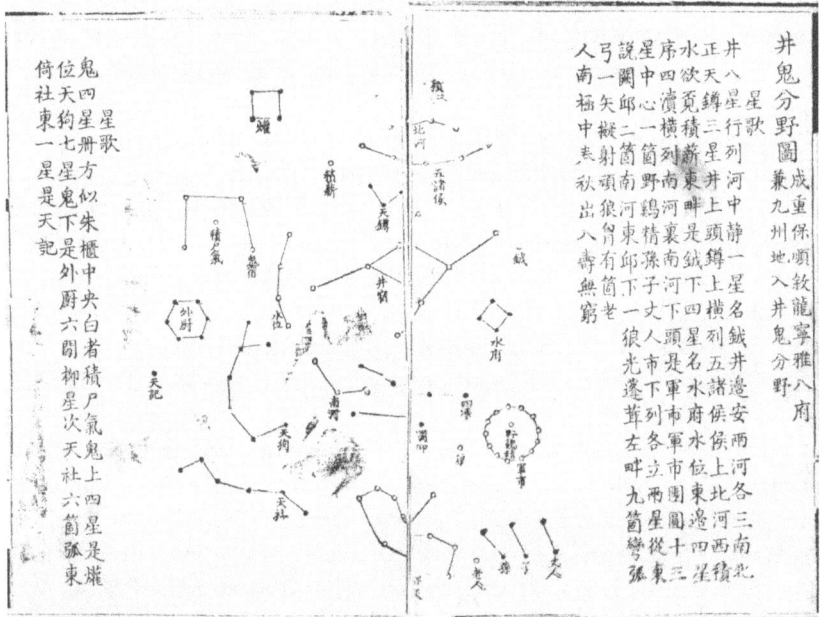

FIGURE 3.7. Ghost and Well Lodges of Guan County (Dujiangyan), Sichuan. This 1786 gazetteer diagram of the "Well" and "Ghost" lunar lodges was reproduced in many of Sichuan's local gazetteers. This image was chosen because it incorporates both constellations within a single frame. The description at the far right lists the prefectural jurisdictions within Sichuan that were covered by these two lodges: Chengdu, Chongqing, Baoning, Shunqing, Xuzhou, Long'an, Ningyuan, and Yazhou. As a matter of fact, Ghost and Well were correlated to a far larger territory by the time of this image's publication. Star map from *Qianlong Guan xian zhi* (1786).

the correlations between the new astral spans of the twenty-eight lodges and the twenty-four geomantic directions for compass calculations:

> Our dynasty's calendar can be clearly investigated, so I used the principle of equinox precession [lit. "annual difference" *suicha*; the gradual change in the orientation of an astral body's axis] to deduce problems in compass calculations.[107] All the dynasties of the past have modified their calendars, and since the Western Calendric System precisely aligns with the Qing Imperial Calendar, the geomantic compass should be updated to match the newly measured placements of astral bodies [*xiudu*]. That way, we may properly analyze mountains for the dispelling of poisonous *qi*, obtain auspiciousness and avoid inauspiciousness, and create fortune for people.[108]

When new knowledge of the natural world entered elite conversation during the dynasty, commercial publishers spread updated understandings of astrology, fengshui, cartography, and the calendar, flooding the market with information people wanted—and used.[109]

The contents of *A Thorough Explanation of the Geomantic Compass* draw attention to the fact that while astrology was part of a Yin-yang officer's career, most of their time was spent practicing fengshui. The market for geomantic consulting services in both the state and private sectors was large, and Yin-yang officers were well equipped to take advantage of the opportunities presented. Yin-yang officers oversaw local sacrifices and opera performances on festival days in accordance with officially sanctioned ritual texts, including the *Imperially Endorsed Treatise on Harmonizing Times and Distinguishing Directions*.

Yin-yang officers also selected the time for beginning construction and the position (i.e., location) and orientation of structures. Some public buildings in Sichuan were built and repositioned under their guidance. In 1862, for example, Li Jixiang presided over the repairs to a county yamen by "selecting its position and picking an auspicious time to begin construction."[110] For that case, no specific outcomes are listed, but some public writings enumerated positive results from the structural changes. In 1690, when the Han bannerman Li Tianzhi was appointed department magistrate of Weizhou, an ethnically diverse area populated by Han Chinese and people now belonging to the Qiang minority group, he summoned a locally esteemed geomancer to analyze the government office and school. After the geomancer's recommended changes were enacted, two provincial degree-holders and three military provincial degree holders appeared in the following decade (during the 1693 and 1696 examinations), leading Li to conclude in an inscribed stele that "there really are reasons to believe in fengshui" (*fengshui xin you ran ye*).[111] This case also reflects the previously discussed theme of Qing officials fostering geomantic fortune along imperial frontiers.

In addition to orienting private and public buildings, supervising private Yin-yang masters, and determining auspicious times for rituals and opera performances, Yin-yang officers in Sichuan were also summoned, alongside state-hired medical officers, to attend the triannual examinations in Chengdu.[112] Sichuan's provincial civil and military examinations were massive events for the city, attracting some fifteen thousand hopeful candidates and their entourages by the closing years of the dynasty. Chengdu Prefecture and the provincial government always needed extra staff to help patrol exam halls, collect exam papers, transcribe essays, check transcriptions, and do other miscellaneous tasks. On top of it all, health emergencies among anxious candidates—understood as extremely negative portents for the afflicted scholars—were common, and it was helpful to have medical and Yin-yang officers on hand.[113] Yin-yang officers could choose either to travel to Chengdu to help administer the examinations or return part of their state stipend for a release from the obligation. Chengdu was far away, and transport and lodgings were comparatively expensive, the latter especially so during exam season. Nanbu's Yin-yang officer declined to attend the 1859 provincial examination offered "by grace" (*enke*) in celebration of the Xianfeng Emperor's impending birthday.[114]

Yin-yang officers in Nanbu, and perhaps elsewhere, expanded their ritual and administrative roles during the nineteenth century. In the years after the White Lotus Rebellion (1796–1804), a Nanbu magistrate gave the serving Yin-yang officer responsibility for overseeing local spirit mediums across the county, in addition to the geomancers they were already managing. Qing officials saw spirit mediums as having spread White Lotus teachings in the decades leading up to the great rebellion, and the role of policing them gave Yin-yang officers considerable influence in rural society.[115] Daoists, who felt spirit mediums belonged under their purview, fought with the county's Yin-yang officers in court several times over this arrangement and related financial tensions.[116]

David Graham (1884–1961) noticed that spirit mediums in Sichuan were overseen by or else associated with Yin-yang masters, which suggests this phenomenon extended across the province.[117] Since Yin-yang officers in some Sichuan counties were already responsible for policing "around 200" local geomancers, the addition of spirit mediums involved even more responsibility.[118] In their policing role, officers were expected to immediately report the slightest appearance of heresy, and they in fact did so during the nineteenth century.[119]

What social factors made Yin-yang officers so administratively relevant? Three factors stand out. First, in wealthy counties with many degree holders who, though not licensed Yin-yang practitioners, knew a lot about the field, the status of Yin-yang officers may not have been especially high. By contrast, in poor counties with a dearth of highly educated people—like Nanbu—Yin-yang officers may well have had more authority. As literate, decently educated men serving in a county with fewer than ten living provincial degree holders in any given year, Nanbu's Yin-yang officers stood out from the crowd and wielded influence accordingly. The protracted regulatory dispute over the monitoring of spirit mediums and policing heresy is one manifestation of that influence.[120]

Second, due to Yin-yang officers' specialized knowledge, their functions sometimes complemented or even overlapped with those of coroners, who analyzed corpses for criminal investigations. Most relevant was the ability of

Yin-yang officers to analyze a corpse for estimating the time of death, which was needed for establishing the proper burial time. In the Ming-era novel *The Plum in the Golden Vase*, the Yin-yang officer Master Xu performs an analysis of a corpse for determining the precise time of death the previous night.[121] Even in relatively cosmopolitan Beijing, Yin-yang masters provided families with official death certificates through the 1930s.[122] Death was big business in Qing China, and Yin-yang officers were close to the center of it.[123]

A third and final reason for their importance probably grew out of social acceptance of their state-licensed expertise. By effectively endorsing building projects by selecting and then announcing appropriate times for their construction, Yin-yang officers served as natural fundraisers. Before offering up hard-earned silver taels or copper cash, donors wanted to see real results in examination rankings—or at least a meaningful change to the physical landscape of the town. A talented Yin-yang officer could help seal a tough investment deal with a convincing argument to donors that better fortune for scholars and merchants alike was just around the corner. With such influence, Sichuan's gentry wanted this position to exist. In one instance, scholars in Ba County petitioned the magistrate to make a quick hire when a vacancy was left unfilled for too long, writing, "... [W]hen constructing buildings and burying the dead there is always the possibility of mistakes, so not having anyone serving as Yin-yang officer in the county for a long time will produce significant errors and omissions."[124]

Naturally, some Yin-yang officers were fired for poor performance. There are few mentions of such cases in gazetteers, which as a genre were more likely to record triumphs than scandals, but authors of vernacular fiction offer some intriguing leads. In the aforementioned Ming-era novel *The Plum in the Golden Vase*, readers are introduced to the character of Wu Dian'en, described as "formerly a Yin-yang master on the staff of the yamen [of Qinghe, a fictional county in Shandong Province] who had been removed from his post 'for cause' and now makes his living by hanging around in front of the yamen guaranteeing loans to the local officials and government staff."[125]

Although the precise reasons for Wu's termination are unstated, the novel implicitly contrasts Wu Dian'en with the county's current Yin-yang officer, Master Xu. Whereas Master Xu is reliable and honest in his readings of the official calendar and almanacs, in selecting times to break ground and start construction, and in presiding over rituals for the dead, Wu is neither reliable nor honest. After obtaining a low-ranking official post with Ximen Qing's help, Wu Dian'en—a literary homonym in Chinese for "Without (*wu*) a bit of (*dian*) grace (*en*)"—even coaxes a runaway page boy to make a false accusation against Ximen Qing's second wife, thus perfectly cementing the character arc of an ex-Yin-yang officer who fails to bring fortune to those around him. In this case, the clue really was in the name.

Fengshui Manuals and the Examinations

Although quality of service varied, the hiring of Yin-yang officers at various administrative levels contributed to the dissemination of knowledge about fengshui. According to the official job specifications, such officers were instructed to consult state-approved geomantic and astrological treatises and the official calendar. Yet the market for private manuals was so large that one assumes they also consulted other sources and, in one capacity or another, shared that information with clients. A mountain of colorful material was available for inquiring readers.

One such manual, *Earthly Principles for Guiding the Original Truth*, was written by Kong Wenxing (1620–1705), also known by his adopted Buddhist name of Che Ying. Kong was a Buddhist monk of Zhejiang Province who became one of the most famous geomancers of the early Qing.[126] *Earthly Principles* not-so-discreetly served as a guide for examination hopefuls. Kong diagrammed the landed estates of prominent examination successes across Jiangnan, particularly those of first-ranked presented scholars. He also profiled many graves that had been inaccurately positioned, detailing the ways he had improved their fengshui and providing descriptions of the positive results.[127] Kong compiled his analyses into a fengshui manual filled with real case studies profiling the success or failure of families across time.

Kong's use of detailed case studies drawn from direct observation provided authority and relevance to specific geomantic theories. By analyzing the waterways of opulent residences, lushly irrigated fields, and the magnificent burial grounds belonging to first-ranked graduates such as Miao Tong (1627–97; 1667 *zhuangyuan*) and Cai Shengyuan (1652–1722; 1682 *zhuanyuan*), Kong provided Qing readers with a tantalizing glimpse into the geomantic secrets of the country's "rich and famous."[128] One can imagine scholar-officials reading Kong's manual as people today scroll through zip codes with expensive real estate on Zillow.

Kong obtained recognition through his work. The literati compilers of the *Provincial Gazetteer of Zhejiang* celebrated Kong for his painstaking fieldwork by including a description of the manual in the literary compositions (*wenhan*) section of the gazetteer.[129] But was his manual influential among geomantic practitioners? No direct link connects Kong's *Earthly Principles* to the Yin-yang officers of western China, but there are some intriguing textual resonances. *Preface to the Origins of Earthly Principles*, an undated geomantic manuscript composed in Sichuan, refers to many of the same geomantic formulas found in Kong's text, such as *mao* facing *you*, as well as some of the same methods for analyzing water configurations (*shuiju*) through the application of specific water methods (*shuifa*). According to the Sichuan manuscript, a grave carefully positioned in the *mao* facing *you* orientation, coupled with water arriving from the *geng* direction, will produce "a civil or military official serving before the emperor."[130] This environmental layout describes with considerable accuracy the Li gravesite at Pig's Trough Pass discussed in the previous chapter. It also echoes the formulae found in a number of Kong's case studies.

These parallels might be explained by the fact that Kong's text and the Sichuan manuscript both ascribe their contents to the geomantic lineage of the great Tang dynasty geomancer, Yang Yunsong (c. 834–900). Yin-yang officers in Sichuan probably consumed Kong's text or others like it. Regardless of whether they did, influence of a broader sort came from the examination system, the national body of cultured elites, and the Qing state's employment of Yin-yang officers. All of these forces shaped consensus regarding effective deployments of geomantic knowledge. As seen previously, for physical structures to hold social prestige, they had to be recognizable through the presence of water and trees, advantageous landforms with auspicious names, and efficacious orientation formulas inscribed onto tombstones. Although there was no empire-wide "standardization" of fengshui beliefs or practices, compelling geomantic claims appealed to recognized precedents, even if those precedents were consistently adjusted to local conditions and new information.[131]

Did scholars and commoners believe that by orienting graves in these directions and along these landforms they could become *jinshi* degree-holders? Here, one might consider by analogy twentieth-century best-selling American self-help books such as *The Science of Getting Rich* (1910) or *The Millionaire Next Door* (1996). These works did not produce scores of financially well-off people, but they did explain and legitimize their respective periods' permutations of capitalism while conveying some practical investment advice. Divinatory manuals may have performed a similar role, helping to explain certain outcomes of the examination system, thus supporting the popular legitimacy of a system for which undisputed statistics offered dire odds against success. Literate consumers of these texts could cite well-known success stories, put some basic principles into practice, and thus participate in the elite national culture. The Qing state did forbid literary works it deemed "false" or subversive, but many popular geomantic manuals freely circulated over the course of the dynasty.[132] The geomantic manuals consulted for this book were hardly seditious. To the contrary, they rationalized the rule of the imperial elite to the audience that most desired to join it.

Manuals also contained guidelines for the construction of academies and schools. Recall the petition of Baoning's gentry for a reoriented screen wall, a Wenqu Star Gate, and a new hall for their temple school. Where did the scholars get the ideas for these improvements to the school? Immediate structural models most likely came from Chengdu's prestigious academies. In providing persuasive advice, Yin-yang officers employed comparison, citing concrete examples of effective layouts from other places against something that was lacking in the county. Officials arriving in Sichuan from high-performing regions may also have been asked by local gentry to offer recommendations for design improvements to academies, schools, and city walls. In any case, local scholars were aware of residential fengshui manuals that identified the layouts of academies and temples from counties that produced successful candidates in the Ming and Qing eras.

Figure 3.8, taken from the fengshui manual *Secret Essentials of Judging Houses*, reveals the structural changes made to the Anqing Prefectural School in Huizhou during the late Ming—after which examination success in the area reportedly improved.[133] This school provides an excellent example of precise geomantic positioning and orientation. It was located at *zi* (north), facing the *wu* (south) direction, with each of its buildings related to the literary and martial astral operators and the five agents of Yin-yang cosmology. A geomancer recommended these specific modifications: adding a screen wall to the south, adding a four-story Library Tower Pavilion to the north, and adding a Wenchang Pagoda to the east of the central complex. In accordance with geomantic principles introduced in the previous chapter, the front screen wall had the lowest elevation of any structure of the academy, while the rearmost building had the highest. These additions added both elevation (the Library and Wenchang Towers) and privacy (the screen wall) to the school, increasing its profile along Anqing's cityscape.

Of note in the modified school is the intentional lack of any "fire" agent symbolism, for fear of inviting conflagrations in one or another of the academy's predominantly wooden buildings. For cosmological balance in the school as a whole, the fire agent was represented by the new red screen wall—the complex's major nonwooden and most southerly (a direction represented by fire in the five agents) located structure—just outside the entrance gate. Beginning with the fire agent in the front (south) and culminating with wood in the rear (north), the layout precisely followed the "mutual generation" cycle of the five agents. The new design strove to foster local talents by arranging the school space so that "wood enters the *kan* palace" (*mu ru kangong*), with the *kan* palace referring to the northernmost central position (water) in the Nine Halls Diagram (*jiugong tu*). By fostering wood on the site of the water agent at

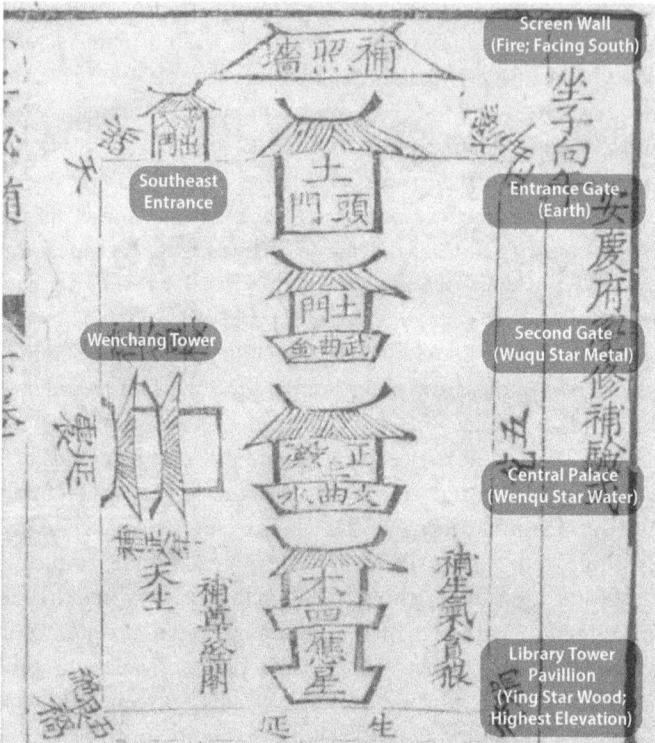

FIGURE 3.8. Geomantic Diagram of Structural Repairs to Anqing Prefectural School from *Secret Essentials of Judging Houses*. Anqing's prefectural school was designed to channel the mutually generative cycle (*xiangsheng*) of the Five Agents: fire produces earth, earth bears metal, metal collects water, and finally, water nourishes wood. Note also the rising elevation denoted with numbers on the school buildings. The "Water" Central Palace has three stories, while the culminating "Wood" Library Tower Pavilion was designed with four. Diagram from *Xinzeng buju misui quanji* (1595). Original manual held at the Library of Congress.

the northern point of the school, the number of examination candidates would—like a well-irrigated tree—grow.

Screen walls (*zhaobi*, lit. reflecting screens) were powerful geomantic structures, providing protection, authority, and privacy to the complexes they shielded while controlling the flow of *qi* into and around a site. Architecturally, elegant screen walls "divided space without completely enclosing it,"[134] thus elevating an institution's import by accentuating the presence of its entrance. Screen walls also warded off malicious spirits who saw their reflection in the screen wall when they considered entering the residence. With much influence on the line, geomancers paid a lot of attention to these structures: sometimes a school needed to add one, and sometimes a school needed to dispense with one. In Anqing, a screen wall was deemed necessary for the local school. In Yunmeng County, however, a geomancer recommended dissembling the local academy's screen wall in 1822 after a few rounds of poor examination performances.[135] Much like strategically placed pagodas, screen walls were sites to watch after provincial examination results were posted.

We can compare the image of Anqing's school to any number of eighteenth- and nineteenth-century temple schools in Sichuan. For illustration, I have chosen the temple school of Hezhou (figure 3.9), located near Ba County.[136] The image is clear, and Nanbu's gazetteer does not contain a detailed image of the county's humble academy. The layout of the Hezhou school resembles Nanbu's temple school (not shown) in several of its features, including a Halberd Gate, a Wenqu (Ling) Star Gate, a Cherishing the Sage Shrine (*Chongsheng ci*), a Central Palace, as well as a *pan* Pond and front screen wall.[137]

In truth, almost any decent gazetteer image from Sichuan would suffice for illustration, as the layouts of schools across China were structurally similar. For instance, the association of halberds with the gates of important buildings is thought to derive from a passage in the *Rites of Zhou* describing how rulers set up "gates" of guards armed with halberds when touring their realms. During the medieval era, halberds became key symbols for the Tang ruling class, with a precise number of halberds permitted to be displayed on state offices, according to rank.[138] By the Qing, the Halberd Gate had lost much of its direct association with weapons displays in becoming used as an architectural term for an outer gate in official, educational, and religious structures.[139]

When we juxtapose the images of the Anqing Prefectural School and the Hezhou Temple School, the similarities between the two structures are unmistakable: the shared screen walls at the front, two twin sets of front gates (Halberd and Wenqu or Wuqu Star), the main entrances to the southeast, the shared orientations facing south, the Wenchang Palace or Pagoda east or southeast of the main structural spine, the Great Perfection Palaces in the back of the complexes, and the culminating Library Tower Pavilions with the highest elevations.

The two complexes were built in different provinces and different dynasties, yet their layouts are strikingly resonant. This impressive level of architectural continuity and congruity could be accomplished only by a congenial combination of influences: the regulatory models and statutes of the imperial state, the universality of the examination system, the advice and assistance of Yin-yang officers, and the invocation of the laws of the land by Qing officials with reference to the cultivation of literary talents.

Some differences between the schools are also noticeable, as one would naturally expect with two structures constructed three hundred years apart.

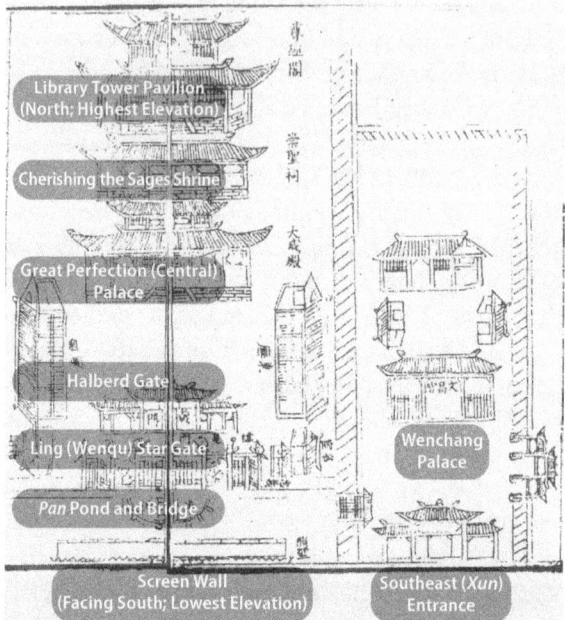

FIGURE 3.9. Layout of Hezhou Confucian Temple School. Image from *Qianlong Hezhou zhi* (1748).

The structural changes to Anqing's temple school rendered the complex's layout in perfect accordance with the generative phases of the five agents.[140] A similar interpretation may have applied to the Qing structure, but it would have necessitated a greater degree of complexity. During the Qing, additional buildings, such as Cherishing the Sages Shrines and Great Perfection Palaces, became part of the expected structural repertoire of Confucian schools and temples.[141] Cherishing the Sages Shrines proliferated in Confucian temples and schools after 1723, when the Yongzheng Emperor ordered that the five generations of Confucius's ancestors be given formal imperial titles as prince (*wang*) and collectively worshiped. By the eighteenth century, there were so many ideal structural elements to a temple school that a simple five agents schema may have become impractical, at least for some prefectures and counties.[142]

Litigating Time, Space, and Sound

In addition to constructing auspicious buildings and hiring Yin-yang officials, county magistrates across Sichuan considered legal petitions concerning the need to protect and foster fengshui. These petitions came from local scholars who demanded various forms of community zoning and landscape protection from the state. These elites, who were often members of a county's gentry, enjoyed certain advantages when it came to pressuring the state to maintain and promote the proper conditions for felicitous examination outcomes. Since they were legally immune from corporal punishment—at least for lesser crimes—such scholars were emboldened to submit sharply worded petitions to Qing officials.[143] As Nanbu and Ba counties saw a good number of such petitions presented during the 1800s, one can surmise that across Sichuan's more than 170 counties, departments, and subprefectures, there were probably many hundreds if not thousands of these petitions presented in the province over the century.

Some gentry petitions were rejected by officials as frivolous or false. Sometimes magistrates accepted a petition about fengshui as legitimate and acted to address the proposed issue, as seen in the cases involving the destruction of allegedly inauspicious mosques. At times, officials suggested a compromise that fell somewhere in the middle. To provide some structure to a complex legal landscape, two specific types of fengshui-related petitions are showcased. The first involves oil-press shops, the zoning for which was contested in Qing Sichuan and in other parts of China at rates that stand out in the historical record. The section that follows concerns bridges, which were also contested in geomantic terms. These sites do not cover the full spectrum of geomantic petitions presented to Qing courts; there were many others, including mining, which the next chapter tackles.

Operating an oil-press shop was a difficult job. Cooking oils were extracted from sesame, peanuts, rapeseed, black soybeans, and castor beans with significant manual labor. Tung oil used in wood finishing was also produced in large amounts through a similar process. An oil-press shop needed to employ three or more men or women to acquire substantial amounts of nuts, seeds, and beans to produce enough oil to have a chance of turning a profit.[144] Production followed seasonal harvests. Tung oil, for instance, was pressed over two winter months through the New Year after the nuts of tung trees ripened, while cooking oils were produced from spring to autumn as rapeseed, sesame, and peanuts came in.[145] Budget margins were tight, and shops operated within time constraints. Rapeseed storage was contingent on sustained control over humidity and temperature that remains difficult even in today's conditions. The handcrafted process involved first stir-frying seeds in a large wok. After softening, the seeds or nuts were ground in a millstone, ideally with the help of an ox. The resultant clay-like powder was then covered and steamed before being hammered with great force by one or two laborers, sometimes in a large wooden log that allowed for the convenient retention of discharged oil in large quantities. Finally, the oil was displayed in jugs for sale at the front of the shop or placed onto carts for transport into urban centers.

Even though oil shops produced commodities that were essential for community life and well-being, legal objections to their activities came from several directions. The constant hammering from within the shops was loud. If livestock were involved, the animals added to the shop's odors and commotion. A seventeenth-century commentary by Xie Zhidao on the medieval geomantic classic, *Snow Heart Rhapsody*—itself recognized in the officially compiled *Correct Doctrines of Fengshui* as an authentic text of fengshui—identified iron furnaces, oil-press shops, water-powered trip hammers, and ox-powered millstones as "objects that make great sound" (*dong xiang zhi wu*).[146] Since these crafts were loud, careful geomantic site selection was necessary to avoid environmental disturbances, general misfortune, and, of course, litigation.

Furthermore, a shop's furnace, which burned for many hours of the day, could also pose fire hazards to the largely wooden structures of walled towns, with billowing smoke cited as a problem in densely populated areas. At least some of the produced oil was flammable, meaning the shops were viewed warily by some as potential tinderboxes. Incendiary commodities like cooking oils did in fact cause fires in Qing-era Hankou.[147] Thus, in addition to noise pollution, the danger of fire also led some to fear that an oil-press shop could be harmful to fengshui. In the *Golden Mirror for Peaceful Living*, the relevant advice on oil-press shops was provided in a section titled "Inspecting an Address" (*xiangzhi*). Indicative of its importance, this section occupies a portion of the first chapter of the manual, where it lists sites that need to be avoided when choosing a residence:

> The fronts of Daoist shrines, the rears of Buddhist temples, old prisons, battlefields, sacrificial alters, discarded residences, smelting furnaces, hulling mills, oil presses, abandoned tombs, sharp ridges, overlapping hills, the spaces between two mountains, banks aligning strong or large water sources (i.e., rivers, lakes, oceans), road intersections, as well as city walls and moats—people's residences must not be near these places.[148]

The list of undesirable addresses combines natural formations, religious structures, and sites of economic activity, two of which—smelting furnaces (*luye*) and oil presses (*youfang*)—involved fire and noise.

There were zoning regulations for these shops across China. One report from the 1920s claimed that during the Qing dynasty oil-press shops had to be established at least forty Chinese miles (around fourteen English miles) away from Beijing's imperial city due to concerns about fengshui.[149] Other cities were said to have had similar regulations. Property too close to a government office, academy, or school was a risky investment for shopkeepers. In Qing-era Hankou, merchants hired fengshui masters to select the locations of shops, a practice that worked as a form of small business insurance against potential litigation. If a reputable Yin-yang master concluded the shop was locationally kosher, there was little room for future protest.[150] Due to the commercial pressures mentioned above, one imagines that some merchants continued to work behind closed doors even after a walled town's night watchman beat a bamboo clapper and announced, "The air is dry, and the land is parched! Be careful with fires and candles!" In walled towns across Sichuan into the early 1900s, a similar announcement was made five times every night, from two hours after sundown to two hours before dawn.

Sichuan's magistrates were occasionally called upon to judge whether an oil-press shop's creation had harmed fengshui. In several instances, officials rejected such petitions. In 1799, for example, when a provincial graduate named Fu Jiayue urged the closure of a merchant's oil-press shop, a Ba County magistrate rejected the claim, saying, "This scholar desires selection in the civil examinations, which can only be achieved by studiously reading books, not by fengshui."[151] Here again, we see a Qing official urging a Sichuan scholar anxious about examination performance to study the classics rather than obsess over fengshui. A similar outcome can be seen in a case from 1836, when another Ba County magistrate concluded that an oil-press shop had been established far enough away from two graves of the He family and thus had avoided summoning misfortune by disturbing an earth vein.[152] In both cases, magistrates were able to protect the shopkeepers by deflecting the geomantic accusations—the first by invoking the need to study the classics and the second by referring to the considerable distance between the shop and the graves.

Yet it would be incorrect to assume that petitions complaining about oil shops always failed. Some magistrates did conclude that an oil-press shop harmed an area's fengshui, and others at least countenanced the possibility by dispatching runners and clerks to investigate a claim.[153] Magistrate Fan's collection of judicial comments contains one such instance from his time serving in Shaanxi Province during the late Qing. In this case, an oil-press shop merchant arrived at the county court with a serious grievance. The merchant had contractually purchased land in a rural village to set up an oil-press shop, and there had been no issue during the contractual negotiations. But after he installed a wooden beam to anchor the grinding millstone, members of the local elite led by a man named Li showed up and destroyed the shop, claiming that the fengshui of the village had been disturbed by its creation. Magistrate Fan took a measured approach to the merchant's plaint, writing, "If the oil-press shop obstructs the fengshui [lit. "vein"] of the village, it is permissible for the villagers to ban it; however, to go so far as to destroy the shop's wooden beam, what kind of cruel intention is behind this act?"[154] Magistrate Fan permitted the accusation to be investigated further, and perhaps he ordered compensation paid to the merchant. As only Fan's initial judicial comment survives, the outcome was not recorded.

Another attempt at mediation appears in a case from Ba County, dated 1838. That year, a merchant surnamed Zhang presented a lawsuit at the county's branch court claiming that local scholars had destroyed his oil-press shop on the pretext that it harmed the fengshui of the town and hindered their examination chances. The official replied to the petition expressing agreement with the gentry:

> Upon investigation, we find that your oil shop was set up at the head of the market town and disturbed the dragon vein of the entire area. This branch yamen urges you to set up your oil shop in another place. On the one hand, this decision is for all the residents of the market town; on the other hand, I fear that, in the future, if the locality is not at peace, there will be many rumors around town regarding your shop and it would produce burdensome litigation. I truly say this with the mind of a heaven and earth parent-official.[155]

Although Zhang had entered a contractual agreement with a local landlord to open his shop at the "head of the town," the gentry refused to recognize the validity of the contract because of its geomantic implication, and it was rendered invalid—a status that the state upheld when it rejected Zhang's plaint.

This property conflict continued for eight months until a formal trial was held at the Ba County court. There, the oil press merchant produced witnesses in the form of local scholars who maintained that the town's fengshui had not been harmed by his shop's establishment. The magistrate then ruled that the merchant could reopen it.[156] The divergence between the branch yamen's resolution and the county court's final verdict is explained by the strategies that officials deemed necessary to prevent further disputes. The branch yamen had endorsed the fengshui understanding of the market town's territoriality in hope that the issue would dissipate through the selection of another site. But since the conflict did not end quickly over the ensuing months, the court intervened to publicly endorse a permanent site for the oil-press shop and in doing so finally close the issue.

Geomantic considerations for the placement of oil presses also extended to other enterprises, with noise pollution being a driving force in fengshui-related petitions to officials. The writings of a county magistrate serving in Sichuan, Shen Shixiu, record one such instance. After Shen, a native of Shandong, arrived in Sichuan during the late Qianlong reign to assume the post of county magistrate in Qingfu County, he was troubled by the scene he encountered. Noodle shops equipped with wooden sifting benches had proliferated throughout the walled town. One of the first steps in making noodles was sieving flour to remove husk, seeds, or clumps of gluten. The shops employed wooden cabinets (or "benches") with a sieve basket for these purposes. Flour was placed into the basket, which was then vertically shaken for straining. The results were excellent, but the process was loud since the basket repeatedly hit the wooden cabinet. Evidence for the cacophonous effects of flour sieving is found in a small episode of *A Dream of Red Mansions*. When the elderly country rustic Grannie Liu visits the Jia family mansion, she has a surprise encounter with a wall clock, and its constant chiming (*ge-dang, ge-dang*) sounds to her like "the banging of a flour sifting bench."[157]

In Qingfu County, noodle shops had opened across town, and they presumably had long lines of hungry customers. The banging noise arising from flour sieving was deafening—audible even in the late evening from inside the sheltered walls of the county's school. After consulting with the gentry, Magistrate Shen learned that academic performance had been poor, as students could not focus on their studies. The scholars were despondent since there had been few local graduates produced in the provincial rankings in recent years. Magistrate Shen needed to act. As a *jinshi* degree-holder representing the Qing government in Sichuan, he knew exactly what to say.

In a public notice issued and inscribed on a stone stele placed at the center of the town, Shen began by tracing the dragon vein of the county. He made sure to mention the county's astrologically affiliated lodges, Well and Ghost, which needed proper attention as they transmitted relevant information to the county from the heavens.[158] Local gentry had presumably drawn upon such astrological information in their initial petition in court. Magistrate Shen then delved into logistical details, observing, "From the beginning of our dynasty, iron furnaces, oil presses, and the benches of noodle shops have all been accorded their proper places by distinguishing directions and harmonizing times. With these regulations, healthy *qi* circulates, the land is efficacious, the people are healthy, and the government school is academically vibrant."[159]

Magistrate Shen certainly was not opposed to a bowl of good noodles, but there was something more important that he identified upon arriving at his post in Sichuan. According to the official, just as iron furnaces had to be operated outside the city walls and oil presses were zoned to a suburban corner of the county town, it was essential for the noodle benches to have their designated place in an area of the county seat. An image of a walled department seat showing oil presses operating outside city walls may be seen in figure 3.10.

From what source did Magistrate Shen derive this regulatory information about iron furnaces, oil presses, and noodle shops? Readers may recall that fengshui manuals and geomantic commentaries, such as Xie Zhidao's

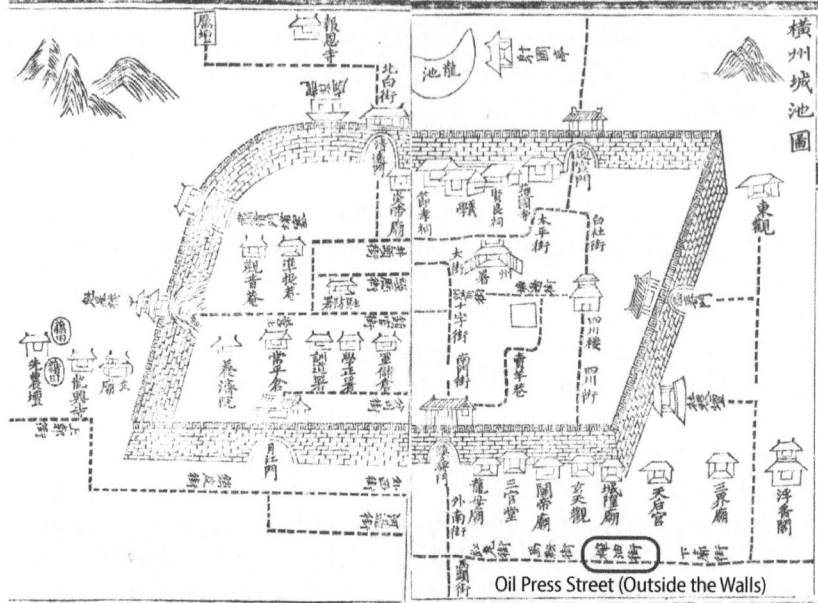

FIGURE 3.10. Oil Press Street Outside the City Walls of Hengzhou. As most gazetteer maps exclusively profiled yamens, schools, and temples, locating an image that depicts oil presses or noodle shops is difficult. Here is a rare image from Hengzhou, Guangxi Province, that was published in an 1899 reprint of a 1746 gazetteer. There, the "Oil Press Street" (*zhayou jie*) was zoned (and drawn) outside of the walled town in accordance with widely recognized geomantic regulations. Image from *Qianlong Hengzhou zhi* (1746/1899).

commentary on *Snow Heart Rhapsody* and *Golden Mirror for Peaceful Living*, urged the careful placement and orientations of related enterprises due to their high levels of noise or other potentially dangerous characteristics. Drawing on that precise principle—which he accorded with imperially recognized regulations—to protect the county's dragon vein and foster fengshui, Magistrate Shen ordered the string of noodle shops to operate outside the city walls. Students could rest assured that dinner remained only a short walk away.

One can only draw tentative conclusions from the five cases about fengshui and local enterprises discussed above, and more research is needed to determine how these negotiations played out across China during the Qing. Commercial alcohol distilleries (*shaojiu*) were subject to similar geomantic objections, which is also a topic worth further exploration.[160] Nonetheless, three observations are warranted. To begin with, these legal cases are not hard to find. To keep the analysis largely limited to Sichuan, I declined to include other cases from eastern provinces, which bear similarities to the ones discussed here.[161] Second, fengshui manuals consistently identified oil press shops as potentially problematic; the legal record demonstrates that those exact principles were invoked in law. Finally, the local gentry's presentation of these petitions can only mean that they knew they might win. From Sichuan's archival record and published judicial collections, it is undeniable they sometimes did, or at least garnered some officials' sympathy. To be clear, most oil-press shops and commercial enterprises opened and operated without protest in nineteenth-century Sichuan. But when there was trouble, officials had to carefully gauge when, where, and how they needed to intervene. If they did not get involved in instances of serious conflict, Sichuan's secret societies would soon be on hand to deliver an offer a shopkeeper could not refuse.

Ceasing Litigation Bridge

Geomancers paid close attention to bridges. They referenced bridges in relation to the examination system and general fortune in ways both positive and negative.[162] On the one hand, a well-placed bridge could increase trade and improve ease of access to previously isolated marketing communities. It would then be celebrated accordingly for heralding good economic fortune. On the other hand, sometimes residents of one settlement did not wish to be physically linked to a community on the other side of the river, or they wanted to limit potential entry points for bandits—the sudden appearance of whom was a telltale sign of poor fengshui.[163] Bridges were also status symbols, tied to the reputations of donors who had funded their creation. Changing the land around these sites through the addition of new water transit or irrigation channels could be contentious. Finally, some bridges held ritual and ornamental significance. Out of a professed desire to honor their ancestors, commoners at times erected high-status stone crescent bridges in front of ancestral halls and tombs. Members of the gentry, keen on guarding the hierarchical social order in relation to the examination system, were seldom impressed.

Instead of repeating cases with similar themes and resolutions, this section examines several cases from Nanbu, Ba, and other Sichuan counties from the second half of the dynasty. Once again, there is no shortage of material to choose from, since Sichuan had approximately 540 large rivers that flowed across the province during the Qing.[164] Local gazetteers in the province often list one, two, or three prominent bridges that were celebrated as improving fengshui, and a good number also chronicle occasional conflicts behind bridge construction. When paired with materials from local archives, one can appreciate the reasons bridges were so important and, occasionally, controversial.

Bridges were tied to fengshui for several reasons. Due to their placement over bodies of water, bridges were associated with the dragon veins of mountain ranges and river systems. Bridges were thought to link the veins of two mountains separated by a water crossing, hence the common appearance in Sichuan's gazetteers of "Arriving Dragon Bridges," "Protecting Dragon Bridges," "Connecting Vein Bridges," and, when located to the east of an ancestral hall or mountain, "Azure Dragon Bridges."[165] The addition of a bridge could link a settlement with the vein of a mountain deemed particularly auspicious to good effect. Sometimes the erection of certain bridges upon the advice of geomancers improved local examination fortunes by channeling the *qi* of a powerful dragon vein. In Jingmen Department, located in Hubei Province, a geomancer of the Form School sought to improve the area's examination results through "restoring the Literary Fortune Bridge in order to elevate the Literary Wind, and constructing the Arriving Dragon Bridge in order to connect the earth vein."[166]

In the context of rural society, ascribing bridges with positive geomantic significance made good sense. Since the imperial state generally did not see rural infrastructure development as its responsibility, most funding for bridges came from private donors.[167] Claiming that the addition of a bridge would improve an area's fengshui provided a compelling reason for people to contribute financially to a local construction effort. Members of the gentry tended to be the people with the most readily available money, so appealing to scholars for donations with a fengshui pitch was generally a sound idea. Echoing the pattern seen in pagoda construction, some members of the gentry might petition the state to publicly endorse a bridge's relocation to a site with better fengshui, presumably to assist with further fundraising.[168] If results were advantageous, which in many cases they were likely to be, donors could obliquely take credit for felicitous local events in the years after the move.

Although multiple interests certainly informed these building efforts, the truth is that better transportation infrastructure from rural settlements to a school, temple, or county seat did, quite literally, improve examination chances for rural residents who otherwise would have little access to those places. Similarly, there were increased opportunities for trade and commerce. It was not empty rhetoric to allege that a well-placed bridge could improve fengshui by bringing fortune to a wide socioeconomic spectrum of individuals. Further, because rivers ran through low-lying valleys between Sichuan's rolling hills and mountains, the link between bridges and dragon veins was natural and intuitively obvious. Geomantic narratives not only drew on commonly accepted knowledge of the land; they also adduced concrete evidence of advantages to be gained that were socially compelling.

During litigation, officials readily understood the geomantic significance of bridges and the desire of local communities to protect them. Returning briefly to the 1838 legal case from Ba County, we can view one instance of such understanding by unpacking a bit more of the story. When members of the gentry presented their petition concerning the activities of the oil-press merchant Zhang, they claimed that the alternate spot he was considering for his shop was beside an important bridge, thus harming the fengshui. Telling, the official commented, "The head of the bridge is where the dragon vein arrives; since you and the others say that this location would hurt the town's earth vein, you should all report in detail and set up a record for this case to protect the locality. Zhang Songsheng should immediately select good land in another location; do not act against the community and engage in wrongdoing."[169] Notable in the official's comment is the blending of observation and opinion. He first recognized that the bridge was in fact the place where the dragon vein arrived, and then said that because the gentry felt the shop would harm that vein, a new site had to be chosen.

A similar dispute ensued over Returning Dragon Bridge, which lay at the border of Han Department and Shifang County to the north of Chengdu. The stone bridge had been constructed at a bend in the river where the Coiled Dragon configurational force (*panlong shi*) was strongest—the same geomantic configuration ascribed to the mountain of Khoja Abd Allāh's shrine in Baoning. The problem was that, by the turn of the nineteenth century, the river was becoming increasingly difficult for boats to navigate due to runoff from intensifying agriculture along its banks. In 1803, when residents from Han Department deepened the channel to make the river navigable once more, the gentry of Shifang, who had donated funds to support the bridge's construction, presented a petition claiming that the fengshui of the county was terribly harmed by the excavation. This sweeping claim about the county's fortune cast a wide net of potential consequences, such as the appearance of droughts, floods, locusts, and examination failures. After investigating the site, a local court issued a compromise, ruling that for every year henceforth, the removal of sediment from the river had to occur at least twenty-six *gong* (c. forty meters) away from the cherished bridge.[170]

In this instance, the fengshui claim of the scholars pitted the sustainability of the expensive bridge's foundation against development that sought to sustain the navigability of the waterway. What should we make of this claim? Qing gazetteers record many well-placed bridges that were specifically constructed near the water exits of towns; these structures were described as having helped counties retain their wealth and fortune.[171] As a matter of fact, many probably did. Such invocations announced the recognized boundaries for the territorial area and common water sources of a settlement, thus preempting efforts to channel water away for private means. In other words, a bridge's publicly celebrated auspicious placement, such as the one in Shifang County discussed above, helped lock a waterway in place, making further alterations to a water source more difficult. Investors in agricultural fields located on either side of such a river or stream, some of whom were undoubtedly members of the gentry, surely benefitted from the extra security of knowing their water sources could not readily be redirected. In instances when water was channeled away from a celebrated stream, litigation over fengshui might easily ensue.[172]

In the two bridge-related cases above from nineteenth-century Sichuan, neither establishing businesses nor deepening waterways were banned because of fengshui. Rather, they were zoned by communities and local governments through fengshui. Once again, we see Sichuan's officials balancing the needs and demands of discrete communities and audiences. But another detail is worth our attention: the "dragon" names of the bridges. By intentionally naming bridges after geomantic terms, local elites established the option of contesting the area around the bridge or presenting litigation over fengshui if the need arose. In other words, as part of the construction process, locals inscribed the land with geomantic information that could later be used as evidence in their communities or in courtrooms. Considerable thought and strategic intention went into the naming of rural infrastructure, where little was left to chance.

Consideration also went into identifying where bridges should not be constructed. People sometimes opposed building bridges on the grounds that such structures offered unwanted access to their communities. Since banditry was common in Sichuan, local elites sought control over the strategic entry and exit points of an area to avoid misfortune. In one stone inscription reproduced in their genealogy, the Shi lineage of Baoning patronized and protected the temple dedicated to Chuanzhu—an important deity in Sichuan—by forbidding anyone from constructing a raised beam bridge around it.[173] This rule, which extended for five Chinese miles (roughly 1.8 English miles) in all directions around the temple, was designed to protect the area's fengshui and limit traffic to pathways that the lineage and allied groups controlled. This early eighteenth-century inscription also served to set the rules of the game for incoming migrants in later decades.

The strategy of the Shis had a natural time limit. As Sichuan filled with arriving migrants over the eighteenth century, new infrastructure became essential, and compromises had to be made. A bridge connecting Rong and Yibin along a major passageway between the two counties was constructed and torn down repeatedly for years because of geomantic disputes. Litigation continued to arrive at the Rong County court until Magistrate Huang, a native of Zhejiang who was said to be talented at fengshui, took on the protracted dispute. With all parties present, the magistrate interpreted the fengshui around the contested area and found a compromise site that anchored the bridge in an outcropping of stones, rather than the riverbed, thus resolving a key component of the previous geomantic objections. Magistrate Huang ordered the resultant structure to be named Ceasing Litigation Bridge (*wusong qiao*).[174] With such a beautifully honest name, one might consider it Qing Sichuan's premier bridge of peace.

Finally, since ornamental bridges expressed social status, they could evoke conflict on those grounds. One such dispute transpired in Nanbu. The petitioner was a member of the gentry named Zhang Zhan'ao, who, for better or worse, was a known entity at the county court. From the 1870s to the 1890s, Zhang presented petitions about many issues, ranging from literal droughts (due to poor rainfall) to "*jinshi* scholar" droughts (due to poor fengshui).[175] On this occasion, Zhang accused a commoner named Yan, who lived more than 180 Chinese miles (about fifty English miles) from the county seat, of harming fengshui by constructing an arched stone bridge said to bear the mark of the previously discussed Kui star. He also alleged that Yan had erected a stone inscription around his ancestral hall of a size and type reserved for officials of at least the fourth rank.[176] And indeed, the inscription exceeded the size permitted to commoners by a considerable margin.

To make sure the court got the point, Zhang brought along a geomancer for the presentation of one of his petitions. The document drew upon the geomancer's witness testimony in the following terms:

> We, members of the gentry, invited a geomancer to accurately analyze our county's geomantic cavern where the *qi* of the dragon vein congregates. Nanbu's dragon vein derives from Wuzi Mountain; it was obstructed and impeded by Yan's overstepping proper boundaries in erecting the stone inscription. This act has obstructed the fengshui of both officials and the people.[177]

Zhang contended that the county had failed to produce successful candidates in the provincial examinations since Yan's construction. To understand the geomancer's logic, we should keep in mind that towns and cities, just like graves and houses, were conceived of as caverns (*xue*) that existed along the dragon veins of mountains. Zhang's petition was backed up by textual sources, as the county gazetteer—conveniently compiled by members of the local gentry—also claimed that the area's dragon vein originated in Wuzi Mountain to the north and followed the Jialing River southward toward Nanbu.[178]

The court accepted the petition and ordered an investigation, which revealed that Yan had come into some money and wanted to show it off. To do so, he constructed an elaborate stone inscription near the Yan ancestral hall that supported several auspicious symbols. Zhang, not to be outdone, alleged that Yan had bribed the runner to alter his report, which did not include a description of the supposedly gaudy crescent bridge. In his verdict, the magistrate took a measured approached, starting by scolding Zhang and dismissing the charge of bribery over the runner's report. Since the construction occurred over fifty English miles away from the walled town, the magistrate ruled that he found it unlikely that the county's dragon vein had been fatally harmed by the creation of one distant bridge or stone inscription. Nonetheless, the court concluded that Yan's construction efforts had violated ritual protocol.[179] The ornate inscription needed to be removed, or at least altered.

Occupying delicate sites of bounded community space, bridges in Sichuan were the loci of status anxiety, landscape protection, and transit control. Two observations may be offered from this sampling of cases. The first concerns class dynamics. Only the degree-holding gentry had the legal standing to present a petition about the fengshui of an entire county or broad area. Commoners were generally restricted to plaints based on the fengshui of their houses and graves, although temples could become important battlegrounds as well.[180] Yet there was a large gray area in Sichuan, where most counties, except for those on the Chengdu Plain, never had particularly large elite gentry bases during the Qing, at least compared with Jiangnan. As the crescent bridge dispute and the Shi lineage's efforts to control a broad territory through temple patronage and fengshui indicate, the social boundaries between literary elites and ambitious commoners could easily become blurred, if not overtly challenged.

Second, it simply did not matter in Sichuan whether fengshui was invoked in law as a "cover" for ulterior motives. Though it certainly could be, fengshui was socially persuasive enough to stand firm as the heart of a legal argument. Reflecting on the legal legacy of fengshui in twentieth-century Hong Kong, the legal scholar Peter Wesley-Smith once observed, "Many fengshui interests are in fact protected through political processes which exploit the susceptibilities of paternalistic British administrators or were pursued so fervently that they simply could not be resisted."[181] That description might be applied to legal proceedings in the Qing dynasty, with an important caveat. During the Qing, the imperial state openly supported and patronized the geomantic arts, employed judges who had memorized the *Classic of Changes*, and issued official guides for time and site selection. The prevailing logic in the Qing was for the government to educate the people in the correct applications of fengshui so there would be no further disputes over it. In all likelihood, the British did not have quite the same idea in mind.

Landscapes of the Imperial Examinations

The deployments of fengshui discussed in this chapter encapsulated much more than one rural county in western China. They comprised something the Qing state engaged on an intimate, rational, and profound level that was hard to express in one sentence, capture in one legal petition, or understand through eavesdropping on one courtroom trial. The average Qing official might invoke related principles once, twice, or even more times in his career—when he was facing down a tough crowd, when he needed to project knowledge and authority in an unfamiliar county with a difficult dialect, or when he just needed to make something happen. In the broader scheme of things, not a few Qing officials felt they needed fengshui more than ever as Sichuan's walled cities and townships exploded with people at the turn of the nineteenth century. Officials needed fengshui to manage and resolve disputes about community space. They needed it to maintain the reputation of the dynasty's examination system, which was bursting at the seams from demographic stress. Blending empirical observation with human narrative and the uncertain future with the recorded past, fengshui was never the subject of a question on the civil examinations. Yet it was often a key to success after a scholar took office.

Over the past decades, historians have evaluated the architectural and spatial histories of the Chinese city, looking for patterns over time to gauge the resiliency of certain cosmic ideals like fengshui. Paul Wheatley identified a strong ritual urban form in the earliest Chinese cities.[182] Arthur Wright saw considerable continuity from antiquity on the level of discourse while observing that fengshui's invocation was applied retroactively rather "than in the actual choice of site."[183] Jeffrey Meyer, in his study of Chang'an and Beijing, noted some shared elements that accorded with geomantic principles but also identified many contradictions in spatial layouts, conflicting interpretations over urban orientations, and a wide range of topographical features.[184] Others have split the difference between form and function by noting that late imperial Beijing was literally composed of "two adjacent cities, one ordered, symbolic, and regular, the other practical, curvilinear, and ragged."[185] This body of scholarship has produced insights on continuities and ruptures in Chinese urban forms, but the broad takeaway is that invocations of fengshui are bound to appear haphazard, discontinuous, and partial when approached exclusively through the assumption of fixed architectural principles.

A fruitful approach for making sense of fengshui's invocations in Qing Sichuan can be found in Sarah Schneewind's discussion of the "minor mandate" of county magistrates and local officials.[186] This concept, which stresses the importance of public opinion for governance during the late imperial era, highlights the "political and cosmic autonomy" that local officials had in county administration.[187] Rather than adhering to a fixed set of political principles bestowed from above, the most striking applications of fengshui in Qing times were local and legal. The cosmology of a city was distinct, and not because of a unilateral projection of the state's political power over the people. Rather, it was distinct through the existence of a wide and diverse range of people who were all invested in that cosmology. Increasingly dire odds for success through the examination system, the need for purchased officeholders to legitimate their positions, the growth and expansion of Sichuan's rural townships, and the constant production and circulation of new fengshui knowledge gave many actors grounded stakes in the landscapes of the imperial examinations, including, in Baoning Prefecture, a highly engaged Muslim community. There was a special kind of power in fengshui that could not be found in commentaries of *The Mencius* or *The Doctrine of the Mean*.

There were some fixed patterns and blueprints behind fengshui, as seen in the relatively set layouts of state-run Confucian temple schools and in the general geomantic principles derived from state-sanctioned and famous geomantic texts. That said, the guidelines of official, classical, or popular geomantic manuals did not constitute fixed regulations but general principles that officials and gentry could invoke during negotiations over local space. Everyone agreed that careful thought had to be given to the placement of oil-press shops and iron furnaces within the wooden structures of urban and village environments. Specific discussions deriving from that principle hinged on grounded details: How far away from a school, a temple, a house, a grave, or a water source should these structures be? When communities could not answer those questions, the legal system did.

We can add geographic and temporal dimensions to this discussion. Along the western frontier of Sichuan during the nineteenth century, how much power did Qing officials have? After all, magistrates served for an average span of less than two to three years in a county. Undoubtedly, some officials did not believe that most geomantic theories worked and were confident they had "made it" solely on their hard work and talent. But if the people in their jurisdictions—especially the gentry—believed that fengshui worked, it was simply not politic to deny its possible ramifications. Officials needed to be receptive to popular opinion, or at least to the demands of local notables who allowed governance to function at the county level. This need was especially applicable in the second half of the dynasty, when dynastic power at the local level increasingly depended on the participation of clerks and runners, militia leaders, tax heads, salt merchants, Yin-yang officers, and other brokers and intermediaries. Resorting to fengshui might mean not resorting to violence—something Qing officials, capable of reading the faces of a courtroom and between the lines of a petition, certainly did not want.

Put aside predictable official rhetoric bemoaning lawsuits, delayed burials, and lavish funerals to grasp the big picture. Fengshui's aspirational logic of family competition rooted in the ancestral cult, its diagnostics of community fortune based on examination performance, and its unabashed celebration of local pride, essential infrastructure, decorative design, and semilegendary histories of conventional success were collectively far better for the imperial state than the treasonous activities of the White Lotus or Sichuan's secret societies. Even with its downsides—such as complex lawsuits and delayed burials—fengshui tied people to the land, ancestors, tax-paying family units, temple-based rural communities, the elusive dream of examination success, and, ultimately, the Qing state. The metropolitan examinations were a faraway ideal for a county that produced zero, one, two, or three "presented scholars" throughout the dynasty. By contrast, fengshui was immediate, familiar, and meaningful. Sichuan's officials were willing to take fengshui into account, and it is doubtful they saw it as a difficult choice when considering the unsavory alternatives on the ground. C. K. Yang's (1911–99) refreshingly blunt comment deserves the final word: "Some Confucians did not believe in geomancy, but they appear to have been in the minority."[188] In Qing Sichuan, the sky was high, the emperor was far away, and C. K. Yang was correct.

There is, of course, the possibility that officials posted to Jiangnan were more confident in dismissing gentry petitions since examination performance was relatively strong there. Perhaps officials serving in Zhejiang believed that scholars living in the wealthy suburbs of Hangzhou enjoyed the privilege of more favorable fengshui and had little reason to complain. It is easy to imagine that frontier zones and poor peripheries may have necessitated more sympathetic posturing from officials than in the more affluent areas of the Qing empire. Then again, one can reasonably assume that elite families in wealthy districts had more resources available for achieving their aims in court.

In the case of Nanbu County, one must recall that the local academy, the examination hall, and the county gazetteer—sites of, or vehicles for, considerable geomantic anxiety—were all constructed in the late eighteenth or nineteenth centuries, as were a good many private ancestral halls. Once the government oversaw the construction of schools and academies in poor rural counties, those institutions had to justify their considerable upkeep costs. They needed real examination successes, which were very hard to come by. Like miraculous tales ascribed to Buddhist temples and Daoist shrines in earlier centuries, compelling geomantic narratives produced through negotiations among officials, scholars, and commoners did the hard work on the ground. As seen in the next chapter, people also mobilized these geomantic narratives to exert control over fortunes that grew beneath the ground.[189]

4
Mining Sichuan

IN 1853, THE XIANFENG EMPEROR was in trouble. War had broken out across southern and northern China, and the Qing needed to rapidly muster resources. Amid the chaos, the western province of Sichuan was a rare oasis of political stability—and the home of valuable mineral wealth. The central government desperately needed minted metals for military expenditures, especially in the wake of tepid results from an effort to enhance mining efforts after the First Opium War (1839–42). The emperor identified the root of the problem as the reluctance of provincial officials to approve new mines. Sichuan's officials, aiming to avoid difficult decisions, regularly claimed that a mine "would disturb fengshui," thereby blocking the extraction of minerals from the soil.[1] In the emperor's view, some officials were overly cautious, while others were corrupt. The emperor compelled Sichuan's governor-general to weigh the advantages and disadvantages of opening mines and to apply caution in all cases. But it was time to double down on mining.

Sichuan was hardly a virgin territory. People had extensively and intensively mined its rocks, minerals, and ores for some two thousand years. The Sichuan Basin occupies what six million years ago was a vast lake, the recession of which left behind thick layers of brine, saline rock, and natural gas. These deposits allowed salt to be mined in the province from the third century BCE onward. During the Qing, merchants enthusiastically mined salt under the government's monopoly on salt distribution.[2] Sichuan also saw the opening of iron, copper, gold, tin, coal, zinc, and sylvite mines with varying degrees of state oversight, the trend being limited regulation.[3] Stone quarrying for the construction of roads, walls, and pagodas was widely practiced across the province. Officials had strong incentives to provide for the people's livelihoods by supporting mining ventures wherever feasible, and the Qing state did not get in the way of well-intended investments. As Madeleine Zelin has argued, the great success of investors in Sichuan's salt industry "belies the long-held beliefs that social structure, the absence of modern banking, and cultural bias against business precluded industrial investment and development in China."[4]

Profit was a strong motivator wherever possible, and the significant number of mines opened in nineteenth-century Sichuan attests to the fact it often was.

With this picture of extensive and intensive extraction, one might ask: Why would an official even consider issuing a ban on mining? A few factors were at play. Sichuan's long history of extraction had implications for Qing-era mining.[5] Apart from salt, some of the province's accessible, attractive, and economic deposits of coal, iron ore, and tin had been exploited well before the nineteenth century. Before the advent of advanced mining technology in the late nineteenth century, some ore and mineral deposits in Sichuan were challenging to reach and thus inaccessible for large-scale exploitation.[6] Terrifying tales of deadly mining accidents reinforced popular views that extraction from certain areas was inauspicious.

Furthermore, due to the population boom of the 1700s, people moved into living and farming along peripheral mountain lands, where many mining and quarrying sites also came to be located. Property conflicts became common, and the presence of Buddhist temples and Daoist shrines in potentially productive mining areas introduced further legal complexities. Some officials identified links between mining and grain prices, unemployment, violent brawls, theft, alcoholism, adultery, and litigation.[7] Officials assigned to Sichuan from faraway provinces wanted to avoid trouble or conflict with local elites, leading them to close unpopular mines. Anxiety over examination performance led scholars to oppose mining in sensitive areas deemed essential for a county's reputation and fortunes.

Also relevant was the primacy of agriculture in the Qing economy. Officials serving in the Chinese provinces of the Qing empire ascribed great importance to the procurement of essential grains for human and animal consumption and timber for fuel and building.[8] Recall that aside from salt and metals minted for its imperial currencies, the Qing did not depend heavily on revenue from mining.[9] For most of the dynasty's 268 years, the state's fiscal foundation was the land tax, around 77 percent of total central government revenue in 1849.[10] Had imperial revenue been more dependent on taxation from mining rather than agriculture, imperial government policies toward the extractive industries may have been different.[11]

Officials cited mining as a threat to agriculture and gravesites, which included consideration of the trees and water sources that surrounded them. For instance, the firm mining advocate and celebrated statesman Chen Hongmou (1696–1771) considered an appeal from the magistrate of Ningyuan County in the 1760s to ban lead, zinc, and tin mining in order to protect the fengshui of a significant mountain and sustain local agriculture.[12] The magistrate's mapped report read, "As for the waters of Jiuyi Mountain, metal, abundant in the mountain, nourishes the river, which is used to irrigate the fields of the people; the water gathers the spirits at the Mausoleum of [Yu] Shun, which is critical for the range's dragon vein.... If we are hasty and permit mines to open, then the metals will be lost and the water exhausted, and the dragon vein chiseled to the point of lost efficacy."[13] Chen agreed with the magistrate's evaluation of the mountain and expressed his support for the ban.[14] In this case, a mining ban overlapped with the imperative to sustain agriculture and protect the fengshui of a significant mountain range in south China.

Some scholarship has addressed the deep ties between fengshui and mining in the Qing period. William Frederick Collins's 1918 *Mineral Enterprise in China* devoted its entire second chapter to "The Chinese Science of Fengshui," "Its Effect on Mining," and "Recent Progress in Opposition to this Superstition."[15] Collins asserted that, due to fengshui's pervasive influence, "the working of mines on a large scale has, from the earliest time, been impossible to individuals, and only practicable by the state."[16] Later historians tempered this appraisal with greater historical finesse. Mark Elvin argues that Chinese cultural practices "seemed to have had no inhibiting effect on economic enterprise, with the exception of geomantic objections to mining."[17] Peter Golas agrees, though he expands that exception to a wider range of religious practices that influenced mining.[18] Nuances aside, historians generally concur that geomantic beliefs and practices had some impact on when and where people pursued mining ventures in China before the twentieth century.

What remains unappreciated is that mining was an arena where the Qing government directly invoked fengshui as a component of its law and regulations. These imperial policies shaped incentives for a range of actors in local society. Due to low levels of governmental regulation, mines were relatively easy to open. However, extraction was relatively easy to oppose if mining occurred near farmland and graves. Accordingly, legal records touching on both fengshui and mining primarily concern arguments for banning extraction, not arguments supporting it, as the latter did not require litigation. The scattered nature of archival cases makes it difficult to quantify fengshui's economic impact, but people—including the young Xianfeng Emperor—understood there was one.

None of these observations imply that fengshui's invocation "prevented" mining in China, at least before industrialization efforts in the late 1800s. Such a contention would mean that fengshui existed apart from, and before, the exploitation of the land. It also imagines fengshui masters and miners opposing each other in an antagonistic struggle over land use. Neither of those things was true. Fengshui was a critical frame for describing and regulating mining processes throughout the nineteenth century.[19] Like any general guidelines, applying geomantic principles in practice required case-by-case interpretation. To open a mine, miners took precautions to ensure that fengshui of an area was not disturbed, and as we shall see, there was a broad spectrum of insurance available to merchants to ensure safe investment. Nonetheless, there were many times when individuals opposed a mining project and cited fengshui in making their case. It was then up to law courts to weigh the available information and make a ruling.

This chapter reveals that officials in Nanbu, Ba, and other Sichuan counties consistently considered fengshui as a factor during the adjudication of mining disputes.[20] Administrative incentives to ban mining came not only from popular pressure from below but also from Qing governing institutions. Two points inform this argument. First, the Qing government at all levels recognized fengshui to be a valid legal reason to ban mining. We cannot know if people always followed mining bans in practice; we can only know that officials in Sichuan sometimes issued them throughout the dynasty and into the latter part of the nineteenth century. Drought, famine, poor examination performance, and worsening security conditions—all of which could be evidence of poor fengshui—sometimes led officials to announce that rocks and minerals needed to be kept in the ground. Second, people across Sichuanese society, including miners and merchants, assumed fengshui's relevance to mining. Fengshui was an expected part of the legal terrain, and merchants incorporated it into investment plans. Over the nineteenth century, when people opened an unprecedented number of mines in Sichuan, various types of mining insurance against geomantic litigation proliferated. Investors needed that insurance.

The sections that follow begin by detailing the Qing regulatory landscape for mining. The analysis then profiles legal invocations of fengshui concerning salt, coal, gold, and stone mining in nineteenth-century Sichuan. Next, focus turns to the types of mining insurance available to merchants seeking to guard against geomantic litigation. Finally, the chapter concludes with a case that brings the abovementioned factors into sharp relief. This case concerns a sweeping ban on excavation issued by Sichuan's governor-general, Wu Tang (1813–76), designed to protect Sichuan's provincial dragon vein in the wake of a famine in the early 1870s. This case, together with the previous sections, indicates that tensions over mining were building in Sichuan throughout the nineteenth century—pressures that were manifest in other parts of China at the time as well.

Qing Mining Regulations

From the beginning of the dynasty through the transitional years of 1898–1902, the Qing government did not intensively regulate most kinds of mining. Metals used in the imperial currencies, namely silver and copper, saw considerable oversight, and some products like gold were sent to Beijing as annual tribute from the provinces that produced it.[21] For most kinds of mining however, the law code, collected administrative compendia, and imperial edicts simply offered general guidelines over when and where mining should occur along with taxation rates for various kinds of mined products.[22] This section considers regulations across these legal sources that were related to fengshui.

The *Great Qing Code* reveals much through the layout of its contents. Specifically, it did not contain a designated statute on mining. As a testament to the primacy of agriculture in dynastic law, the code situated punishments for illicit excavation (*daowa*) under the statute of stealing grains from people's fields.[23] An additional substatute under coinage (*qianfa*) first adopted in *Substatutes of the Board of Punishments in Current Use* (*Xingbu xianxing zeli*; 1680) forbade copper and lead mining "in any places with gravesites" without exception.[24] Since many places were home to gravesites, on-sight appraising was required before mining could commence. The opening of a mine technically required merchants to apply for permission by registering with a governmental office for the purposes of taxation. Nonetheless, legal cases make it clear that private exploitation was widespread and generally tolerated in the absence of protest.

Qing law also designated certain areas where mining was forbidden due to geomantic considerations. Naturally, excavation around the imperial tombs was strictly controlled, but the law also set some limits on excavation around the imperial capital. The *Imperially Endorsed Collected Statutes and Regulations of the Great Qing* offers a number of such places, including the Dayu Mountains, which were identified as "connected to the fengshui of *jingcheng* [i.e., Beijing]."[25] A similar regulation in the *Great Qing Code* forbade stone quarrying in the Western Hills (*Xishan*) of Beijing and the unauthorized establishment of kilns for selling coal there.[26] As described by Susan Naquin, the Western Hills offered Beijing's residents "scenic vistas, uninhabited mountain trails, wintry landscapes, dark caves, crystalline springs, and tree-covered hillsides."[27] The Qianlong Emperor authorized some coal mines in the Western Hills to meet the growing fuel needs of capital's population, but Qing law formally extended protection to the area.[28]

Resonant regulations applied beyond the imperial capital. The *Draft History of the Qing*, a semiofficial history of the dynasty composed in the early twentieth century, summarized the imperial government's mining policies in the following terms: "If mining disturbs the fengshui of a forbidden mountain, harms the field or graves of the people, or instigates a mob that disturbs the people, or if there is crop failure and the price of grain increases, then mining should be banned in that area."[29] This passage offers common reasons mining was banned by Qing officials. Indeed, because these reasons were cited so often, emperors and provincial governors at times wondered whether the rules were being sincerely or duplicitously invoked, as indicated by the imperial edict at the start of this chapter.[30]

Whether an official or commoner wanted a mine to open or wanted one to close, the reasons above constituted the key boxes that needed to be checked— in writing. For most cases where the government approved mining in the *Imperially Endorsed Collected Statutes and Precedents of the Great Qing*, authorities explicitly stated that no harm would come to farming, graves, or fengshui.[31] One memorial composed by the lieutenant governor of Guangxi in 1743 stated, "For any mountains found with exploitable metals, when mining does not harm farmland or the fengshui of private buildings or graves, we should approve the merchants' requests for exploiting, so that it can yield a profit from nature."[32] The crux of the sentence was not contingent on grave destruction, which was already identified as a serious crime in the law code, but on the principle that the environment around a grave (i.e., its fengshui) could not be disturbed by mining.

The government's policies toward mining naturally shifted over the course of the dynasty's history. To summarize a complex topic, it is often held that the reigns of Kangxi and Yongzheng during the early Qing presided over relatively conservative policies toward extraction.[33] Historians have pointed to the controversies surrounding mining tax disputes in the late Ming as a key frame of reference for the early Qing state's policies.[34] The two abovementioned emperors rejected a fair number of mining and excavation proposals, with Yongzheng once commenting in his characteristic style on the plan of the governor-general of Zhili, Li Weijun (d. 1727), to dig irrigation ditches around Beijing, "This matter is something that can wait, but you act with such impatience, which is truly a big mistake! If someone brings an accusation that you have hindered fengshui, it will be very hard for me to pardon you."[35] The imperial comment cannot be taken as timeless state policy, but it does indicate that harming fengshui through activities akin to mining could be understood as a criminal offense.

A relaxation of mining bans occurred during the reign of the Qianlong Emperor, who in 1740 began consenting to requests from provincial governors to open mines for increasing state revenue, employing landless laborers, and expanding fuel sources. That year, Grand Academician and Minister of Rites Zhao Guolin (1673–1751) memorialized the emperor to recommend opening coal mines wherever possible "as long as they do not disturb the city moats, the dragon vein of a town, or the tombs and mausoleums of past rulers and sagely worthies, and as long as it does not affect the places where thoroughfares and embankments have been constructed."[36] Another memorial composed by Zhang Tingyu (1672–1755) in 1743 summarized the voices of mining proponents in the government by rhetorically asking, "If a mine injures the dragon vein of an area, the people themselves will not mine there; if a dragon vein is not disturbed, why should we leave minerals in the ground?"[37] It is striking that these memorials, both composed by mid-eighteenth-century advocates for increasing mining, invoked fengshui. One must assume that these elite officials found the idea that local people could themselves decide if fengshui had been harmed to be a powerful way to frame their arguments to the emperor.

The liberalization of mining under Qianlong had long-term consequences. One important effect was that the central government's policies placed more power in the hands of provincial officials and resident administrators, who henceforth had greater autonomy to determine whether mining should be pursued or banned. The result was more mining, with the central government retaining limited control over many ventures. Further complicating matters, provincial authorities often had incentives to underreport their revenue to the central government, and Sichuan was an especially notorious case during the Qing period. Focusing specifically on Sichuan, William Skinner has clearly exposed the province's tradition of irregular reporting and cheating the central government with respect to population figures.[38] The revenue side of these trends accelerated with the levying of new taxes in the three decades (1864–94) after the Taiping Civil War, when at least 40 percent of all revenue accrued in Qing provinces went unreported to Beijing.[39]

Nonetheless, pressing information about important mines, mine closures, and taxation were still reported to Beijing through the end of the dynasty. When officials proposed the opening of a large mine, they usually affirmed that the selected site "will not harm farmland nor the fengshui of private buildings or graves."[40] Although the line's inclusion was perfunctory in most cases, it remained standard to include in memorials concerning mining. When mining was alleged to harm fengshui and cause significant trouble, provincial officials composed memorials to report such conflicts to the emperor. One such memorial was composed in 1819 by the investigating censor of the Jiangnan Circuit, Wang Yunhui (d.u.). Wang described concerns about quarrying and mining around Shandong's provincial capital at Jinan, where resident administrators had failed to halt fengshui-harming extraction to the ire of the region's examination-taking gentry.[41]

One final source of imperial regulation is easy to overlook but deserves serious consideration. The *Imperially Endorsed Treatise on Harmonizing Times and Distinguishing Directions* influenced many activities that were akin to mining through setting out recommendations suitable to the dictates of ritual time. This authoritative compendium laid out year tables (*nianbiao*), month tables (*yuebiao*), and day tables (*ribiao*) that cosmologically analyzed discrete segments of time for their inclusion in the annual imperial calendar. The result was much richer than anything people would recognize in a calendar today.

Consider an example from the month tables. Within a given month, days were identified as either auspicious or inauspicious for engaging a particular action based on the sexagenary cycle and the five agents (wood, fire, earth, metal, and water). When *bingyin*, the third day of the sixty-day cycle that is closely associated with the fire (*bing*) and wood (*yin*) agents, fell three days after New Year, many activities related to altering the land needed to be avoided.[42] Constructing a dyke (*zhudi*), moving the earth for the building of a residence (*dongtu*), breaking the land for creating a grave (*potu*), opening a water channel or dredging a well (*kaiqu chuanjing*), cutting a tree for wood (*famu*), and constructing a road (*pingzhi daotu*) were all taken as inauspicious actions on that day. It was nonetheless an ideal day for inquiring after a prospective bride's name and date of birth (*wenming*), making a proposal of marriage (*nacai*), or writing a contract (*liquan*). The calendar always kept people busy.

The imperial calendar did not object to dredging wells or opening mines as a matter of principle. There were many days over the course of a given year that were suitable for opening a well or building a road, so development plans were never delayed indefinitely. The key point is that the imperial calendar lent gravitas to a wide range of social and economic activities, ritualizing them for people. Moving the earth for building a residence was placed in the same "bundle" of important actions that also included marriage—as was cutting down a tree. These were actions that mattered in rural society, and the government knew it. Because the actions covered by the calendar also produced the most litigation at the county level, it made sense for the imperial state to provide ample advice about them. In this way, the calendar regulated family and community behavior much like law, with "punishments" for violations self-induced through the suffering of inauspicious repercussions.

In brief, Qing law recognized that mining could be pursued legally if (1) local security was ensured, (2) agriculture was not harmed, and (3) the fengshui of gravesites was not disturbed.[43] Interpretation and negotiation were central to the dynasty's decision-making process, and sometimes officials leaned one way or another for or against a specific project. Over time, administrative pressures from both above and below resulted in the opening of more mines, but these three conditions were the laws of the land—and people knew it. By formally recognizing that mining could be inauspicious, Qing law affirmed and reinforced widely recognized knowledge of land's features and principles, providing all the logic needed to make a persuasive claim in court.

"Salt for the Highs, Graves on the Side"

Among all products mined in Sichuan, none was more prominent than salt. Prior to the twentieth century, brine was first raised in drilled wells through long bamboo pipes and then evaporated into salt by burning natural gas, coal, or timber. The product then went to market. The wells of Zigong were the province's largest during the nineteenth century, when they produced around 60 percent of all Sichuan's salt.[44] By 1849, Nanbu registered over five thousand small- to medium-sized wells, which produced most of the salt consumed in Baoning Prefecture.[45] Across Sichuan, salt mining was unambiguously celebrated. The benefits of the industry were immense to both the Qing state and the people—not only because salt was essential as a preservative, a medicinal substance, and a food additive, but also because wells, furnaces, and pipes offered stable employment to a large pool of Sichuan's skilled and unskilled workers.[46]

One insight into salt mining in Northern Sichuan was captured by the scholar Liu Xianting (1648–95). Drilling a single well into Sichuan's sandstone and carboniferous rock could take years, he wrote. Wells extended deep into the ground—down to a kilometer or more—with drilling that might be "as little as an inch or as much as a foot a day."[47] Large sums of money were needed even to consider beginning the process, and no one was certain where salt brine might be found. Liu observed that because finding a good spot could make the difference between fortune and bankruptcy, the merchants of Northern Sichuan employed geomancers to locate the sites of potential salt wells.[48] Some of those geomancers ended up in the salt business themselves.

Salt mine divination was a Sichuan specialty, recorded in the provincial gazetteer and echoed by observers into the early twentieth century, when it was still common to invite a geomancer to endorse a prospective drilling site in the presence of potential mine investors.[49] Once brine was struck, other merchants rushed in, competing to create new wells around the "auspicious" spot—in just the way that people competed to add graves along mountains proven to be geomantically efficacious through the evidence offered by wealth accumulation and examination outcomes. Sichuan's salt merchants and fengshui masters quite literally settled the west.

Though much of Sichuan salt's history with fengshui is tied to legend, some links were preserved in both scholarly and down-to-earth practical writings. Qing-era technical treatises of salt production made ample use of *qi* and the five agents to explain the subterranean dynamics of brine formation.[50] Geomantic applications were thus intuitive to participants in the industry. One of the wealthiest salt merchants of early twentieth-century Sichuan, Xiong Zuozhou (1892–1958), was renowned for his mastery of fengshui texts such as

Jiang Dahong's *Distinguishing Correct Earthly Principles*, which he drew on to meet high popular demand for his supplementary geomantic consulting.[51] A proven record of investments in successful brine drills—the Sichuan salt merchant's equivalent of passing the civil examinations—served as powerful evidence that a person understood the ways of fortune and could be trusted to provide competent advice.

Geomantic terminology also frequently appeared in salt well deeds. Across Sichuan, the term earth vein (*dimai*) referred to the prospective site of subterranean salt brine and hence the site where a well was drilled, just as burial plot was dug along an auspicious vein.[52] Houses of the living (*yangzhai*) and houses of the dead (*yinzhai*)—two terms readers of this book will be familiar with—were employed in reference to topsoil and subterranean (brine) rights respectively, alongside their original geomantic meanings of "house" and "grave."[53]

Some salt contracts, which were often long and highly descriptive, identified dragon veins, or subterranean channels of powerful *qi* that coagulated into valuable brine:

> Since the well was closed for many years, in the thirty-first year of the Guangxu Emperor's reign (1905) the dragon vein recovered, and Jing Haiquan invited his business partners to again exploit the lower vein (of salt brine) while changing the well's name to "Overflowing River" Well.[54]

By allowing the brine to rest without exploitation for a few decades, merchants could allow the earth to recover and regenerate the brine's dragon vein. Powerful dragon veins needed attention and protection: the act of rushing in to drill new wells around a site that had been proven to be profitable was called "riding the dragon and severing the vein" (*qilong jiemai*).[55] Similar terms appear in contracts for burial sites in Sichuan. In one such contract, a landowner—fearful that the addition of more graves in an area found to be especially auspicious might disturb the delicate balance of *qi*—forbade the placement of extra graves on the spot:

> Above the grave, it is not permissible to "ride the dragon" [add another grave along the dragon vein], below the grave, it is not permissible to "cut the cavern" [harm the grave with the addition of an extra grave in the cavern]. Cows and sheep may not trample on this area.[56]

The persistent use of geomantic terminology in Sichuan's salt industry indicates that Liu Xianting's observation about selecting salt well sites in the early Qing remained accurate into later eras.

Legal complaints about fengshui and salt extraction occasionally arose in Nanbu and other salt-producing counties in Sichuan. Most cases concerned small private wells that were allegedly created too close to gravesites. In 1887, Liao Daren, the head of a local surveillance unit, accused four residents of his market town, Beiyuan Temple, of drilling a salt well behind the Liao cemetery, thus severing an earth vein.[57] Liao also mentioned that the men were privately evaporating salt for smuggling. The court dispatched a runner to investigate the claim, but no resolution survives in the archive. To avoid potential disputes like this one, salt well contracts included detailed information about nearby graves and trees, setting out clear expectations for their management upon drilling:

> No matter whether the graves are of the Wang family, of other families, or the one ancient grave within the sold boundary [for drilling]—none of these graves have "forbidden land." As for the trees and grass growing on top of the grave, only the managers of the original grave sites may cut them. If the grave is moved, then the buyer may open the land for building, and reburial is not permitted.[58]

Disputes of this kind were uncommon, at least in Nanbu. Out of the 1,530 litigation files concerning salt wells that survive in the archive from the last century of Qing rule, only a small handful of cases invokes the alleged detrimental effects of fengshui. Concerns over salt smuggling notwithstanding, the major objections to salt production were related to the management of natural resources needed to sustain local industries. Mianzhu County, located north of Chengdu, produced some salt in its landlocked northwestern mountains, where it was difficult to import coal. Since significant amounts of timber were needed to evaporate salt brine there, a magistrate declared that the well-being of the area had been "obstructed" (*ai*) through the overexploitation of local forests, and he banned the further opening of wells in the area.[59] Again, the term *ai* in relation to a forested mountain suggests that considerations of fengshui may have influenced the magistrate's decision.

In counties where salt was produced in considerable quantities, several factors contributed to the security and reputation of salt-extraction enterprises: the profitability of the wells, the industry's associations with the Qing state—at least in terms of salt distribution—and the fact that fengshui was usually examined by reputable experts before the land was drilled. Considering the considerable social capital of salt merchants, who donated generously for the building of schools, temples, academies, and pagodas, it was probably difficult to find a respectable geomancer willing to legally testify against these powerful tradesmen.[60] Indeed, in some cases, the most celebrated geomancers were found among the salt merchants. Salt was a very special kind of fortune in Sichuan, as the people of the province clearly recognized.

Contesting Coal

Patterns in coal mining and consumption in Qing Sichuan were naturally dictated by geography.[61] Ferdinand von Richthofen (1833–1905), who passed through several coal-producing areas in the 1870s, noted that the coal reserves of eastern Sichuan "would probably be found to exceed in size the total area" of coal reserves in every other province of China.[62] He was correct that Sichuan's deposits were substantial and spread out over a large area, as he was regarding the difficulty of exploiting most of those reserves because they were buried in the Sichuan Basin's deep strata of red sandstone.[63] In general, coal mining was pursued wherever it was profitable, and most of the province's salt works outside of Zigong, which had ample deposits of natural gas, depended on burning the bituminous form of the rock that was scattered across northern, southern, and, most importantly, eastern Sichuan.

And yet, because many coal deposits were located near farmland, which, by the Qing period, extended into peripheral mountainous areas, coal extraction could be contentious.[64] Parcels of land for coal mining had to be acquired or leased "from many landowners, one small piece at a time," which pitted coal mining against agriculture on the ground.[65] Unlike salt mining, where it was in theory possible to hit "savory gold" and become suddenly wealthy, the vertical and horizontal fragmentation of Sichuan's coal industry offered modest livelihoods for miners who typically pursued small-scale ventures.

Coal mining also came with considerable risk and danger. Gazetteers printed in counties that produced coal are replete with stories of mine collapses and drowning deaths, which contributed to popular perceptions of coal-related misfortune.[66] The association of some mining sites with poisonous qi in the Ming and Qing eras was borne from the fact that invisible killers like carbon monoxide and methane suddenly ended many miners' lives.[67] The potential misfortunes associated with coal mining were never relegated to mere articles of faith. Plenty of hard evidence backed up those impressions.

There was also a long-standing tradition, found in the writings of Song Yingxing (1587–1666), which held that a coal seam should not be exploited down to the core so as to allow the earth's qi to regenerate the rock.[68] Arthur H. Smith (1845–1932), who spent more than fifty years in China as a missionary, cited a related understanding of that tradition as present in nineteenth-century Sichuan, but I have never seen it explicitly invoked in a legal case over coal mining.[69] The previously discussed "regenerated" dragon vein of salt brine might be an extension of that principle.

By the 1700s, in counties where coal could be mined locally or imported cheaply by boat along rivers, coal largely replaced or supplemented timber as a core energy source.⁷⁰ Coal was plentiful in Ba County and eastern Sichuan from the beginning of the Qing, with coal fields becoming "the single most important asset" besides agriculture in the Chongqing region during the nineteenth century.⁷¹ Daning County's saltworks, located near Ba County, transitioned to coal from wood as the major local fuel source around 1800.⁷² By contrast, in landlocked An County, located north of Chengdu, timber continued to supply over 90 percent of fuel needs into the early 1900s.⁷³ Though coal was generally cheap around Chongqing, moving it could be difficult across the vast province of Sichuan, and consumption levels depended on local transportation and market conditions. Though Sichuan's navigable rivers provided the province with an impressive waterway network, boat transport was more expensive than in other regions due to the rapids on most major rivers and a lack of artificial canals.⁷⁴ Coal was most profitably transported downstream with the current.

Nanbu, as a non-coal-producing county, provides an intriguing middle-ground case. The county's location aligning the Jialing River meant that it imported coal for its local saltworks and sericulture industries over 770 Chinese miles (239 English miles) by boat from Guangyuan County, the largest producer of coal in Northern Sichuan. Prices are difficult to measure across time, but the price of coal appears to have been reasonably affordable since Nanbu was able to manage a large saltworks without locally sourcing the fuel. That said, prices are always relative to one's means, and some counties in Baoning Prefecture reported considerable price fluctuations at various points in the 1800s.⁷⁵ The many hundreds of legal disputes over trees from Nanbu—with few recorded disputes over coal transactions—suggest that timber remained a vital part of local energy consumption through the early twentieth century. As we saw in chapter 1, Nanbu magistrates serving in the county were especially protective of community commons in the form of fengshui groves.

Rather than attempt to gauge the precise economic or environmental effects of coal mining bans, it is more feasible, and more useful, for our purposes, to consider their legal and political implications. Counties that saw geomantic objections to coal mining were those that produced considerable amounts of coal and, accordingly, relatively cheap local retail prices. From the perspective of a magistrate serving in these counties, a targeted ban on mining in a contentious area generally would not affect the local price of coal, meaning that the downsides of a ban were slim. On the other hand, allowing a contentious coal mine to stay open could pose risks for officials if doing so alienated local elites, spurred litigation, or elevated the chance of violence. Fengshui offered flexibility to officials in that a ban could be zoned to a specific area, settlement, or temple. In other words, by citing fengshui, magistrates did not have to ban mining from an entire county, but rather from a limited space in a county.

There were also rationales for banning coal mining that could be unrelated to fengshui, such as a mine's effects on agriculture or an outbreak of violence due to ethnic conflict or criminal activity. This section focuses on cases concerning fengshui, but those other reasons for banning mining are found in legal files too.[76]

The Qing state's revenue incentives may have also contributed to litigation trends. Coal's patchwork, small-batch production meant that, unlike silk or salt, it was not in the purview of large transport merchants.[77] After 1855, items that were transported in bulk fell under the new *lijin* tax on transported commercial goods, which formed an increasingly important part of provincial and national revenues during the late 1800s.[78] For most of the dynasty, taxes on coal were light, if not absent, in practice; a tax specifically on coal production in Sichuan was not even attempted until 1909.[79] In other words, the government did not lose much revenue by issuing occasional bans over contentious coal mining sites. As we will see, Sichuan's officials understood coal as essential for the province's salt works, and they could mobilize local bureaucracies to increase production in the event of a price spike. Officials also surely considered the importance of coal mining for people's livelihoods. Yet the general picture is one of minimal government involvement in coal mining and distribution for most of the Qing period.

Geomantic objections to coal mining found voice from two major groups in nineteenth-century Sichuan. The first came from local gentry, who sometimes objected to the exploitation of coal in a significant place out of concern for examination fortune. The second came from families who cited the disturbance of graves and cemeteries in opposing coal mining. Sichuan's local gazetteers provide many glimpses of their petitions to the state. Consider Rong County, which had no shortage of locally mined coal in the 1800s.[80] During the Tongzhi era (1862–75), the gentry of Rong presented a petition for the banning of coal mining around a cherished natural landform in the county due to concerns over fengshui. The landform, named Heavenly Generated Bridge (*tiansheng qiao*), was a stone overhang that connected two cliffs in the shape of a land bridge.[81] The magistrate accepted the request and issued a ban on coal mining for five Chinese miles (roughly one and a half English miles) around the natural formation. A similar ban was issued by a Rong magistrate in 1904 for an area around a historically important stone fort in the county, though fengshui was not explicitly mentioned in the surviving record of that case.[82]

The act of identifying a significant landform or mountain and linking its conservation to examination performance was often pursued by Sichuan's gentry in court. Since members of the gentry typically did not own property in the areas immediately adjacent to the mining activity, appeals to examination fortunes were the best means available for lodging a protest, though we must remember that, in many instances, members of the gentry appear to have been genuinely concerned. In 1870, for instance, a provincial degree holder in Ba County urged a ban on coal mining on Zhenwu Mountain on the basis that the entire county's fengshui was obstructed by the extraction; unfortunately, the result of that specific petition is not included in the archive.[83]

When a site was accorded geomantic significance due to its appearance, location, or heritage, members of Sichuan's gentry could target mining as posing deleterious effects for an entire county, thus morally pitting the contractual terms of a mining lease against the well-being of the broader community. Consider a protest lodged by a lower-ranking Ba County gentry member over mining at Crimson Silk Cloud Temple in 1838 that was met by a complete ban on coal mining after the magistrate weighed all arguments offered by the competing interest groups. There, members of the gentry had contended that "this temple is a mountain of immortals—one of the 'nine great peaks and eight sceneries'—it is not only connected to the literary wind [fengshui] of both Ba and Bishan counties—but moreover is part of the dragon vein of Yucheng [Chongqing]."[84] Similar bans on coal mining were issued in Ba County for Tiger Peak Transcendent Temple in 1810 and 1818.[85] From only six scattered cases, we cannot conclude that every gentry petition against coal mining was successful, but they clearly could be.

Lawsuits presented by families over coal mining near their farmlands were even more numerous than gentry petitions. From a survey of fifty coal mining contracts from Ba County, it is easy to see how disputes could emerge. In some ways, contractual terms favored investors in coal mining rights, with many contracts stipulating that payment of rent would commence only after a mine began producing coal and that a mountain would be leased for an unspecified amount of time.[86] Nonetheless, the deeds reveal some latent concerns of mining investors. While Qing land deeds often include a generic line justifying a contract's composition due to the capriciousness of human nature, nine of the mining contracts explicitly included terms related to the potential blocking of mining, such as "[the lessor] is forbidden to violate this contract with alternative terms or attempt to block the mining."[87] Attempts at obstructing mining were evidently not rare.

Nine of the fifty contracts included stipulations for trees or graves around the mining site, which, at least in some cases, were designed to limit the lessor's ability to protest extraction because of concerns about fengshui.[88] For instance, in an 1831 mining contract for the sale of a waste mountain holding the Zhao family graves, the purchasing Pengs were accorded the right to farm, cut down trees, and dig for coal. While the Zhao family retained the ritual right to sweep the graves on the mountain, the contract stipulated that "the old, the young, the present, and the yet to be born of the Zhao family are not permitted to block the mining, and if anyone uses the graves as a pretext to extort money, it is up to the Zhao family to handle the issue without involvement by the Pengs."[89] The tricky dimension of arrangements that involved "selling the land but keeping the graves" was that one side retained ritual access to the site but not its environment, while the other side obtained control over the mountain's environment while pledging to not destroy the existing graves. When disputes over these arrangements arose, magistrates attempted to identify a border between the grave and surrounding lands. In this specific contract, "not an inch of land" was retained by the Zhaos. The contract was composed in a way that would preclude extortion through an appeal to fengshui, which once again suggests attempts to do so were not rare in Ba County.

As a matter of fact, they were not. Coal miners sought out the state's intervention against claims that their extraction had "severed an earth vein and frightened a grave" (*jiemai jingzhong*).[90] In a case from 1870, two men adopted false identities as the descendants of temple patrons to accuse the resident monks of Lu Family Temple of harming fengshui by leasing temple land for coal mining. Essentially, the men accused the monks of harming the fengshui of their pseudo-ancestral graves around Lu Family Temple and demanded compensation. The monks identified this artifice as extortionary in intent and sought the state's protection through the courts.[91] The county court granted that protection by ordering the two men who adopted false identities to be beaten with a heavy stick while banning the monks from allowing further coal mining around the temple. All such mining had to cease immediately. We might recall the 1847 Nanbu tax evasion case from chapter 1, wherein a group of Deng descendants faked identities as Lis to surreptitiously destroy their own lineage's valuable fengshui. In the 1870 Lu Family Temple case from Ba County, two men faked identities to claim someone else's fengshui. In Qing Sichuan, human identities were apparently easier to fake than excellent fengshui.

Protests by landowners against nearby coal mines could be deemed legitimate or extortionary by county courts, and sometimes officials banned mining to avoid future trouble. For locals opposed to mining, citing the existence of an older state-issued ban, or alleging mining contracts were forged, served as common strategies alongside detailed descriptions of geomantic repercussions.[92] These pleas could be in good faith, and courts responded as if they might be legitimate by dispatching clerks and runners to investigate the allegations.[93] Regardless of intentions in bringing a claim about fengshui, wealth and social class certainly influenced one's ability to engage in litigation. For those with the means to afford lengthy legal battles in court, persistence could pay off, as magistrates sought to avoid burdensome litigation. In Changshou

County, the Xiang lineage shut down coal mining around Yellow Dragon Temple by employing such a strategy in the nineteenth century.[94]

In the absence of statistically significant data from Ba County, it is difficult to ascertain exactly how common geomantic protests over coal mining were. However, one case overseen by Sichuan's provincial salt authorities offers intriguing clues into broader legal trends. In the winter of 1879, provincial officials expressed concern about the rising price of salt produced at the Qianwei and Leshan salt yards. Leshan's saltworks burned great quantities of coal daily, linking the price of salt with the cost of coal.[95] Qianwei and Leshan both sourced significant amounts of coal locally, but because the 1870s had seen production drop and demand increase for coal in Chengdu, the price of salt production was rising alongside the price of coal.[96] Sichuan's Office of Official Salt Transport instructed the prefect of Jiading to dispatch surveyors to look for exploitable coal seams closer to the salt yards. Prefectural instructions to county magistrates were clear:

> In any areas with rich coal veins, as long as the mining sites do not obstruct the fields or graves of the people, issue a notice inviting merchants and commoners to exploit the seams while strictly forbidding local bullies from extorting money; further, make it impermissible to employ vague talk about fengshui as an excuse to willfully obstruct mining, so that merchants and commoners can be allowed to pursue these investments without fear.[97]

In short, Sichuan's provincial authorities identified fengshui objections, whether substantial or "empty," to be the core hindrance to official efforts at expanding coal mining in the late 1870s.

A year later, the officials were proved prescient. The Jiading prefect dispatched surveyors in early 1880 to scope out potential sites, and a report issued in the spring of 1881 captured the results of their efforts. A promising coal seam, deemed by surveyors far enough away from existing gravesites, was identified in Qianwei County. However, shortly after mining began, a property owner rallied the area's families together to halt coal extraction due to concerns about fengshui. Upon hearing about the stalled production, provincial authorities ordered the prefect of Jiading and the county magistrates of both Leshan and Qianwei to travel together to the site to inspect the fengshui of the area to determine if the allegation had merit:

> The corrupt elites and local bullies of this place are relying on local lineages and protesting this matter on the basis that it hurts fengshui, thereby repeatedly blocking the mining and halting of extraction.... The officer should at once go forward to this site, and first take counsel with Prefect Yu of Jiading, and then immediately go together with Magistrate Zhang of Leshan

County and Magistrate Han of Qianwei County along with the deputy officials of each county to personally inspect the site. There, they should put up a notice announcing the original merchants shall continue to mine as before, and then map out and label the site with an accompanying explanation, reporting back whether the fengshui of the site in contention is obstructed. If they continue to dare relying on the crowd to unlawfully resist, immediately arrest and transport them to Chengdu, so that they can be transferred to a detaining area and punished as a warning to future offenders; do not disobey this command.[98]

As indicated in the quotation above, the officials were explicitly instructed to draw up a map of the mining site with annotations for nearby tombs.

In no uncertain terms, the officials were tasked by the Sichuan provincial authorities to cartographically document whether fengshui had been harmed by the mining, just as they did regularly in Nanbu County during legal disputes over houses, graves, and trees. The provincial instructions to officials left open the possibility that fengshui could be threatened, meaning that the dispatched county functionaries had to analyze the contested area carefully. The results of that analysis mattered, and it is noteworthy that provincial authorities were willing to pursue it even when they wanted the mining to take place. If the state-commissioned map verified that fengshui was not disturbed, but locals continued to insist it was, officials were ordered to arrest them and transport them to Chengdu for punishment. The communications do not say whether monetary compensation was rejected or even offered to the protesting parties, and it would be intriguing if the protesters had refused a generous payoff. But the key point is that the mining had been effectively blocked for an extended period because of concerns over fengshui.

As suggested above, the legal landscape of coal mining in Sichuan was influenced by many factors. Population growth in Sichuan pushed agriculture into the mountainous peripheries of the province and drove high demand for fuel, thus pitting coal mining against farmlands and graves. Of the legal conflicts that emerged, many involved contractual issues unrelated to fengshui. Yet, enough of them involved fengshui for provincial authorities to single it out as a significant factor in plans for new coal mines in the 1870s and 1880s. It is a fact that many people in Sichuan sold gravesites, willfully harmed fengshui, and welcomed miners onto their property in the pursuit of profit. Fengshui was never believed to the point of mechanically dictating human behavior. Still, it took only a few families' protests to delay one government-backed coal initiative for a year, with a plan devised by officials in late 1879 remaining unaccomplished by early 1881, when officials were required to inspect fengshui before taking any further action.

Within this landscape of resource management in Sichuan, a few things were true at the same time. Members of the gentry sincerely cared about the natural landscapes of their home counties and sometimes sought to protect them by invoking fengshui. Many people invoked fengshui as a way of extorting money from miners and merchants. Others invoked fengshui in lawsuits as a last-ditch effort against what they perceived to be the unlawful exploitation of land. All these things could be true at the same time because fengshui was a language of power over the land, and power is never confined to simplistic black and white morality. Accordingly, the Qing state appraised related claims by verifying the status of an area's fengshui—even those levied against its own proposed mining sites.

Gold and Stone

Although not used as a domestic currency in the ways silver and copper were, gold had many uses in Qing times, ranging from fire-gilding artisanal objects to engaging foreign trade.[99] Gold has been found in one form or another in every province of China, and Sichuan had quite a bit of it. For its part, Nanbu County had placer and vein deposits of gold, the local exploitation of which begun in the Ming dynasty.[100] Alluvial placer deposits containing flakes, dust, or grains of gold were relatively easy to exploit through panning because of their visible concentrations along Sichuan's many watercourses.

The 1870 edition of the county's gazetteer records residents along the Jialing River mining "sand gold" (*shajin*) for a part of their livelihoods.[101] Gold panning, though easy compared to more invasive kinds of mining, still involved dredging deep into riverbanks, thus drawing attention to the ownership of the affected lands. The gazetteer's brief description is revealing for its emphasis on gold's singular exploitation by residents who lived along the river. The subtext was clear: persons not local to the Jialing riverbed area of Nanbu did not engage in gold mining there. As in other parts of China, prominent families in Sichuan invoked fengshui to prevent others from mining within their ancestral territories.[102] County magistrates knew the score, and the Qing state did not even attempt to get involved in the local industry's expansion prior to the 1900s.[103]

One local resident's far-reaching claim over a gold-producing area can be seen in the following case. In 1857, a member of Nanbu's gentry, Zhang Bing, accused a nonlocal named Chen of panning for gold near his land, thereby severing the rear dragon vein of the protective mountain that shielded his ancestral graves. While processing Zhang's case, no contracts were presented or referenced, and the county yamen apparently did not ask for them. A clerk was dispatched to inspect the scene, where he found Chen employing more than ten individuals to pan for gold along the river. The clerk's report to the yamen indicated that the rear vein of the Zhang cemetery extended for one Chinese mile (around one-third of an English mile) until it met the riverbed, effectively endorsing Zhang's own description of his ancestral properties.[104] The subsequent trial concluded with a verdict ruling that the Zhang property had been obstructed by the panning, thus forbidding Chen from further mining in the area:

> Now the court has determined that I [Chen Shaoxun] have dredged a hole in the earth for mining gold and the estate of the Zhangs is obstructed; the court orders that after today it is not permissible to privately mine in this area again. I [Chen Shaoxun] will follow this resolution as ordered.[105]

It is hard to discern a person's motivations for making a certain argument, either today or in the past. Maybe Zhang sincerely felt that the gold mining had disturbed fengshui. Perhaps he was economically motivated to guard his lineage's private wealth. Most likely, these two possibilities were not in tension, since one's fengshui was, quite literally, one's fortune. Zhang's successful mobilization of fengshui may have been related to his status as a member of the gentry and a resident living on valuable waterfront property alongside the Jialing River. One should not overlook the fact that the dispatched clerk did not map the mining site, as was common practice with fengshui allegations. The clerk may have even known the Zhangs.

While graves were accorded a zone of protection in Qing times, such customary arrangements typically did not extend for more than one *li* (one-third of an English mile). Only a gentry or salt merchant family of considerable standing would have been able to pull off the abovementioned claim in court. On the other hand, considering that gold placer mining required ample space to collect enough surface sediment for profitable returns, Zhang's claims may have appeared reasonable to the county magistrate. Regardless of the intentions involved, one may assume that the county magistrate had not read the Xianfeng Emperor's edict about Sichuan's officials and gold mining bans issued four years prior. Perhaps he simply chose to ignore it.

Another target of geomantic concern extended to stone quarrying. Two major kinds of quarrying were pursued in Nanbu. The first involved the acquisition of red stone (*chishi*; cinnabar stone), which, in addition to its links to mercury production, was quarried to produce inkstone.[106] As with many other kinds of mining, inkstone quarrying was intertwined with geomantic concerns—both because of the relevance of geological terminology and the potentially deadly risks of the extraction involved. Miners in Guangdong discussed the raw material for inkstones as submerged veins requiring careful removal from the body of the earth. Dorothy Ko writes that "the miner intuits the trajectory of the hidden vein and moves with it: a little to the left; a little further up. The waste rock, called 'the bones' (*shigu*), has to be left undisturbed or the tunnel may collapse."[107] Although legal disputes over fengshui and inkstone quarrying existed during the Qing, I have not seen such a case in Nanbu, where extraction appears to have been modest relative to other regions.[108]

The second, more common form of quarrying involved rocks for construction or road building. As with the mining of coal, rocks were sometimes acquired from temple, tower, or grave-adorned mountains, which led to litigation. To give an impression of the frequency of this kind of dispute, three quarrying cases involving fengshui were presented by different families at Nanbu's court over March 10, 11, and 12 of 1890 by three different kinship groups: the Zhaos, the Zengs, and the Dengs.[109] Here again, petitioners can be broken down into two major groups: gentry who opposed quarrying for reasons of examination performance, and commoners who cited the presence of gravesites and ancestral lands. Magistrates adopted different strategies for resolving these conflicts. In response to gentry petitions concerning the protection of a culturally significant mountain, officials ordered notices forbidding quarrying to protect a county's fengshui.[110] If quarrying occurred too close to a family cemetery, magistrates ordered stone inscriptions to be erected that forbade further exploitation.[111] Other accusations were deemed extortionary by courts and subsequently dismissed.[112]

When stones were quarried from a mountain owned in common by a lineage, verdicts often echoed resolutions for fengshui grove disputes by retroactively condoning the action and mandating protection of the site going forward.[113] When a member of the Deng lineage accused other relatives of quarrying stones near commonly maintained ritual fields for the construction of a road, thus causing inauspicious events to befall his family, the court ordered a clerk to map the mountain in contention. The clerk found the claim to be true but noted that the quarrying had occurred several years before the lawsuit's submission in court. In his verdict, the magistrate emphasized that the quarrying had occurred "more than three years earlier," drawing on the statute of limitations for cosmic repercussions that we have seen elsewhere in Qing legal settings:

> The court rules ... even if Deng Yulong claims that, in the past, stones were quarried from the Azure Dragon Corner behind his house, thereby obstructing fengshui—three years have already passed since the quarrying and since the two sides in this dispute constitute a single vein, they should not lose peaceful relations and engage in litigation. Now I order Deng Zhongchun (a witness and relation) to privately meet with each party to resolve the conflict, so that such incidents do not occur again. Both sides have accepted this ruling and the case is over.[114]

Although the small-scale quarrying of stones was less economically consequential than the felling of trees from a community grove, the act of doing so set a strong precedent for control over those communally owned lineage properties. Accordingly, the county's court tended to accept related plaints for the preservation of harmony among kin.

Mining Insurance

The market for state-issued mining permits and other forms of ritual insurance exploded across Sichuan during the nineteenth century. Official mining notices, which always bore the surname and title of the official as well as the date of issue, announced in advance that lawsuits about fengshui would not be accepted in relation to a project deemed to be of importance by the state. We have already seen one of these notices from the 1879–81 Leshan coal mining case, but there are others that will be considered below. Another kind of insurance involved the hiring of ritual specialists such as geomancers and Daoist priests to be present at the prospecting or opening of a mine. We previously saw geomancers involved in the salt drilling process, but Sichuan's Daoists were also regularly summoned for the opening of mines. Both forms of mining insurance reveal merchants' strategies to avoid litigation and extortion.

One collection of copied announcements and communications with resident administrators reveals how often state notices were issued for important development projects. *A Brief Account of Taming the Rapids* contains the legal files of Li Benzhong (c. 1760–1840).[115] Li, a native of Hubei Province, was a man of exceptional wealth who devoted his later years to funding and overseeing engineering projects in Hubei and Sichuan. Li's philanthropic focus was the treacherous rapids along the Yangzi, where violent waters crashed against sharp rocks as they flowed swiftly through the Three Gorges. The rapids and rocks of the river made the passage out of Sichuan dangerous for cargo ships, while the passage into Sichuan required the manual pulling of boats against the current. Deforestation and the expansion of agriculture along the river's banks contributed to sedimentary runoff, which created sudden surges and drops in water levels, adding to the risks of ship navigation. As the volume of cargo shipped down the Yangzi increased with Sichuan's population recovery over the 1700s, Li Benzhong, whose grandfather died in a shipwreck along the rapids, resolved to use his personal wealth for charitable infrastructure work on the river early in the nineteenth century.

Li Benzhong collaborated with resident administrators in Sichuan and Hubei between 1823 and 1840 to spend a large fortune, of two-hundred thousand taels, to improve forty-eight dangerous locations along the river.[116] He hired dozens of stone masons and other laborers for these jobs. Projects were undertaken during the winter and early spring, avoiding the monsoon season, when flooding added to the work's danger. Among the most difficult projects Li and his team undertook was the removal of gigantic fifty- to one-hundred-meter-long rocks that jutted into the riverbed. His primary methods for removal were chiseling and fire setting, a process involving heating rocks and then dousing them with water, causing fracture by thermal shock.[117] The absence of references to saltpeter or gunpowder from Li's reports may indicate that none were used, or that explosives were employed but were omitted in writing due to strict state regulations on their private, nonmilitary use.

Officials in Sichuan and Hubei appear to have been generally supportive of Li's commitments to infrastructure improvements, and by midcentury his projects had temporarily improved navigational safety along the rapids of the Upper Yangzi. In addition to the benefits of safe transit and the facilitation of commercial networks, river shipping came to employ millions of Sichuanese men in the late Qing, so there were plenty of incentives for officials to support these efforts.[118] In recognition of his life and career, a stone inscription celebrating Li Benzhong's charitable deeds was erected in 1838 at the Palace of King Yu in Ba County.[119]

The official communications copied into *A Brief Account of Taming the Rapids* reveal that Li Benzhong imposed conditions of his own. Before embarking on ambitious projects, the merchant-philanthropist requested that officials issue notices announcing that lawsuits based on fengshui would not be accepted by courts during the duration of his team's demolitions. The demolition work along the river was too dangerous for abrupt pauses, and Li did not want his team subjected to expensive, lengthy litigation. Though unstated in the collection, geomantic objections obviously could have arisen from a few directions. The methods of chiseling or setting the fires could have been perceived by property owners as harming the fengshui of nearby graves, houses, or temples. Another possibility is the rocks, which were often given proper names in Li's collection, may have had religious or geomantic significance by virtue of their unique physical characteristics. By altering the flow and current of water along sections of the Yangzi, there also could have been winners and losers among local farmers and property owners.

Regardless of their motivations, such allegations were common in Sichuan—and notoriously effective. An 1889 government order regarding salt transport observed that when the crews of cargo ships attempted to dig out of impassable waterways following severe flooding, people pretended to be local property owners and invoked fengshui to extort money. Such incidents were identified by Sichuan's provincial government as "difficult to enumerate":

Previously, when boats became stranded in shallow water, crews attempted to dig themselves out. There were invariably unlawful local bullies who pretended to be property owners, saying that fengshui had been disturbed, willfully obstructing the boats, and causing various kinds of damage. These instances are difficult to enumerate![120]

Notice that the order's contention was that the accusations were levied by people who pretended to be property owners. In other words, if they did own the land, those people may well have had valid legal claims on the fengshui of the area.

Because resident administrators wanted the work completed, and since two-hundred thousand taels of labor and equipment were on the table, they gave Li exactly what he needed. Between 1823 and 1832, a string of magistrates and prefects in the Upper and Middle Yangzi issued at least twelve government notices concerning fengshui and Li's demolition work.[121] These notices all contained a core sentence that read, "If there are wicked people who falsely claim this work affects fengshui in order to block the action or extort money, it is permissible for Li Benzhong to present a petition at the court [with this notice as evidence] for the apprehension and questioning of the accusers."[122] The consistency of the wording in these notices suggests Li specifically petitioned for these precise terms. The accompanying administrative documents do not reveal if Li invoked them in court, but their rate of issue and number suggest that he needed them. Akin to government permits for zoned construction, state notices announcing that fengshui allegations would be automatically deemed false served to establish rules of the game for significant, sensitive projects.

In addition to soliciting government-issued notices, merchants sought out ritual specialists for commencing important mining projects. Temple sacrifices and opera performances were often considered essential for the opening and operation of significant mines.[123] A range of specialists, including geomancers, Daoists, and perhaps even Buddhists were summoned for the opening of mines. In addition to prospecting for salt wells, geomancers aided miners in locating the sites of other valuable minerals.[124] Buddhist abbots in Ba County presiding over valuable monastic lands and temple endowments were well known for leasing out coal mines with ruthlessly high rents.[125]

Local gazetteers contain many references to Daoist rituals related to mining. Ming and Qing-era gazetteers from Sichuan, Shanxi, Gansu, Hunan, Henan, and Guizhou cite Daoist ritual texts and sacrifices performed upon the opening of coal mines specifically, though texts for other kinds of mines may have been created as well.[126] These coal rituals allayed fears of inauspicious repercussions and announced to a community that the extraction had been thoughtfully planned. Thus, in addition to the religious dimensions of these rituals, an important incentive for holding them was preventing litigation over the mining.

One example of ritual insurance came from Liu Yuan's (1768–1856) *Compendium of Ritual Words*, a collection of liturgies compiled between 1821 and 1844. Volker Olles has provided a thorough overview of the text and its social contexts, which I'll expand on here.[127] Liu Yuan, a native of Shuangliu County near Chengdu, was a provincial degree holder well-versed in classical Confucian learning; he was also receptive to various Daoist traditions that had proliferated in the province during the Qing.[128] In his *Compendium of Ritual Words*, which he compiled with disciples based on earlier Daoist ritual texts, Liu Yuan recorded a liturgy for opening a mine. This mining ritual was intended to assuage fears of inauspicious consequences and prevent litigation of the type seen throughout this chapter. Though the Daoist lineage that once claimed this specific ritual remains prominent in Sichuan today, the mining liturgy is no longer regularly performed, and thus it is difficult to reconstruct precisely how it was executed. The text of the liturgy suggests that it was newly conceived in the early nineteenth century. It also circulated beyond Sichuan. In the list of deities invoked toward the end of the mining liturgy, a note clarifies that Chuanzhu, a popular deity in Sichuan, should be replaced with a provincial deity as appropriate.[129]

In the script for the liturgy's performance, the significance of the ritual was explained. Reflecting on the growing population of Sichuan, the author contended that mines provided employment to poor people without land so they could feed their families. Even the central text for the liturgy's performance incorporates the theme of Sichuan's growing population, with one line of the liturgy reading, "Now, there is a Sage Ruler on the Throne [the Daoguang Emperor, r. 1820–50], and he takes the prosperity of the people as his concern; as a result, officials are elevated to good conduct, and everyone knows that not being greedy is treasured. But the large population increases daily; everywhere people clamor for food, and it is hard to fulfill their daily needs."[130] The liturgy also claimed that mine creation helped fill the state's coffers, which presaged the concerns voiced by the Xianfeng Emperor a few decades later.[131]

The *Compendium of Ritual Words* explains the reason for offering the mining liturgy. The problem with mining was that people feared heavenly punishment for penetrating the earth. Accordingly, it was essential to perform a ritual to assuage these fears, which could not be allowed to hinder socially beneficial mining development.[132] Pivoting away from geomantic concerns such as the severing of dragon veins, the liturgy suggests that the real danger of mining lies in not properly respecting and thanking the gods for the use of the minerals.

FIGURE 4.1. Offering to a Mountain Deity for the Extraction of Coal. This print is held in the Département des Manuscrits at the Bibliothèque nationale in Paris under the listing OE-117-4 (Recueil. Extraction de la houille). The image here is reproduced from Peter Golas's *Science and Civilisation in China: Mining* with permission from Cambridge University Press through PLSclear.

As John Lagerwey and Gil Raz remind us, mountains were identified as the homes of the gods, and early Daoist interest in gold, jade, cinnabar, and other minerals was primarily motivated by their role in the creation of elixirs.[133] In this liturgy, that esoteric interest was channeled for the broader social welfare. The image in figure 4.1, printed in a nineteenth-century pictographic album on mining, displays a sacrificial offering to a mountain deity prior to the extraction of coal.

Was ritual insurance effective for protecting miners and mining investors? Since Daoist liturgies for opening mines proliferated during the Qing, one suspects they helped prevent some litigation. Their deployment followed patterns seen elsewhere in the legal landscape. Specifically, mining liturgies paralleled court-mandated *jiao* sacrifices to ancestors whose graves had been disturbed. In cases involving real harm done to gravesites, magistrates might seek to redirect social anxieties away from intractable geomantic harm and toward appeasing restless spirits through ritual sacrifices. Likewise, mining liturgies did not deny geomantic harm arising from extraction, but redirected attention to the improvement of conditions under which mining inevitably would happen.

Nonetheless, mining liturgies did not and could not counteract the imperial law specifying that mining could not disturb gravesites. Mining's potentially negative consequences were recognized, and there were legitimate vehicles and venues for concerns to be voiced to, and considered by, Qing officials. At the same time, geomantic extortion was a serious issue in Sichuanese society, and mining investors sought formal guarantees against extortion and lawsuits by looking directly to the state and specialists from the religious realm. During the nineteenth century, resident administrators, merchants, geomancers, and Daoists sought creative ways to make mining more acceptable, legitimize specific ventures, and support their own careers within Sichuan's competitive landscape for ritual consulting.

The Dragon Vein of Sichuan

Question: Which are the most important methods for seeking the dragon, observing the configurational force, and examining caverns?

Answer: . . . In general, where there is a great dragon [vein], *qi* is abundant. For much of the year, clouds and mists cover its summit. On the top of the great mountain peaks are pools and springs, pure and clear, which do not dry up even during a great drought. These pools are called "nourishing shade." This water is present because of the abundance of *qi*. *Qi* is the mother of water so that where there is *qi*, there will be water. Through examining the depth of water, one can divine whether *qi* is flourishing or declining.[134]

—TWENTY-FOUR DIFFICULT PROBLEMS OF FENGSHUI

The passage above, which derives from an early sixteenth century geomantic text, was not composed by Wu Tang, but he likely was familiar with it. Wu, governor-general of Sichuan from 1867 to 1875, needed such knowledge, for he arrived in the provincial capital just in time for a major crisis. Wu's political career, like that of many men of his generation, had been shaped by the troubles of his era. Born in Huizhou, Wu received the provincial degree in 1835, but never obtained the *jinshi* degree.[135] Wu served as a county magistrate in Hunan in the 1840s and 1850s and would have probably fallen into historical obscurity if he had not attracted the palace's attention for his exemplary performance suppressing Nian rebels. A close political ally of the Empress Dowager Cixi (1835–1908), Wu was promoted rapidly in the 1860s, becoming governor-general of grain transport in 1863 and governor-general of Guangdong and Guangxi the following year. But even with formative governmental experience under his belt, Sichuan's droughts of the late 1860s and early 1870s posed daunting challenges.

By the mid-1800s, Sichuan's reputation as a breadbasket was increasingly belied by the limitations wrought by food supply, coupled with the province's geography and transportation infrastructure. Sending grain out of Sichuan via the Yangzi River was easier than getting it into the province. As seen from the demolition work of Li Benzhong, even getting ships out through the Yangzi's rapids was not without risk. While major grain imports had been largely unnecessary during the first half of the dynasty, at a time when Sichuan consistently exported grain, the province's resource equilibrium become increasingly unfavorable over the 1800s, making rural farms more vulnerable to aberrant weather patterns.[136]

Sichuan's disaggregated data for the 1700s reveals many nondrought years as well as several years with only a handful of affected counties—with one dramatic spike in 1778. In general, the reliable fertility of Chengdu Prefecture rendered it less prone to famine than Chongqing and Baoning prefectures, which respectively endured seventy-five and forty drought years during the Qing, compared to Chengdu's twenty-one.[137] That said, drought and famine became more common across the province in the mid-1800s. In 1864, drought conditions were even reported on the Chengdu Plain, and in 1871, all three prefectures of Chengdu, Chongqing, and Baoning experienced dry spells.[138] That year, widespread summer drought was followed by heavy autumn rains, resulting in crop failures lasting into the spring of 1872. Rice prices more than doubled in Chengdu, and famine relief took weeks to reach Chongqing. Richthofen, who was present in Sichuan during the drought, described scenes of provincial authorities passing out food to twenty thousand people who gathered outside the walls of Chengdu.[139]

In responding to famine, Qing officials employed an all-of-the-above strategy, beginning with an effort to secure necessary resources. Accordingly, in March 1872, Governor-General Wu composed a memorial asking the central government to remit the province's surcharge and *lijin* tax revenues for famine relief.[140] That request was granted by the emperor, who also directed the governors of neighboring provinces to instruct merchants to ship rice into Sichuan. There were also actions taken by officials that spoke to Buddhist and Daoist linking the killing of animals with the appearance of famine.[141]

Richthofen noted that community fasts structured by prohibitions on the consumption of meat, fish, and eggs were issued by Sichuan's authorities during the drought.[142] Meat prohibitions were almost certainly followed by public rain-making rituals conducted by county magistrates.[143] Government messaging was critical, and officials needed to demonstrate to the people that they were doing everything possible to resolve the situation, especially in the light of an impending provincial-level examination in Chengdu.

It is against the backdrop of this famine that we should consider the following fengshui-related policy decision. In the summer of 1870, prior to the failed harvests of 1871, Sichuan's most eminent living scholar-officials—including former Vice-Minister of the Board of Works Xue Huan (1815–80), former Governor-General of Zhejiang Bao Chao (1828–86), former Governor of Hubei Yan Shusen (1814–76), former Zhili Judicial Administrator Sun Zhi (1811–76), compiler for the Ministry of the Interior Chen Shouzun (d.u.), Supervising Secretary Wu Fuxiang (1810–83), and junior compiler in the Hanlin Academy Wu Zhaoling (1826–1915), among others—presented Governor-General Wu Tang with an urgent petition.[144] They alleged that the mining operations of tile works and brick kilns had injured the dragon vein of the province, thus heralding imminent misfortune. The officials urged their immediate closure.

Artisans employed at the kilns operating in Chengdu, Guan, Chongning, and Pi counties dug deep into Sichuan's red earth for the excavation of illitic and kaolinite clays. Tiles and bricks were made by excavating clay, mixing it with water, trampling it into a thick paste, scooping the paste into standardized frames, smoothing the surfaces, and then baking them in kilns. As described by Timothy Brook, the process involved "hot, filthy" work and was perceived as low status in late imperial society.[145] There was also precedent for targeted, temporary bans on kiln firing. During the nineteenth century in parts of south China, including the famed porcelain production center of Jingdezhen, kilns were regulated through the observation of the spring kiln ban (*jin chunyao*), whereby kiln firing was forbidden from the New Year to the Qingming Festival. The ban allowed the owners of large kilns to store up enough firewood at the beginning of the year to protect operations from price gouging during the resource-intensive months of production.[146] Yet, in this specific instance, the pressing issue for the scholar-officials was not the kilns themselves. The problem was that the clay's mining took place along the irrigational arteries at the heart of the Chengdu Plain: the tributaries of the Min River.

Although Sichuan's authorities had consented to the request for a ban in 1870, by 1872, after drought conditions extended to the counties of Guan, Pi, and Shuangliu around Chengdu, rumors were circulating that the kilns were still operating in violation of the law.[147] Adding to official concerns was the fact that 1873 was a provincial examination year, meaning that well over ten thousand individuals were expected to arrive in Chengdu to sit for the exams. With dangerously high food prices, Chengdu's supplies would be put under immense strain, and the county, prefectural, and provincial governments would have to manage the difficult logistics. As pressure mounted, Governor-General Wu reissued the order in August 1872, dispatching officials to rigorously inspect the affected counties to ensure compliance. Adding severity to the original 1870 ban, Wu also banned the sale of private farmland to the kiln operators for the purposes of clay excavation:

> An order for forever strictly forbidding excavation associated with harm to earth veins, and especially for the protection of the Arriving Dragon Vein. The provincial capital of Chengdu is surrounded by great rivers on its north, south, and east sides; to the west, only a single embankment connects the capital to the mountains of Guan County. The two tributaries of the Min River flow through the counties of Chongning, Pi, and Chengdu before reaching the provincial seat. . . . Henceforth, in the entire area from Chongning, Guan, and Pi counties and extending to Xipu, Tuqiao, and Wu Family Ridge, it is forever forbidden to open kilns to burn clay for tiles. If any of the old kilns remain, they are to be immediately destroyed and forbidden from reopening. Among the people, it is forbidden to buy or sell wet fields to any of the kiln operators for clay mining. If unworthy troublemakers illicitly open a kiln for the burning of clay, or if any commoners sell their land to the kiln operators, then the gentry, militia heads, community leaders, and heads of surveillance units should notify resident administrators, and they will be individually detained, interrogated, and punished as a clear warning. This order will extend protection to the earth veins and offer great benefit to the advantageous terrain of the provincial capital. Everyone should follow this order. Do not trespass the law.[148]

A confluence of factors informed Wu Tang's decision to ban clay mining and close the kilns of the Chengdu Plain. Many prominent Qing officials felt strongly that the mining posed risks for the province. Their high-profile petition may have been impossible for Wu Tang to reject. There were also potential economic dimensions informing the ban. As Pierre-Étienne Will and Kathryn Edgerton-Tarpley have shown at length, farmers tended to sell or lease their temporarily worthless farmland for needed money or food in the wake of famine conditions.[149] Wu Tang's forbidding the sale of wet fields for clay mining on the Chengdu Plain would ensure those properties remained in use for agricultural production.

Moreover, famine conditions tended to hurt landless artisans, laborers, and miners most grievously, since the market for their products typically collapsed as everyone concentrated on buying grain. We cannot dismiss the possibility that shaping a political narrative by assigning blame for inauspicious events on the miners and the kilns was appealing to Sichuan's officials and gentry. Politically, low-status groups would have been safer for officials to blame for disasters in the 1860s and 1870s than the growing number of foreign missionaries active in the province or a weakened Heavenly Mandate.

To unpack the geomantic dimensions of Wu's order, it is helpful to look at a map. As for any serious fengshui claim worth its Sichuan salt, Governor-General Wu or members of his staff commissioned one. The key to using the map to understand Wu's geomantic logic is the watershed of the Min River. Sichuan's provincial dragon vein was defined as the dynamic cosmic interactions between the central mountains and the rivers that constituted the Min watershed. Readers should recall that in fengshui, mountains, like rivers, have a directional flow. The Min River, which flows north to south, originates to the west of the Min Mountains in Northern Sichuan. According to the map's annotation, the provincial dragon vein originated in the Min Mountains, near the valley of Jiuzhaigou in the far north of Sichuan. In the petition over the crescent bridge discussed in chapter 3, the scholars of Nanbu had also contended that the dragon vein of the county emerged from the north.

The provincial dragon vein continued southward until it reached Guan County, home of the famed Daoist peak, Mount Qingcheng. There, the vein also passed the historic Dujiangyan Irrigation System, which had been constructed in the third century BCE and has remained in operation ever since. In the space between Guan and Chongning counties, a vein broke off from the west and headed directly toward Chengdu. That vein was shepherded by two tributaries of the Min on both sides: the Zouma (Qingshui) below and the Youzi above. These rivers were the lifeblood of the Chengdu Plain, and between them was a narrow chain of temple-dotted hills that extended down toward the west gate of Chengdu's city wall. The place-names where illegal clay excavation was occurring were labeled at the center-right of the map, situated along the connective vein just before it reached the walled city of Chengdu.

The map of Sichuan's dragon vein invoked a blend of landforms and compass directions. The mappers of the Sichuan dragon vein, for instance, labeled the two geomantic "table mountains" (*anshan*) of Chengdu, the Longquan Mountains. In fengshui, table mountains are identified below the landscape's central cavern (*xue*), which in this case was the walled city of Chengdu. Positioned in the southeast, the Longquan Mountains directly faced the dragon vein arriving from the northwest. Geologically, the mountains do indeed form a topographical barrier between the Chengdu Plain and the rest of the Sichuan Basin, making the geomantic description accurate. The mappers also employed compass directions to signal the flow of the vein across the land. At the

FIGURE 4.2. Map of Sichuan's Dragon Vein Composed Around the Time of the 1872 Excavation Ban. This map depicts the provincial dragon vein of Sichuan in the wake of Governor-General Wu Tang's excavation ban on clay mining. The walled provincial seat of Chengdu is at the far right of the image. The mappers labeled the following county seats (right to left) within square cartouches: Chengdu, Shuangliu, Wenjiang, Pi, Chongning, and Guan. The provincial dragon vein arrived from the northwest, trailing the Min River down to Chengdu, with one branch extending east to envelop the city by way of the Longquan Mountains (far right). On the far left, the Min River can be seen flowing below the famous Daoist peak Mount Qingcheng, where it splits off into the Youzi (upper) and Zouma (Qingshui; lower) rivers. Clay excavation was strictly forbidden in the demarcated areas along the vein. Map from *Tongzhi Chengdu xian zhi* (1873).

point where the dragon vein jutted toward Chengdu, an annotation reads, "The protective dragon ends here; a vein emerges from the *you* (west) direction and continues to Luzhou (southeastern Sichuan), where it stops." As discussed in chapter 2, some Qing-era maps, including those created for administrative purposes, included geomantic directionals.

Readers have already encountered cases resonant with Governor-General Wu's mining ban from nineteenth-century Sichuan: a petition from the gentry of Shifang against dredging a waterway in chapter 2, petitions from the gentry of Ba County against coal mining earlier in this chapter, and implicitly through the many official notices extending legal protections to the demolition activities of Li Benzhong along the Yangzi. Those gentry pleas were met by government action—either to enforce an understanding of fengshui or to reject one. In this instance, it was important for Governor-General Wu to send a message to the farmers of the Sichuan Basin that they needed to keep plowing their fields, and to alert the province's scholars that all would be well for the examinations the following year.

While it is impossible to identify the exact reason for any given political decision, what stands out most is that geomantic knowledge—channeled into this narrative, leveraged at this moment—mattered to Sichuan's government. This decision was not taken in 1672. It was not made in 1772. It was issued in 1872. Officials needed to shape narratives based on what was happening in the material world that people saw around them. With extractive activities intensifying across Sichuan against the backdrop of rural unrest, droughts, crop failures, and famines, that need became more acute.

Mining: The Good, the Bad, and the Unlucky

The Qing dynasty institutionally recognized fengshui's relevance to mining. No one contested that point in the abstract. Instead, officials, gentry, and commoners discussed its implementation on the ground. Since Qing law confronted a rapidly changing social and economic landscape over the 1800s, there was much to discuss. Sichuan saw substantial and growing amounts of mining each year—much of which was unregistered, untaxed, and hence illicit from a strictly legal standpoint.[150] Within this context of intensive extraction and low regulation, property owners sought to exploit miners, and scholars sought to maintain the integrity of threatened natural sites. Above all, officials serving on the ground wanted governing options, and fengshui gave them one. People across Qing society—official, gentry, and commoner—were involved in discussions over fengshui as it related to mining in their communities.

To understand why this was so, recall David Faure's observation that "stability was maintained in late imperial China not by strict enforcement of law but by connivance between the officialdom and the local leadership to maintain the propriety of rituals."[151] This chapter showed that connivance in two ways. First, to keep the peace, resident administrators weighed incentives to raise revenue through taxing commercial activity, which played an increasingly large role in government finances during the second half of the century.[152] While selectively approving new mines, officials also hid extractive operations from provincial superiors to keep tax revenue in their jurisdictions. Other officials blocked nonlocals from exploiting resources deemed connected to the fengshui of local elites. As the Xianfeng Emperor was all too aware, fengshui found roles among officials pursuing clever endeavors related to the strategic concealment of locally accrued wealth and the appeasement of local power-brokers.

Second, Qing officials also weighed arguments that sincerely sought to keep ores and minerals in the ground. They faced demands from influential gentry, complicated property disputes concerning peripheral lands never recorded on the government's tax rolls, and the increasing appearance of drought and natural disasters leading to faltering agricultural production. No official wanted the blame for the latest drought, the latest murder, or the latest frightening rumor. Even when Sichuan's provincial government endorsed a much-needed coal mining site for the salt industry in the 1880s, officials delayed the project for months to carefully map and analyze the area's fengshui, without doubt aiming to influence public opinion. The costs of doing otherwise were too high. From Sichuan's archives, one gets the sense that resident administrators knew how far they could push and the dangers of going too far.

Although impossible to measure, Qing governing policies rooted in agrarian paternalism and prevailing knowledge systems shaped mining development during the dynasty. Merchants and miners operated in a regulatory context where they could be legally liable for geomantic damages. This observation is a far cry from concluding that the dynasty's political and economic institutions were not conducive to market expansion, abundant trade, or high levels of commercialization. Yet this much is indisputable: late into the 1800s, county magistrates and even higher-ranking officials issued extraction bans over places where mining was possible with the available technologies. In other words, officials were presented with clear choices, and they sometimes ruled to keep mining products in the ground. This book lacks the quantitative data to make sweeping conclusions regarding the Qing's developmental trajectory in comparative terms, which has been a subject of considerable debate over the past few decades.[153] What it can show, however, is that fengshui was an arena where today's categories of law, religion, science, and economy seamlessly intersected in China. Further studies on divergences, great or small, might do well to look not at one of those elements in isolation but all of them in tandem.

For Sichuan and beyond, the reality was simple. Commoners could destroy fengshui. Those in desperate circumstances did so without hesitation. They could sell their gravesites and cut down ancient trees to obtain money when needed. But many people in Sichuan, especially the landholding classes and gentry, held their government and officials to a higher standard. People looked to the government for aid, protection, and upholding the social order, i.e., the propriety of rituals. For its part, the imperial state needed to maintain the reputation of its courts and its examination system while preserving a delicate balance of competing interests.

Ambitious infrastructure projects were possible and many precedents existed for them in China's history. However, considering the patchwork system of lineage property holdings that prevailed during the dynasty, the practical limitations of existing financial institutions, and the imperial state's well-established relationship to the kinds of legal claims presented throughout this chapter, such projects required meticulous planning and careful, tailored messaging.

In the late nineteenth century, geopolitical circumstances forced the Manchu dynasty to attempt those projects under the intense pressure of Western imperialism and in the face of numerous internal crises. Just when national, provincial, and community fortunes appeared to be most actively declining, the imperial state was called upon to take actions that threatened to undermine the fengshui of its people. Official reactions would involve excruciating political choices, new winners and losers, and a fierce bureaucratic battle over the letter of the law. A component of Qing governance that had helped Sichuan's officials manage disorienting commercial and demographic changes over the eighteenth and nineteenth centuries was about to become the subject of heated international controversies and profound misunderstandings between China and "the West." Pressures rising in the province long before the Opium War of 1839–42 would suddenly erupt into the open as the bureaucracy debated the implications of breaking the land—and breaking the law.

5

Breaking the Land

THE SCHOLAR AND STATESMAN Li Hongzhang (1823–1901) stood at the center of many of the most important political events of the late Qing dynasty. Li came to serve as one of the towering figures of the Self-Strengthening Movement, an ambitious effort from 1861 to 1895 that sought to address the crises precipitated by Western imperialism. As early as 1862, he employed technology from the West for his Anhui army to suppress the Taiping Civil War.[1] Nevertheless, through much of the 1860s, Li expressed caution over the introduction of railways and telegraph lines in China. His late 1867 memorial to capital authorities while serving as governor-general of Hubei and Hunan captures his objections at the time:

> Railways will chisel into the mountains and rivers of our country, they will harm the people's fields and houses, they will obstruct fengshui, and they will undermine the livelihoods of merchants; the common people will certainly rise to tear them down, but officials will not be able to punish these wrongs or order compensation for losses because then the people would rebel.[2]

A few years later, Li Hongzhang had a dramatic change of heart. Political fallout from the Tianjin Massacre of 1870, coupled with Prussia's strategic use of railways in its 1871 military victory over France and the 1872 opening of Japan's first rail track, convinced Li that the development of industrial infrastructure could be a solution to the unprecedented crisis facing China.[3] Li's move toward openly advocating for railway and telegraph construction was an important moment for the broader Self-Strengthening Movement. Still, Li's journey was a difficult one, full of roadblocks and hard compromises. Not every Qing official agreed with Li's program—and many had serious reservations.

The struggles that defined the Self-Strengthening era have become a lightning rod for debate about the "success" of China's efforts to "modernize." One school of interpretation considers the Self-Strengthening Movement to be a failure, using as its ultimate benchmark China's crushing defeat by Japan in the

Sino-Japanese War of 1894–95. In this narrative, a "conservative Confucian" China resisted a "Western" model of modernity, with disastrous consequences. According to this view, the hindrances to China's embrace of newly introduced Western ideas, institutions, and technologies included "superstition," technophobia, and blind adherence to timeless Confucian traditions.[4] Not surprisingly, in this interpretation, fengshui, as a pernicious and widespread "superstition," played a significant role in hindering commercial modernization by blocking the rollout of railways, mining, and telegraph lines across the country. Even the strong advocate for industrial development Zheng Guanying (1842–1922) expressed a version of this view in his appraisal of the country's mining industry.[5]

A second historical narrative has sought to revise the first. This narrative focuses on the "modernizing" efforts of self-strengtheners such as Li Hongzhang, who promoted the acquisition of Western arms and ships, the reform of China's military and naval forces, the building of shipyards and arsenals, and the adoption of new manufacturing industries, mining technologies, telegraphs, and railways. Naturally, this interpretive perspective downplays the role of fengshui as a meaningful impediment to economic development. When these revisionist accounts acknowledge obstacles to self-strengthening, they draw attention to factors such as officials' concerns over foreign espionage via telegraphy or the dynasty's desire to avoid widespread unemployment in the labor market.[6]

Historically revisionist accounts of the Self-Strengthening Movement have greatly improved our understanding of these decades. According to Benjamin Elman, "The view that Qing China was irrevocably weak and backward, in contrast to a powerful and industrialized Europe and a rapidly industrializing Japan, is an artifact of the impact of the Sino-Japanese War after 1895 on international and domestic opinion."[7] To rectify what he sees as a damaging misperception, Elman documents a process by which the Qing state thoughtfully and strategically pursued new technologies in the last decades of the nineteenth century. Other historians have now provided similar documentation. Peter Lavelle demonstrates, for example, that Qing statesmen like Zuo Zongtang (1812–85) embarked on intensive campaigns to develop natural resources along previously untapped frontier regions.[8] Stephen Halsey maintains that the swift formation of a military-fiscal state in China after 1850 enabled the country to avoid wholesale colonization.[9] Seen in this light, the Self-Strengthening Movement of the late Qing was at least a qualified success.

But the facts remain that most late Qing officials and literati were ambivalent about Westernization, and that before the Sino-Japanese War of 1894–95, "the voices of the Self-Strengtheners were very much in the minority."[10] How else, then, did local elites and Qing officials resist foreign encroachment? And how did fengshui fit into that picture? One finds a fascinating example of an effort to reconcile new Western knowledge and fengshui in the writings of Song Gengping (d. c. 1910).[11] Although details of Song's life are sketchy, he claimed to have traveled widely in India, Europe, and the United States while cultivating mining and mineral surveying expertise. Upon returning to China, Song became one of the leading mining experts in Sichuan during the 1890s and 1900s, when he taught at the newly established School of Mining Affairs in Chongqing. Though Song enthusiastically supported the pursuit of Western learning as it applied to Chinese mining, he was also sincerely devoted to the study of fengshui, and in a 1902 work titled *New Edition of the Core Essentials on Mining Studies*, he argued for a creative synthesis of the two approaches to managing industrial extraction.[12]

This chapter focuses less on intellectual approaches to late Qing reform efforts and more on the strategic role of fengshui in contending with foreign pressure and allaying widespread domestic fears. It is structured to show that what was previously written off by Westerners as Chinese superstition or as an ancient custom, was, actually, an expression of officials and scholars strategically and intelligently applying Qing law to new problems. This chapter also reveals that anxiety over a perceived decline in fortune due to harmed fengshui had tangible political, social, and economic consequences during the dynasty's final years. Following decades of sustained state management over geomantic tensions and anxieties, the imperial government faced tough choices.

The following section reveals how members of Sichuan's gentry responded to perceptions of declining fengshui by lodging complaints in local courts. The analysis then turns to memorials written by late Qing officials on the need to protect local fengshui in the wake of new industrial projects, and then an analysis of legal and regulatory reforms for mining and development in the late 1890s. Finally, the chapter describes the impact of these changes on the ground in Nanbu County, where the faltering imperial order in the 1900s further exposed fengshui's relevance to rural society and local administration.

Sichuan's Scholars on Fortune's Decline

By the time the Guangxu Emperor ascended the throne in 1875, the scholars of Sichuan had already begun to sense that the province's fortunes were in trouble. As seen in the last chapter, it was well known among many of the province's most esteemed officials that the provincial dragon vein had been injured by mining in the early 1870s. From that point on, further inauspicious events in the province and other parts of China continued to portend worsening fortune. Turning to the county-level, the trend is apparent. Over the Guangxu era (1875–1909), severe drought or flooding hit Nanbu County in 1877, 1878, 1881, 1889, 1899, 1903, and 1904, with an average of around two significant environmental crises per decade.[13] These years do not include instances of milder drought or flooding, which also affected food production, livelihoods, and local government revenues.

Amid these natural disasters, Christian churches were constructed along sensitive urban landscapes with seemingly less local consent than Khoja Abd Allāh's disciples had sought and obtained during the Muslim community's acquisition of Coiled Dragon Mountain earlier in the dynasty (see chapter 3). To top it all off, taxes were rising on a range of commercial goods, including Sichuan's most famous commodity, salt. To the provincial gentry, these signs constituted compelling evidence that Sichuan's fortunes were in decline, and they leveraged that evidence in legal petitions to the state.

Missionary activity—in particular, the building of churches—was a source of geomantic conflict in nineteenth-century Sichuan and in other parts of China. With at least 127 antimissionary incidents recorded by 1910, Sichuan registered more "missionary cases" than any province in the late Qing period.[14] Ba County and Chongqing were sites of considerable geomantic conflict created by missionaries even before the city's opening as a treaty port in the 1890s. The first serious incident in Chongqing, in 1863, originated from a French bishop's determination to reclaim what had once been church property near a prominent pagoda.[15] A later and more unsettling incident, the Chongqing Uprising of 1886, was described by an American minister as "the most serious riot that has occurred in China for many years."[16] It was sparked by a geomantic controversy over stone quarrying on the highest promontory in the area. This activity, which coincided with the hosting of the local military examinations, threatened to undermine the prospects of the candidates taking the exams. The previous chapters of this book make it abundantly clear why these actions were so threatening.

Other parts of Sichuan saw their share of geomantic conflicts into the 1890s, with both urban cores and rural peripheries affected. Kristin Stapleton cites one incident in Chengdu from 1893, but there were surely others considering the city's importance as the site of the provincial examinations.[17] Nanbu witnessed a missionary incident in 1885, when a military provincial graduate named Yang Shuda led a group of men to destroy a building being used as a Catholic church. This incident was referenced in a petition concerning fengshui that will be discussed in detail below. Naturally, not every missionary case involved the fengshui of houses, gravesites, or the examinations, but a good number did. Despite rising protests, Qing officials had limited options for resolving tensions in favor of local elites because stringent treaty obligations demanded the strict protection of foreign property and lives.[18]

Fiscal problems, like natural disasters and unsettling missionary cases, were viewed by scholars and commoners alike as indications of declining fortune resulting from actions that undermined local fengshui. One growing target of concern was the salt *lijin* tax, which affected Sichuan's merchants and gentry with ties to the industry. A bit of background on this tax is necessary for understanding the geomantic petition discussed below.

The salt *lijin* tax first appeared in Nanbu in 1863 under Magistrate Huang, the official celebrated for constructing the county's premier fengshui pagoda.[19] For sixteen years, collection of the tax was overseen by members of the gentry and brine furnace operators.[20] These interested parties collected the tax, but some salt mysteriously disappeared in the process and payment delays were common. During a provincial effort to tackle corruption, tax collection was reformed in 1879 with the creation of a county Salt Tax Bureau and the establishment of twenty-eight levying branches across Nanbu to ensure reliable intake and to prevent salt smuggling. Headed by a state-appointed officer, the county salt commissioner (*yanwei*), the bureau managed to streamline levying, but it also weakened the influence of local elites—including members of the county's gentry—who were previously responsible for collecting the tax.

A few years later, the presiding commissioner of the Salt Tax Bureau proposed to purchase the building that housed Nanbu's Confucian academy, Aofeng Academy, as the location of its central levying office.[21] The site's location near the county yamen was ideal for coordinating matters related to salt administration, which oversaw an important and growing portion of the province's revenue. The salt commissioner justified the decision by alleging that Aofeng Academy welcomed scholars who took pleasure in presenting meddlesome petitions at the adjacent yamen court. Described as an unfit place for cultivating true literary talents, the academy had evidently earned the nickname Turtle Peak Tavern, with the diminutive word tavern (*zhan*) suggesting a kind of unscholarly guest house for drinking and socializing.[22] Nanbu's magistrate agreed to oversee the bureau's purchase while relocating the academy to Towering Cloud Cave, outside of the walled town and considerably farther from the county yamen.

In response to the site's purchase, a long list of members from Nanbu's gentry, led by a provincial degree holder, filed a petition in 1887 accusing the salt commissioner, a man named Zhao, of engaging in corruption and bribery.[23] The following year, the group presented a second petition declaring that the county's fengshui had been gravely damaged. In that petition, the scholars claimed that previously the academy's Ling (Wenqu) Star Pavilion was perfectly positioned between the Confucian temple and the shrine of the City God—the celestial counterpart of the county magistrate—physically affirming the cosmic link between preparing for the examinations and assuming office.[24] All buildings were oriented toward the south, channeling the *qi* of Nanbu's dragon vein that flowed south toward Mount Aofeng, where a treasured fengshui pagoda had once delivered so much good fortune for the county.

The scholars specifically claimed that by purchasing Aofeng Academy's building, the salt commissioner had "forgotten righteousness at the sight of profit."[25] The cosmic consequences of the move were manifested through the recurring misfortunes of resident administrators and literati, high turnover for official posts, and the untimely deaths of several elite degree holders. Among the grievances mentioned was the fate of military provincial graduate Yang Shuda, who had been detained and punished by Qing officials after his role in burning down the French church in 1885. Events related to recent judicial verdict in the county were offered as evidence for compromised fengshui:

> As for the present inauspicious effects on the county's gentry, there was the death of Provincial Degree Holder Lin by the geomantic *sha*, the detaining of Military Provincial Degree Holder Yang despite his innocence, the beating of *xiucai* Ren even though he was not guilty of a crime. Case after case bears evidence that the arrival of the Salt Tax Bureau has not benefitted officials or gentry.[26]

The gentry petition tied the fallout from the 1885 missionary conflict—specifically, the legal punishments of esteemed members of the country's gentry class—to the county's declining fengshui. The petition then identified the Salt Tax Bureau's move as a step too far for comfort.[27]

As discussed in chapter 3, the gentry class held legal privileges in Qing society. Members of the gentry could not be humiliated in public or insulted by commoners, who were forbidden to serve as witnesses in routine lawsuits involving the former.[28] When members of the gentry were involved in local litigation, they were not required to personally attend trial and instead could dispatch servants on their behalf to court. Of course, gentry-scholars also committed offenses that required punishments, but, in such instances, they had to first be officially stripped of their titles so that any punishment would not reflect poorly on members of the class. Since a judicial punishment against a member of the gentry was seen as an offense to all members of that special class, public beatings mandated by officials against scholars were understood by all as significant political acts. This cultural context is essential for understanding the message of the gentry petition protesting the move of Nanbu's Confucian academy.

On paper, the Salt Tax Bureau went on to achieve many of the provincial government's aims for extracting revenue.[29] Yet, for the gentry, perceptions of declining fengshui did not go away. Nanbu never produced another palace graduate following the bureau's creation, and provincial degrees remained out of reach for most. Poor examination performance and continuing natural disasters reinforced the perception that the county's fortunes were declining, with the prestige of the local gentry descending in tandem. Even the government appeared concerned. The year before the salt bureau protest, 1887, the Baoning prefect informed counties under his jurisdictional authority, including Nanbu, of new, stricter empire-wide legal regulations (*zhangcheng*) for punishing acts of grave destruction and tomb robbery, which capital authorities cited as a growing trend.[30]

It is easy to find heightened troubles from this point onward. One consequence that followed the gentry's fengshui protest came in 1895, when brine furnace operators looted and destroyed salt tax collection posts across the county following rumors of another rate hike.[31] Over the 1890s, a new magistrate oversaw the completion of repairs to the fengshui pagoda on Mount Aofeng to keep spirits up, though one suspects that already a higher degree of cynicism than usual had set in among the scholars.[32] Among many reasons for their disapproval was continued displeasure with the site of the new academy, Towering Cloud Cave. In a local gazetteer composed not long after significant flooding and drought hit the county, the compilers alleged that the fengshui of the new site was terrible:

> Aofeng Academy was originally in front of the county school and government yamen; it was built in the *wuxu* year [1778] in the Qianlong reign by the county magistrate Li Yuanfen. After its establishment, the literary achievements of the county splendidly rose, and the gentry and the commoners valued it. The academy held erected tablets and altars honoring the prior worthies of the county. Then, in the *dinghai* year [1887] of the Guangxu reign, the academy was met with disaster when the salt commissioner Zhao Huicheng connived with local gentry to transform the academy into the Salt Tax Bureau! The county's academy was moved to the foot of Cloud Soaring Peak, where a cemetery was leveled to create its foundation; the *qi* of this place was abominable, and the county's scholars denounced the site.[33]

In an ominous sign of troubled times, the local literati were willing to critique the local government in writing, with an acerbic tone not as evident in the earlier 1849 and 1870 editions of the county gazetteer. The compilers of the 1906 gazetteer condemned the Salt Tax Bureau for having stolen the authentic Aofeng Academy—and the county's good fortune along with it.

Following a short burst of good fortune for Nanbu around the Taiping Civil War—symbolized by the unexpected appearance of two rare *jinshi*-degree holders and a new pagoda in the county—there was a growing perception among many scholars that public fortunes were in trouble from the 1880s onward. Anxieties apparently accelerated in the 1890s and continued into the early twentieth century. Missionary activity was one important source of anxiety, but it was not the only one. Natural disasters contributed to a growing sense of crisis, as did rising taxation on Sichuan's key industries. Government reports regarding a troubling trend of tomb destruction probably reflected public moods more generally. The perception of declining public fortunes related to fengshui in Ba County has been well documented by Mark Driscoll, but it also applied to Nanbu County, which was a far more remote place.[34] If Ba and Nanbu both saw local scholars sounding the alarm about disturbed fengshui in late Qing Sichuan, it can only mean that similar protests were being lodged in other parts of the empire.

One might also speculate that the perception of harmed regional fengshui may have set in later in Sichuan compared to other provinces due to its relative isolation along the western frontier and its relative lack of violent unrest over the 1850s and 1860s. Chuck Wooldridge, for instance, cites related anxiety over regional fengshui from just after the First Opium War in Nanjing, the capital of coastal Jiangsu Province.[35] Regardless of when and where these anxieties began to be voiced, it is important to recognize that resident administrators found themselves increasingly hard pressed to offer satisfying solutions to a growing number of complaints.

Late Qing Officials on Protecting Fengshui

During the period from the 1860s to the 1890s, as reports of resident administrators and gentry over declining fengshui grew in number and volume throughout China proper, provincial governors and palace officials became more involved in Sino-foreign disputes involving the law and the land. Although Western diplomats repeatedly dismissed fengshui as a trivial concern, it was decidedly not.[36] A wide and diverse spectrum of high-ranking Qing officials took a stand on fengshui-related issues, proposing various solutions to deal with the potentially harmful effects of new forms of economic development.

Most Qing officials agreed that fengshui was a significant and legitimate issue in the 1860s, especially during the diplomatic negotiations over revisions to the Treaty of Tianjin.[37] As readers will recall from earlier, that group initially included Li Hongzhang, who later championed telegraph and railway development in China. But what about the Qing officials who disagreed with these policies? In historical hindsight, their strategy was obvious. The opponents of

Li Hongzhang's industrial proposals in the 1870s, 1880s, and 1890s repeated or expounded upon the objections he had raised in the late 1860s in their own memorials to the emperor. In these documents, fengshui was a commonly cited concern.

Telegraphs were introduced into China in 1852, but their acceptance and development took around thirty years because of the state's wariness of alien technologies and the ways they might be used to China's disadvantage.[38] In the early 1860s, Jiangxi governor Shen Baozhen (1820–79) criticized the use of telegraphs in China on the grounds that they would enable Westerners to gain advantageous positions in the transfer of military and commercial information. Far from being ignorant of global affairs, Qing officials like Shen were aware that telegraphs and railways had opened India to accelerated Western military and economic penetration.[39] And, as previously mentioned, geomantic objections to telegraph lines had been raised by Qing bureaucrats in the 1860s.[40] Some concerns carried over through 1880, when the imperial court gave its explicit approval for the construction of the Tianjin-Shanghai line.

Some Qing officials identified methods for protecting fengshui in the event of telegraph construction, while others insisted that the risks of construction were too high. In drawing on existing practices in Qing law, Shen Baozhen suggested in 1867 that the areas around gravesites be mapped and annotated to preemptively investigate fengshui and prevent popular protest from arising should the court choose to construct telegraph lines.[41] Such a view proved prescient in the ensuing decade. In the wake of a political "fiasco" over abandoned line construction in Fujian in 1875, the imperial censor for the Office of Scrutiny for Works, Chen Yi (c. 1827–96), raised objections to further development.[42] Chen's memorial is notable for its specific focus on geomantic consequences:

> The setting up of lightning wires [*dianxian*, i.e., telegraph lines] penetrates deep into the ground. Running unimpeded both vertically and horizontally, they extend into all directions; the veins of the earth are severed by them, wind invades the land and water pours in [idiomatic for "bad fengshui"]. With such influences that must eventually manifest themselves, how can the minds of children and grandchildren be at peace? The classics say: "One should find a loyal minister in the home of a filial son." If we encourage the people of China not to care about their ancestors and their ancestral graves, when they hear of the establishment of these copper wires, will they still look up to their emperor and their superiors? The matter of defying one's superiors and fomenting rebellion would be hard to avoid.[43]

Pointing to the geomantic conditions of burial grounds, Chen opposed telegraphs on the assumption that telegrams, believed to be transmitted underground, would harm the earth veins around cemeteries and gravesites by redirecting the flow of *qi* in the soil. He leveraged that technical dimension of geomantic surveying to connect the adoption of Western technologies to moral decay. Chen Yi emphasized that the geomantic problems associated with telegraphs related directly to the imperial state's role in building them. In his mind, it was not a good idea for the Qing to be seen publicly undermining the fengshui of its people.

Chen Yi's polemic must also be understood within the context of Qing law. At the time of its writing, the heights of private and public buildings were monitored by communities in Sichuan and elsewhere. Qing authorities, as we have seen, recognized these rules and upheld them when appropriate in legal disputes. The erection of telegraph poles in a county where residents had donated funds over many decades to construct fengshui pagodas or other protective structures could be understandably controversial if the lines were thought to threaten those hard-won investments. Counties with not many local examination successes to celebrate, or those suffering from devastating natural disasters, could be especially sensitive to these additions.

By the same logic, allegedly inauspicious structures like pagodas, temples, and mosques might be destroyed, altered, or moved by Qing officials—a process that took place throughout the entire nineteenth century. Recall the words of Magistrate Fan responding to a late Qing dispute over an oil-press shop: "If the oil-press shop obstructs the fengshui of the village, it is permissible for the villagers to ban it." With the creation of permanent telegraph lines under government sponsorship, there could be no such option. Chen Yi's memorial argued that by exerting control over rural landscapes and skylines through the construction of telegraph lines, the government would relinquish power to shape local narratives in its own interests. For the imperial censor, the medium was literally the message.

After the Qing government resolved to build telegraph lines in the 1880s and beyond, fengshui was no longer a major obstacle to their construction.[44] Likewise, once the central authorities approved the construction of a telegraph network, rollout across and between major cities was relatively swift, with 45,448 kilometers of lines constructed by 1908.[45] When construction began on these lines, surveyors, engineers, and laborers took care to avoid trespassing "graveyards, groves of trees, and private homesteads," just as development contractors do today.[46] Historians have compellingly interpreted these developments as signaling the sophistication and adaptability of the late Qing state.[47]

Yet, lest we dismiss Chen Yi's memorial as impractical, naïve, or reactionary, one should ask if his contention that telegraph lines would lead the people to "defy superiors" proved to have merit. It did. Excluding the notorious events of the Boxer Uprising or early incidents that saw fengshui cited in relation to telegraphs in Fujian and Jiangsu, telegraph lines were torn down in Guizhou in 1887, in Hunan in 1892, in a string of rural counties in Shanxi, Shaanxi, and Gansu in 1892, and in Sichuan in 1898.[48] The Sichuan and Hunan incidents appear to have been primarily tied to antimissionary or anti-Western mobilization, while the others were linked to high food prices and droughts, which increased in frequency during these years.[49] Undoubtedly many incidents resulted from a combination of all these factors.

Fengshui-related anxieties about telegraphs did not go away, even after the government's decision to build up a communication network.[50] The Shanghai-based *Dianshizhai Pictorial* ran an image in 1896 concerning disturbed ancestral spirits and telegraph wires, revealing that the assumptions Imperial Censor Chen Yi made about the technology twenty years earlier still held sway among some people. In counties where foreign or domestic political and business interests did not deem telegraph construction essential, people could mobilize fengshui to delay construction for some time. The gentry of Xinghua, in Jiangsu Province, invoked fengshui to deter the construction of telegraph lines in and around the walled county seat, presumably in relation to examination anxieties. With financial backing hard to obtain due to a lack of support from the local elite, Xinghua's first telegraph station was not fully operational until 1915.[51]

For its part, the Qing government responded to troubles over telegraphs by adopting security procedures and adding amendments to the law code. The government hired guards to protect lines from destruction from the opening of the first national telegraph line in late 1881. Their annual salaries were not an inconsiderable expense.[52] Later, in 1892, the Board of Punishments, citing a memorial from Li Hongzhang concerning an upswing in the destruction of telegraphs, introduced punishments for anyone caught damaging such lines, which were identified as government property.[53] Mandated punishments included sixty blows with the heavy stick and one year of penal servitude. To encourage magistrates to strictly uphold the law regardless of popular opinion, provincial officials punished resident administrators in the event of a serious incident in their districts.

Such regulations were emphasized and enforced through the end of the dynasty, including in Sichuan. By the early twentieth century, Sichuan had fifty telegraph stations spread out over 163 counties, with a ratio of less than one station per three counties.[54] Chengdu and Chongqing were well connected to the broader network, as were the major commercial hubs of the province. For peripheral areas and hinterlands, however, the situation was different. The nearest telegraph station to the Northern Sichuan counties of Nanbu, Tongjiang, Nanjiang, Cangxi, and Bazhou was in Baoning, around two hundred kilometers away from some of the rural areas it served. Yet, as late as 1910, Nanbu's magistrate received orders from the Sichuan provincial government to strongly prohibit "the cutting or theft" of telegraph wires, even though the county lacked a physical telegraph station.[55] Service coverage of telegraph lines—and their popular reception by Sichuan residents—were uneven to say the least.

There were, of course, many factors behind the controversies involving telegraph wires in China, and fengshui's influence should not be exaggerated. Fengshui was not the only reason telegraph lines were vandalized in the late Qing. In addition to popular expressions of anti-Western sentiment, bandits stole wires and components for black market sale as scrap metal.[56] And even in cases of antiforeign collective action, we can assume that the torn-down metal wiring was put to profitable, albeit illegal, use. Nonetheless, the important point to keep in mind is that—regardless of the motives behind the actions of the central government, resident administrators, gentry, and commoners—Qing law was the final arbiter of planning and development, even if it took time for the legal system to adjust to the needs spawned by new technologies.

Unlike telegraph lines, which were relatively cheap to construct and could be flexibly placed along the ground, railways posed greater problems for Qing administrators. While proponents of telegraphs could plausibly claim that they could be constructed without harming fengshui, it was more difficult to make the same claim for rail lines. Constructing rail lines invariably involved the destruction of graves and their surrounding foliage, not to mention the leveling of entire hillscapes and mountains. As we shall see, even a relatively short rail line could involve the uprooting of tens of thousands of gravesites and trees.

One of the best-known arguments against railways was offered by the head of the Office of Transmission and a Qing diplomat by the name of Liu Xihong (d. 1891). Liu had spent considerable time in Western countries, where he had personally observed new transit technologies.[57] With more tangible experience with the technology than most Qing officials, Liu was reasonably well-informed. Although mid-twentieth-century depictions of Liu typecast him as a "xenophobe *par excellence*," his 1881 memorial listing the many reasons railways would harm the country expanded on many of the same arguments Li Hongzhang had offered in 1867, including concerns about fengshui.[58] Liu wrote, "If we suddenly burn and chisel away at the mountains and rivers, I fear the people will be frightened and shocked, and the masses will view these actions as highly inauspicious. The gods of the mountains and rivers will not be at peace, and disasters concerning droughts and floods will easily be summoned."[59]

Rather than viewing Liu Xihong as a Confucian "luddite" or "ignoramus" in the face of Western power, we would do well to situate his thought in the context of China's severe agricultural and environmental crises of the nineteenth century.[60] Liu's connecting of droughts with harmed fengshui resonates with Governor-General Wu Tang's order to protect Sichuan's dragon vein in the wake of a provincial drought in 1872. Further, as Mark Elvin reminds us, Liu Xihong's political positions were "probably near the norm among the well-educated" in the 1880s.[61] He was hardly a lone voice in the wilderness. Some officials, notably Yu Yue (1821–1907), a retired scholar-official of the Hanlin Academy, argued against the importation of new technologies such as railways on the grounds that they would consume large amounts of China's dwindling natural resources like timber and coal—which proved to be the case.[62] Qing officials like Yu Yue were aware of the fragility of the country's vast agricultural sector. They also knew that, for reasons unrelated to newly imported technologies, officials had fewer options to combat food shortages than their counterparts had earlier in the dynasty, when the state-civilian granary system was still relatively functional.

Other Qing officials underlined the economic and cosmological consequences of railways in tandem. In 1885, fifteen years before the Boxer Uprising, Academician of the Grand Secretariat Xu Zhixiang (1838–99) railed against railway construction in north China on the grounds that the resultant unemployed laborers would become bandits while emphasizing that, "the entire area is connected to the earth veins of the capital and is a place where royal *qi* congregates; carelessly excavating this area is bound to summon inauspicious phenomena."[63] As documented in further detail below, reports of banditry involving grave robberies were becoming widespread, especially around the imperial capital. Such reports lent credence to Xu's concerns.

Turning to an 1897 image of railway construction in Shanghai's *Dianshizhai Pictorial* titled, "The Dragon's Lair Is Already Broken" (*Longxue yipo*), one can view a visual depiction of the landscape that Liu Xihong and Xu Zhixiang had described in their memorials from the previous decade. In 1896, the first railroad tracks were constructed within the vicinity of the imperial capital, and the following year, they were extended from suburban Beijing to a place named Majiapu. This second route passed through Nine Dragon Mountain, which had been purchased by a British businessman for tree-denuding, demolition, and, eventually, track construction. The image depicts snakes emerging from the disturbed landscape of the mountain, driving the terrified railway construction workers away. Unlike dragons, snakes cannot fly, but here they are lifted into the air via the branches of the disturbed mountain's vengeful tree cover. The Majiapu station and rail tracks south of Beijing were attacked by Boxer rebels three years after the image's publication.[64]

FIGURE 5.1. Illustration of the Geomantic Consequences of Railway Construction (1897) This illustration depicts the transformation of Nine Dragon Mountain (Zhili Province) as its tree cover was cleared for rail construction. Rail tracks can be seen in the lower right corner. Image reproduced from Wu Youru, *Dianshizhai huabao: Dake tang ban*, vol. 14 (1897/2001).

The annotated image reads, in part, "This area's fengshui has already been ruined; one fears that in the future, the hills and valleys will be transformed, and it is impossible to know what the landscape will look like." The text's focus on the mountain's forest cover should not be overlooked. As previously discussed, gravesites and cemeteries, particularly those of the gentry and moderately wealthy lineages, often had collections of trees surrounding them. The railways planned in North China during the late 1800s traversed natural environments where timber was especially scarce, which accordingly drew considerable attention to potential damage to the fengshui groves of gravesites and cemeteries.[65] Lineage cemeteries on the North China Plain (see, for instance, figure 1.3) produced some of the only timber available for fuel into the early 1900s, and as Myron Cohen reminds us, "The importance of such graveyard products (trees) in a north China countryside otherwise largely denuded of trees and other overgrowth should not be underestimated."[66] None of these observations captures the entirety of skepticism about railways among the elite or common people, but they help explain the persuasive logic behind it.

One strategy of Qing officials to delay or cancel projects was to cite a dramatic number of disputes and lawsuits about disturbed gravesites in palace memorials. In 1888, an opponent of railway development, the Minister of Rites Kuirun (1829–90), a Manchu of the Plain Blue Banner and member of the imperial Aisin Gioro clan, claimed that among the two hundred to three hundred lawsuits concerning land along the planned route of the Tianjin-Tongzhou Railway, "there are over a thousand gravesites mentioned."

> It is rumored that the people living along the Tianjin-Tongzhou Route have filed no less than two hundred to three hundred lawsuits at the Tongyong Circuit Yamen; the circuit intendant has not yet responded, so the people are now appealing directly to the governor-general. . . . The contents of these petitions all concern the difficulty of moving houses and gravesites. Among these two to three hundred petitions, there are over a thousand graves mentioned. Most of them are ancient tombs with decayed coffins. Will descendants not be heartbroken to see the desiccated bones of their grandfathers? [67]

Kuirun's core message was that the development project was causing elevated levels of grave litigation and should be shelved. Whether for this reason or because of other complaints concerning restless laborers along the Grand Canal and the security of the capital, the plan was abandoned.[68] In an instance from 1897, the same year the dragon's lair image was published in the *Dianshizhai Pictorial*, rumors over grave disturbances during the planning of a Hangzhou railway led the investigating censor, Chen Qizhang, to appeal to the emperor for the cessation of the project:

> However, the present situation is not reasonable. As soon as the land is chiseled for the railway, people abandon farmlands and destroy homes, and there are tens of thousands of graves in the demarcated area of the rail path. In recent days, I heard that the city of Hangzhou is filled with public notices, and that the gentry and commoners wish to make things difficult for the railway builders. There is a saying among the people that if one's house is destroyed, they will move into the house of the builder; if one's grave is removed, then the family will place the coffin on the tomb of the builders' grandfather.[69]

Using a strategy like Kuirun's a decade earlier, Chen leveraged rumors about fengshui as evidence of an impending crisis, painting a picture for capital authorities of imminent revolt due to disturbed gravesites, which this time numbered in the "tens of thousands." Here again, construction on the proposed railway was halted, and the line was canceled.[70]

The situation on the ground may have been more complicated than these officials conveyed in their memorials. Poorer residents were sometimes willing to part with their ancestral lands and modest gravesites for the relatively generous price obtained by selling them to foreign companies.[71] As seen in chapter 1, the practice of selling the land around graves was common in rural China over the course of the dynasty and especially in the nineteenth century, when the practice was legalized for people in poverty. Elisabeth Köll finds a similar situation during the planning of the Jin-Pu Railway in the 1900s.[72] The class dimension identified by Köll is worth highlighting, because some of the strongest geomantic objections to railways and telegraphs came from Qing officials and gentry rather than commoners. It is also true, however, that commoners left fewer written documents than members of the elite, and thus we have less direct evidence of their opinions and experiences.

The importance of fengshui to the political debates of the late Qing is further revealed by the fact that advocates of railway construction drew on geomantic logic to support their positions. Liu Mingchuan (1836–96), for instance, argued in 1889 that just as fengshui masters advocated the moving of graves to more auspicious sites, the Qing government could pay generous compensation for the reburial of corpses to better sites away from planned rail lines.[73] That is, rail construction could be facilitated by appealing to existing geomantic practices surrounding reburial.

In the same year, Li Hongzhang mobilized those exact principles, citing in a memorial the example of a charitable cemetery in north China with poor fengshui. Recalling the inauspicious grave with poor drainage in chapter 1, Li described how the cemetery faced perennial flooding (*changnian jishui*) and was overcrowded with unburied coffins.[74] Compensation from the railway for its relocation away from a planned railway would help bring better fortune to the descendants of people buried there. Liu and Li's appeal to consider the "fortunes of the less fortunate" in relation to burials and grave lands certainly speaks to the class-based nuances discussed above.

Some officials also appealed to the "fortunes of the more fortunate" in advocating for industrial development. In one letter to Li Hongzhang, the Qing diplomat Guo Songtao (1818–91) cited the examples of two districts in his native province of Hunan that had produced a good number of elite degree-holders despite intensive coal and iron ore mining. Guo's defense of new technologies, made in the late 1870s, contended that railway and telegraph development would not harm fengshui for scholars seeking success in the examination system since there was "nothing to dig up or to destroy" in their creation.[75] Arguments like these would never have been offered by proponents of rail construction unless fengshui was a serious and meaningful issue for actors across the Qing political spectrum.

From the 1860s to the 1890s, a good number of Qing officials drew on narratives of declining fengshui and grave disturbances for political advantage, out of genuine conviction, or both. Well after Li Hongzhang changed his position on railway construction, officials continued to emphasize the need to protect fengshui in the wake of industrial development. One issue shared by Li Hongzhang's 1867 memorial, Chen Yi's 1875 memorial, and Liu Xihong's 1881 memorial was the identification of a direct connection between fengshui and the Qing dynasty's legitimacy—specifically, the idea that undermining fengshui would lead to unrest and rebellion. All three men at various stages of their careers claimed that the Qing government needed to protect fengshui, even though they may have held somewhat different personal beliefs about it. This shared understanding of fengshui and law enabled high-ranking Qing officials to oppose industrial projects on those grounds into the 1890s, during which time the palace bureaucracy carefully weighed the spectrum of official opinions as it attempted to steer the course.

Reforming the Laws of the Land

By the 1890s, it was evident that officials in Sichuan could not protect fengshui to the degree they had in the past and that the central government increasingly could not do so either. The decade also witnessed particularly strident calls for a dramatic shift in the state's position regarding fengshui. Some members of the emerging urban gentry-merchant class, exemplified by people such as Zheng Guanying, targeted fengshui for blocking mining and development in the country, thus weakening the government's ability to confront Western technological and economic power.[76] And even persons fully conversant with the principles of fengshui, like the Sichuan mining expert Song Gengping, called for changes in popular applications of the practice to permit industrial levels of extraction for the country's benefit. Over the decade, the Qing government began to compromise on matters related to grave relocation and compensation, and soon after the turn of the century, new sets of provincial and national regulations sought to write fengshui out of Qing law by formally forbidding its consideration in legal settings.

To appreciate the dramatic changes that took place in the second half of the 1890s, we should first consider the Qing government's position on fengshui prior to that time. As Qing officials didn't give public interviews about fengshui, one might imagine reconstructing that position would be difficult. However, there is archival evidence revealing a Chinese perspective on fengshui from inside the Qing government from precisely this time. Six years before the Chongqing Uprising, the German minister, Maximilian August Scipio von Brandt (1835–1920), objected to the rejection of a lease on Fuzhou's Wushi Hill because of concerns about fengshui, which he found to be "such a vague concept that it could be applied to any case of land acquisition."[77] The Zongli Yamen, which was the dynasty's office for foreign affairs from 1861 to 1901, provided the minister with the following response:

> Let me explain the meaning of "fengshui" because You, the Honorable Minister, have asked for a detailed explanation. This Zongli Yamen will specifically cite the clearly stated writings from the Treaty of Tianjin, such as "Do not hinder the dwelling places of the people and their orientations . . ." to explain the meaning of fengshui. When Chinese people speak of fengshui, we separate the grave and the house. Many people believe in house fengshui. But a belief in grave fengshui is unbreakable among our people. For you, the Honorable Minister, this is absurd, remote, and groundless. But Chinese people who believe in fengshui take it as undeniable and provable: people will, because of contestations of fengshui, engage in lawsuits for years![78]

The Zongli Yamen's response explained fengshui as a Chinese cultural practice related to houses and graves—with consequences for the legal system. The Qing government even invoked the idea of fengshui as law, arguing that it had been written into the Treaty of Tianjin in 1858 as a stipulation for property relations between foreigners and Chinese.

The Zongli Yamen's response continued: "If a foreign country rents or purchases a landed foundation and does not permit our people to believe in fengshui, it will surely bring on quarrels. This is not the way for peace between China and foreign peoples. This is really the situation in China; I do not say this as a mere excuse. If your settlements do not hinder the dwellings of our people and their orientations, our people are prohibited from finding any pretext to cause confusion."[79] The logic of this government statement is also revealing. Even though fengshui caused headaches for Qing officials presiding over geomantic litigation, the idea of ceasing to practice fengshui was considered so ludicrous that it was not even considered in conversation. The government's message was unambiguous: foreigners needed to respect the laws of the land.

Likewise, through the 1880s, the Qing state did not support large-scale grave relocations for railway construction, even for generous monetary compensation. Up to that point, rail lines had to be planned around graves, rather than the other way around. Li Hongzhang acknowledged this reality in an 1881 written reply to Prince Chun, also known as Yixuan (1840–91), stating, "As you have ordered, the fields and houses of the people can be moved if necessary for compensation, but graves and tombs cannot be moved."[80] Yixuan referenced related communications in 1889, reporting, "As we have started discussions, I have repeatedly emphasized to Li Hongzhang the point that avoiding disturbing the houses and graves of the people is the most important issue [*zui yao zhi duan*] in constructing railways."[81] The officials participating in these discussions never doubted that some commoners would willingly depart with their grave lands for compensation. Yet, as a matter of policy, the imperial state did not want them to—at least through the 1880s.

The imperial state's reluctance to move graves for compensation prior to the 1890s derived from the fact that its moral authority was related to the protection of the resting places of the dead.[82] Grave removal involving compensation also posed risks for litigation over stolen money, fraternal fraud, fake graves, and a host of other issues related to the sustainability of kinship units in the regions that planned lines traversed.[83] The governing paradox for Qing officials was that the government's public destruction of gravesites encouraged unsavory actors to take advantage of the situation by also destroying them, thus contributing to a growing sense of unrest across society. New technologies promised to help imperial governance in the long term, but what of the immediate costs for officials serving on the ground? As previously mentioned, a growing number of provincial reports on rampant tomb destruction galvanized the central government to continue amending the official law code, mandating stricter punishments for uncovering graves.[84] Related reports continued through the end of the dynasty. The powerful official Yuan Shikai (1859–1916) reported many occurrences of grave destruction by "bandits" in Zhili Province around the turn of the century (1903), not long after the Boxer Uprising.[85]

Despite the challenges mentioned above, new national laws and regulations for mining were adopted between 1898 and 1902. As Shellen Wu has shown, these new laws sought to make Qing laws over mining broadly congruent with Western laws.[86] These moves were especially important given the rising involvement of foreign companies in China. The laws also revealed what Qing officials' priorities were, or were becoming. The law of 1902 enshrined the principle of monetary compensation for grave removal if needed—a stark change from the government's position just a decade earlier.[87] In other words, the 1902 law extended state protection to gravesites, but it did not extend official recognition to fengshui. High-level official documents concerning mining thereafter seldom invoked "popular" fengshui sympathetically.[88]

These national trends were echoed at the provincial level in Sichuan. Prior to 1898, Sichuan lacked a mining bureau, which was a type of government office during the late Qing responsible for surveying lands for new mining projects. The province received imperial consent to establish one in 1898, and shortly thereafter two mining schools were established to train local Chinese in the techniques of surveying. Foreign companies wanting to exploit Sichuan's resources negotiated with the mining bureau and its affiliated company, the Huayi Corporation, to establish the terms of mining leases in specific areas. From this point onward, international mining contracts typically contained a statute limiting locals' abilities to protest extraction because of fengshui. In one agreement drawn up in 1899, resident administrators were obliged to protect "with a sincere effort" the interests of the company from such litigation.[89]

Other proposed regulations in Sichuan further sought to diminish fengshui's relevance. A series of communications, dated 1901, involving the governor-general of Sichuan, the Zongli Yamen, and the official overseeing Sichuan's mining and commercial affairs—a person named Li Zhengyong (1848–1902)—explicated new provincial mining regulations in considerable detail.[90] Li was aware that mining bans had proliferated across Sichuan in earlier decades. To deal with these proscriptions, he proposed the opening of all previously forbidden sites to mining on the condition that merchants first obtain a license and pay the appropriate taxes. Considering the number of local bans that were issued over the prior decades by county magistrates in Sichuan, this proposal signaled a radical change.

Li Zhengyong continued by noting that if a mine did not directly "invade the graves of people, the action would not be taken as illegal."[91] Here again, a new distinction was being made between graves and grave environments. Li also wrote that any members of the local gentry who attempted to obstruct mining by claiming that fengshui was obstructed were to be severely punished—a warning that this behavior would henceforth not be tolerated. Finally, Li proposed that mining around temples, communally owned properties, and governmental land could be encouraged under the condition that part of the profits went to charitable deeds, such as the establishment of new schools. This way, extraction around these sensitive sites could "avoid local controversy."[92]

The degree to which Li Zhengyong's proposals were accepted, amended, and enforced in Sichuan during the 1900s is difficult to measure. By the end of the 1890s, Ba County saw the relaxation of coal mining bans on sensitive sites, such as Mountain of the Perfected Warrior (*Zhenwu shan*), with a local gazetteer from the early twentieth century citing 1898 as an important milestone for the relaxation of local mining regulations.[93] Facing foreign encroachment into the interior, powerful actors within the provincial government wanted to open Sichuan up for an unprecedented degree of extraction and drafted regulations to facilitate that change. Both at the national and provincial levels, proposed regulations sought to contain or diminish fengshui's legal relevance for mining and other forms of development. Yet, as demonstrated below, local archives reveal that targeted mining bans in sensitive areas did not completely disappear during the last decade of the dynasty.

Cutting Down Fortune

From 1901 to 1911, in response to the humiliations occasioned by the Boxer Uprising of 1899–1900—which included the occupation of the Qing capital by foreign troops, and severe economic and other sanctions imposed by the Western powers and Japan—the dynasty inaugurated a series of far-reaching cultural, economic, educational, military, and political reforms designed to keep the regime in power. These reforms, known collectively as the New Policies (*Xinzheng*), were felt across the country, even in remote Nanbu County.

Two major changes are evident: the first involving the provincial mining bureau's efforts to expand gold mining, the second concerning initiatives to establish new-style schools in rural temples. Both acts coincided with the legal changes mentioned in the previous section. Undoubtedly, some people profited in this new climate, and through the 1920s, cultivators in Northern Sichuan took advantage of expanding markets for tung oil, opium, and mined goods.[94] However, in fostering new kinds of development, the New Policies revealed fengshui's continued importance to legal practice. As illicit mining and tree-felling increased across Nanbu County over the decade, locals resorted to protests based on fengshui. By 1911, rumors that the state had sanctioned the felling of all fengshui trees in the county led the local government to order a halt to the further clearing of timber on rural commons.

In the early 1900s, surveyors backed by Sichuan's provincial government set out to locate potentially lucrative mining sites that could be exploited with imported technologies and new expertise. As discussed previously, Nanbu had seen the local exploitation of placer gold deposits alongside the banks of the Jialing River over the nineteenth century, with occasional disputes finding their way to court. As in other parts of Sichuan, however, disputes over mining rose sharply in the 1900s.[95] From 1894 to 1905—a decade for which many hundreds of archival legal case files survive—only two recorded disputes over gold mining reached the county's court.[96] But in the mere five years from 1906 to 1911, at least twenty-one lawsuits and petitions concerning gold mining were presented. This number of cases was greater than any previous five-year period, including the earliest preserved archives from the eighteenth century. Of these twenty-one cases, twelve involved graves, fengshui, or disturbances to farmland.[97] The remaining nine cases alleged illicit mining that evaded proper registration and taxation.

In a few instances, fengshui proved consequential prior to extraction. In a 1910 survey of twenty proposed mining sites in the county, two charted locations (10 percent of surveyed sites) had been shelved due to geomantic objections. The first involved an area of three square Chinese miles (one square English mile) around Guoqing Temple. The second involved a smaller area adjacent to the Wang Family Embankment and the Lord Lao Temple. For the Guoqing Temple site, the surveyor recorded that "the people here are obsessed with fengshui, therefore the mine has not opened."[98] For the Wang Family Embankment and the Lord Lao Temple site, the surveyor commented that "the education levels of the people are not high; they are obsessed with fengshui."[99] Although the surveyor's notations denigrated the locals involved for their obsession with fengshui, the protests apparently succeeded in blocking the mining. It is also worth noting that residents in these two places were not interested in monetary compensation for the use of lands adjoining their properties.

In addition to the abovementioned locations, the county court continued to issue bans on mining around sensitive sites. One magistrate issued a gold mining ban around Baoben Temple when a member of the gentry appealed for one. It may be relevant that the petitioner bore the prestigious Xianyu surname—an old Sichuan lineage that claimed a string of ancient gravesites in Nanbu reportedly dating back to the Tang dynasty.[100] Around this time, opportunists even attempted to quarry Mount Aofeng, the site of Nanbu's most famous fengshui tower. This act was protested with a geomantic petition at the county court and ordered halted immediately by the presiding magistrate.[101] Incidents of this sort reveal that people were aware that the rules of the game regarding development were changing and that some locals were more willing to push their luck in exploiting previously off-limits areas. Yet, despite the proposed lifting of all mining bans in the province, resident administrators still felt compelled to occasionally issue restrictions in response to the rising number of conflicts.

Disruption of fengshui was not the only objection to increased mining during the period of the New Policies, but its appearance in government documents from the era is striking considering the burst of new provincial and national regulations aiming to restrict its legal invocation. With illicit mining increasing in the county, resident administrators could not simply sideline fengshui with a few brushstrokes on paper or telegrams to Chengdu. People still recognized fengshui as a valid, compelling reason to object to mining development, and new laws and regulations had not penetrated rural areas to the extent that they could systematically replace deeply held beliefs and practices.

Rising concerns about fengshui were also expressed in the wake of news announcing the 1905 abolition of the civil service examinations, which marked a significant shift in the relations between the state and the gentry class. No longer would an examination system serve as the primary vehicle for obtaining degrees that immediately conferred gentry status and could lead to governmental office. In place of older Confucian centers of learning, there was an effort to construct new schools—another prominent feature of the New Policies. Dimensions of the "building schools with temple property" movement have been discussed at length by Paul Katz and Vincent Goossaert, who draw attention to the destruction of temple institutions and the deleterious effect on popular religious cults.[102] The Nanbu Archive offers an additional side-effect. Because so many temple and lineage assets were tied up with forest groves and mineral wealth, as discussed above, popular reactions to the campaign to establish new-style schools in Sichuan took a distinctly geomantic turn. And yet in the larger picture, as one of the many ironic twists and turns of this tumultuous period of reform, the abolition of the imperial examinations took at least one major source of inspiration for geomantic legal petitions off the table.

As discussed previously, for a county lacking local coal, like Nanbu, lineage woodlots and temple forests held great significance, not only for reasons related to fengshui but also as sources of fuel. County magistrates projected their authority by strictly protecting grave and temple groves, not least because any attempts to modify them often ended in lengthy legal disputes. Harmful cutting could be retroactively "condoned" with an order to pay compensation, to perform sacrifices, or to host a funerary opera to remind the community of the ritual gravity of a violation over lineage commons, but official permission to cut fengshui trees down in advance was seldom given in the 1800s. The rural social order had been sustained in large measure because the government maintained a high bar for the protection of collective fengshui and a low bar for forgiveness—one that always allowed for family and community reconciliation.

Although there had always been local conflicts over fengshui groves in nineteenth-century Nanbu, the persistent threat of legal punishment prevented some corruption and opportunistic behavior. But in the early 1900s, the Qing state altered its imperial-style social contract by encouraging rural development around the temples, with limited official oversight. The results were messy.[103] Under the pretext of raising money for schools, school headmasters and other self-interested actors began to steal other people's fortunes. Thereupon, lawsuits and petitions poured into the county court.

Troubles were felt across the province, but they were more pronounced in some places than others.[104] As historian Xu Yue has shown, in the well-off counties of the Chengdu Plain, there was often enough money available to establish schools without relying too much on extensively felling temple groves or leasing lands for mining.[105] Even when cutting or mining was done by temple stewards, strong lineages and religious communities in wealthy counties carefully oversaw the process and the distribution of profits. The issue for Sichuan's poorer counties, like Nanbu, was that there was little surplus available to establish the schools, meaning that one of the only choices available was the sale of assets. Furthermore, because the county's modest gentry base was thin on the ground, there were fewer impartial scholars of moral standing to mediate conflicts and check abuses.

Around the same time that legal disputes over gold mining increased, disputes over the fengshui groves of temple and lineage estates rose markedly. Popular requests for official notices of protection for temple fengshui picked up around 1902, when the county's court issued a general notice about protecting monastic trees.[106] By the end of 1907, there were ten open school-related cases in Nanbu that revolved around the harming of fengshui and the cutting of trees. These cases reflect only instances in which litigants had the time, means, and the will to travel to court.[107] Some petitions had ten or more signatories, affirming a foreboding sense of cosmic disaster or evidence that one had already arrived.[108] Magistrates, who could never be fully certain where all the money from tree sales went, were subsequently compelled at trials to condone prior sales to support the new schools while urging the cutting to cease going forward. The result was a markedly mixed message from county officials.

In the summer of 1911, the county's newly established Deliberative Assembly (mandated in each county as part of the New Policies) reviewed thirteen lawsuits alleging unlawful tree-cutting since the beginning of the lunar year. In a written recommendation to the Nanbu magistrate, the members of the assembly denounced rumors circulated by opportunists who sought to justify the cutting of community groves and the selling of lineage common spaces or properties collectively owned by a group of related families. "Recently," they wrote, "many unworthy residents have fabricated rumors and incited the minds of the people regarding the building of new schools; these rumors claim that since the government has solicited funds from felled trees, it is thereby requiring the felling of all ancient trees around temples, monastic estates, and cemeteries."[109]

The assembly's recommendation to the county government was to dispatch the head of the Agricultural Branch Bureau to travel across the countryside and forbid the further cutting of trees, thereby silencing the malicious rumors. It also recommended a campaign to encourage "Western" techniques for commercial tree planting, including one technique identified as the "planting trees to summon rain method" (*zhongshu zhiyu zhi fa*), which was said to ameliorate drought conditions. As seen in chapter 1, in the wake of drought—one of which hit Nanbu in 1904—desperate communities appealed to the county yamen for permission to cut community groves in order to raise money for food purchases.

By pairing two recommendations—protecting existing trees and planting more trees to summon rain—the Deliberative Assembly conveyed concern about damaging rumors in the wake of recent events. Rumors were driven by mutually reinforcing religious and economic factors. The perceived links between appearance of drought and harmed fengshui was obvious to the assembly and needed immediate addressing. In terms of the timber market, once the government mandated—or allegedly mandated—the cutting of fengshui trees, rural communities had little remaining bargaining power over their worth. Timber merchants could acquire wood at cheap prices, knowing that the trees had to be sold.

In a longer view, these actions removed medium- and long-term sources of collective emergency funds for rural kinship and temple communities. Reliable data for year-by-year pricing is difficult to find, but a local gazetteer from Nanchong, which neighbored Nanbu, ascribed fuel price hikes in the 1920s to the overcutting of local forests in the preceding decade.[110] Norman Shaw (1878–1955), who worked for Imperial Maritime Customs of China in the early twentieth century, offers a complementary observation for the Baoning region in his 1914 book, *Chinese Forest Trees and Timber Supply*, just three years after the crisis in Nanbu:

> The local woods near Paoning [Baoning] have almost entirely disappeared; in fact, the only remains of the old cypress forest near there is the fine avenue lining the post-road for a great distance, and preserved by official command. Even near the sources of the rivers in this part of Szechwan the forests are disappearing; the cypress logs available at Paoning are now of an average diameter of only from 8 to 12 inches, and a log of 40 feet in length and 1 ft. 3 in. diameter is considered a very good specimen.[111]

As seen from Shaw's testimony, by the early twentieth century most local cypresses—essential for building—were harvested long before their point of maturity. Cypresses, common fengshui trees in Sichuan, require thirty to forty years to grow to maturity, and readers will recall that some Nanbu families owned trees that were considerably older.[112] Shaw understood, of course, that forests can be regenerated with time, but was nevertheless thoroughly shaken by the turn of events in Northern Sichuan, adding that "the only replanting of trees done is that of mulberry and tung, and the timber question may before long become acute here also."[113]

The submission of the Deliberative Assembly's recommendation to the government in 1911 and its description of chaotic county events were unprecedented. Prior to the 1900s there had never been a local rumor that the imperial state had sanctioned the cutting down of all groves around temples and graves, and, in fact, such a rumor would not have even made sense to people before then. To be sure, there were always borderline cases regarding fengshui that required a magistrate's personal judgment and interpretation, and complex property questions involving multiple contracts, genealogies, and inscriptions might go either way. But Qing officials had been remarkably consistent on the question of community groves. Everyone had generally known what to expect from officials—until, rather suddenly, they could not.

Reports of tree felling were not unique to the poorer corners of Sichuan.[114] Consider railway construction in North China during the same decade. In a record from Yuanshi County, Zhili Province, construction on the Beijing-Hankou Railway after 1900 was said to have precipitated the mass felling of family and temple groves. In 1901, workers on the rail line allegedly colluded with local ruffians to extort villagers of their groves. Claiming that their trees were to be requisitioned by the government without market compensation due to their placement near the planned rail lines, the schemers exploited local fears, encouraging villagers to cut and sell valuable woodlots en masse for low prices.

Not long after, the trees that had once grown near the planned route across counties, villages, and towns had been completely cut down.[115] The droughts that followed in the ensuing years inspired rumors that underlined the popularly perceived causes and effects of such actions, and the local government was compelled to launch new tree-planting efforts to stem the tide—just as Nanbu's authorities had to do in 1911. Not every part of China witnessed these economic changes in the wake of school or railway development, and one surmises that places with ready access to ample fuel sources had a smoother transition.[116] Yet, at least for some corners of China, material outcomes uncannily coincided with the reservations that Qing officials had expressed over new development in previous decades.

In addition to tree felling, the confiscation of popular temple and shrine buildings during the New Policies era added to anxieties about fengshui. Reflecting a common theme seen across Sichuan, local gazetteers offered droughts as evidence that fengshui had been harmed. Some residential administrators handled these anxieties reasonably well, but that was not always the case. In Chongqing Department near Chengdu (not to be confused with Chongqing Prefecture and Ba County), a police training center was established in the City God Temple and its adjacent old examination hall in 1909. Many residents opposed those changes and protested to the prefect, who alleged that popular "superstition"—invoking the post-1900 neologism *mixin*—was hindering the enactment of the New Policies in Sichuan.[117] The prefect rejected the request to halt construction of the training center in those structures.

Later that year, the appearance of drought conditions led residents to contend that Chongqing's fengshui had been disturbed. After calling for the shutting of the city wall's south gate to protect fengshui and demanding the performance of a rain-making ritual, a mob destroyed the newly established police training center. A 1926 retrospective account of the surrounding events from 1909, which involved the embezzlement of public funds by yamen functionaries during the fighting between the mob and local authorities, found fault with both resident administrators and the crowd involved.[118] Specifically, the passage observed that the officials should have been more cautious in dismantling the City God Temple since doing so removed a key mechanism for handling public relations during the drought. In other words, the local officials had given the crowd all the evidence it needed to contend that fengshui had been disturbed.

The decade in which the Qing state attempted to sideline fengshui's influence most actively revealed how deep and essential its engagement with it had been in prior times. This observation supports Prasenjit Duara's notion of a "cultural nexus of power," wherein popular religion played essential roles in certain important features of Qing rural governance.[119] Fengshui played many roles within that nexus, including ones that distinctly extended into law. Thus, when the Qing state decided to disengage from its relationship with fengshui, the process was a difficult, gradual, and geographically based one. Change could not, and did not, occur all at once across China, and by 1912, there were still considerable divergences between urban and rural areas within Sichuan Province, not to mention other parts of the empire.

In Nanbu County, fengshui continued to be invoked in law well into the Republican era (1912–49). Readers caught a glimpse of its continued relevance in chapter 2's discussion of the Liu house dispute following the dissolution of the Northern Sichuan Soviet. Many other examples exist, however. In 1917, for instance, a widow and her sons presented a string of lawsuits concerning stolen fengshui trees, contending that "Ren Shun last year felled and sold one fengshui cypress tree from my property, which has affected four members of my family and resulted in the deaths of two of my grandchildren. Now, he again felled trees by force; with so many trees cut, how could these activities not have broken our earth vein and threatened our lives?"[120] After a series of county court rejections, the widow appealed to the new superior court established at former prefectural seat of Baoning, where her geomantic claims were ultimately recognized as valid.[121]

The Republican Revolution of 1911–12 brought a new political system to China, but fengshui remained a prominent feature of Chinese rural life. For decades afterward, provincial land surveyors made note of fengshui's widespread appearance in contracts, household division registers, and other written records—inclusions that made "scientific" land registration especially challenging.[122] Though the laws of the country had changed with the collapse of the imperial state, the laws of the land proved difficult to uproot.

A Century of Fengshui

Chinese attitudes toward fengshui in the late nineteenth century involved far more than questions of cultural preference or "the following of ancient customs," as Western critics at the time too often incorrectly assumed. Fengshui mattered to Qing officials, scholars, and commoners because it was deeply enmeshed in political, social, economic, and legal life. It had its principles and logic, which might be subject to debate but could not be dismissed out of hand. Fears over the disruption or weakening of fengshui were not irrational, and those who expressed anxieties either in palace memorials or in county lawsuits were not necessarily naïve, uninformed, or uniformly xenophobic or technophobic. Their arguments reflected understandable, albeit often conflicting, responses to the rapid changes in politics and law, the forces of global imperialism, and the emergence of industrial capitalism.

What seems possible in retrospect is that fengshui-related issues contributed to the crises of the late Qing by undermining the legitimacy of the Manchus in the eyes of Han Chinese officials, scholars, and commoners. In the first place, the country's welfare was in serious trouble in the decades leading up to the Boxer Uprising. Rivers were flooding, the land was drying up, famines were breaking out, fuel prices were increasing, taxes were rising, and church steeples were piercing the sky.[123] Considering legal cases from across the nineteenth century, it is unsurprising that people linked those trends with damaged fengshui. Such trends left resident administrators managing an increasingly treacherous public relations landscape—a reality that made some wary of church-building, telegraphs, and industrialization in general.

Historically, the Qing state had accorded a space within legal practice for addressing credible geomantic claims. Its officials protected fengshui, albeit with uneven success. This position, firmly rooted in imperial law and legal precedent, likely helped the Qing navigate several difficult decades of foreign penetration into its territories. Even the man of the moment, Li Hongzhang, saw a need to invoke fengshui against new technologies from the West in the years following the Taiping Civil War. But as time wore on, the tide shifted in consequential ways. The Qing government ultimately could not protect the resting places of the dead, secure advantageous examination conditions for aspiring candidates to officialdom, or consistently secure the well-being of community commons and lineage properties from illicit sales. By the dynasty's final decade, Qing law formally distinguished grave environments from gravesites, the government altogether canceled the examination system, and community spaces like temples and shrines were confiscated and sold to construct schools and other enterprises.

Although not uniform in application, these arrangements required a very different kind of political and legal infrastructure from the one that the Qing dynasty had presided over successfully from its inception. People across classes undoubtedly believed in and cared about fengshui, but that was never the paramount point. The key factor was that imperial law had been written and practiced with fengshui's relevance assumed. To render fengshui's influence defunct did not only require people to stop believing in it. It also necessitated a transformation of mapping practices, mining regulations, property relations, burial customs, and the civil examination system, not to mention a rewriting of Qing law.

The national implications of those changes were stark. During land appropriation for the Shanghai-Nanjing (Hu-Ning) Railway, which took place between 1904 and 1908, surveyors relocated 73,561 gravesites and cut over two million trees along the planned route. Those numbers applied to a single line that extended for just a few hundred English miles.[124] Even if one assumes that most of the affected families along the route generally welcomed the railway—a bold and perhaps unwarranted assumption—the government's limited land ownership records ensured there would be conflict over the distribution of compensatory funds. In hindsight, it is not hard to understand why pernicious rumors about the detrimental effects of railways circulated in the fuel-and-timber-scarce North China Plain, just as Qing officials had repeatedly warned from the 1870s to the 1890s.

In Nanbu County, where no railways or telegraph stations were even planned, let alone operated, rumors about harmed fengshui resulting from far more modest developmental efforts also circulated during the dynasty's final years. If Qing officials even noted those rumors in Nanbu, it can only mean they were widespread across many counties. It is unnecessary to rely on missionary accounts, Western diplomatic writings, or treaty port newspapers from the era to appreciate the extent of the challenges facing Qing officials and rural communities. Sichuan's legal archives, palace memorials, imperial edicts, the records of the Zongli Yamen, and local gazetteers point in at least two directions: first, to a kaleidoscope of seemingly impossible policy choices that the dynasty was forced by circumstance to make; and second, to a longer process of disorienting changes in the forces of rural production that burst asunder existing economic relations and were fought out at the intersection of law and religion.

The political, legal, and economic shifts that arrived at the turn of the twentieth century provoked different reactions, depending on where people lived, their social class, educational background, and other variables. The example

FIGURE 5.2. Boxers Destroying Railway (Image Dated 1905). Image from Li Di, *Quan huo ji* (1905).

of Sichuan mining expert Song Gengping, who sought to update and apply fengshui to a shifting legal landscape for the extractive industries, offers one fascinating response. Still, there were many others.[125] Above all, reactions varied because of the material outcomes that emerged from the dramatic political, social, economic, and legal transformations of the late Qing. As the mines were opened, the lines were raised, the schools were built, and the tracks were laid; some people gained while others lost—increasingly, it seemed, not because of "fate," but because of the Qing government's actions.

The early 1900s saw the inglorious end of a century-long chapter in the history of fengshui in China, with its conclusion evident in notorious crises like the Boxer attack on railways and telegraphs, and little-known events like the rampant clearing of timber on rural commons. Those alive during this final century of the imperial era witnessed "changes in the land" that pushed the human world to its material limits.[126] They experienced the ways that the trees of the dead shaped the fates of the living. They observed jurists of the realm weigh past action and future consequence in judging not only what was legal but what was law. And some lived long enough to see the laws and machines of a new kind of empire forcibly remake the land, permanently changing the connections between time, people, and place that had once been common sense—like the directions of a map or the commands of a calendar.

Concluding Remarks

LAWS OF THE LAND is motivated by a desire to take a fresh look at Qing law through the lens of fengshui, an important yet largely misunderstood historical phenomenon. It invites readers to rethink the nature of law in the late imperial era, the inner workings of the Qing dynasty's administration, and the complex social and economic forces that drove people to record and analyze changes in natural and human landscapes. It explores facets of Chinese life in the Qing dynasty, including the culturally constructed boundaries between law and nature, science and religion, rationality and superstition, as well as "modernity" and "tradition"—all categories that remain challenging to parse in our present time.[1] This book will hopefully also shed valuable light on many other facets of Chinese culture, including politics, cosmology, examinations, ethics, architecture, aesthetics, ancestor worship, and city planning.

This book uses documents produced by officials, yamen functionaries, and literati to argue that fengshui played a critical role in Qing legal culture and was interwoven with the dynasty's governance at all levels. The imperial state accorded a space for fengshui in its administration of land, law, examinations, and the other activities it oversaw. Fengshui's importance to the core imperial institutions of family, society, and state supported the continued production of ritual specialists, technical manuals, gazetteer narratives, and, last but certainly not least, lawsuits. Each of these attested to fengshui's relevance and, quite often, its effectiveness. Not everyone in Qing China thought fengshui always worked, but the smart money knew it sometimes did.

Although neither Nanbu nor Sichuan is fully representative of all of China, both provide local insight into the broader features of imperial law and governance.[2] Sichuan's archives are especially useful in this regard. One might expect to see a monograph discussing fengshui based on materials from Singapore, Taiwan, or Hong Kong, where most research on the topic has been conducted.[3] Even among places in the Chinese mainland, Nanbu is hardly an obvious candidate because it lacked celebrated fengshui, famous scholars, and highly productive land. In this instance, the exception proves the rule, with the rarely heard testimonies of people living along the western edge of the empire helping clarify the challenges, pressures, and remedies that transformed Chinese society during the Qing dynasty.

In a broader sense, by giving a powerful voice to pervasive anxieties over health, land, wealth, and family in the Qing period, fengshui helped to preserve the imperial order in Sichuan and, most certainly, in other parts of China. Fengshui gave stakes to filial piety, allocated natural resources, checked unwanted development, extended property claims, and provided explanations for the prosperity or decline of individuals, lineages, and entire communities. It addressed fortune and misfortune in an environmental context where one seasonal harvest could determine the life or death of a family. Fengshui projected the power of emperors and officials, gave influence to scholars, and lent rural families a precious avenue for legal recourse and protest. Bridging heaven and earth, emperor and commoner, official and scholar, and the living and the dead, fengshui's principles served in countless ways as laws of the land in the last of China's imperial dynasties.

A Boundary with No Border

In theory and practice, fengshui served as a constantly evolving understanding of the natural and built environment. A wide range of people, groups, and institutions participated in the creation of knowledge about fengshui. At the top was the Qing state, which hired geomantic professionals, printed and circulated *Imperially Endorsed Treatise on Harmonizing Times and Distinguishing Directions*, and annually issued the imperial calendar. Below were commercially produced geomantic handbooks and almanacs that offered additional practical advice concerning irrigation, building, and business to farmers, artisans, and merchants. These manuals often contained several if not dozens of case studies or model examples detailing inauspicious sources of litigation. On the ground were scholars and commoners who strategically named their communities' mountains, bridges, residences, gardens, and shrines with eyes to the future. In doing so, they made astute observations about the fengshui of dwellings that belonged to the celebrated locals who managed to pass the higher-level examinations or struck sudden wealth through salt mining and related activities.

In Sichuan, as elsewhere in the empire, county magistrates and higher-ranking officials mediated between imperial and local interests. Many officials hailed from the southeast, famous across the realm for the intensive practice of fengshui among competitive counties, aspiring scholars, and elite lineages. These men arrived at their posts with the experience, status, and knowledge prized by the literati class and revered by the common people. Calling to mind the work of James Watson on orthodoxy versus orthopraxy, these administrators perceived local people as *believing* in fengshui but *practicing* it incorrectly due to a lack of proper knowledge and limited education.[4] Officials used that distinction to the government's temporary advantage. In court, officials examined the petitions of litigants, listened to their complaints, and drew on their knowledge of fengshui to help a widow, disarm a bully, or resolve a long-standing property dispute. The path to the county court was a two-way street, as the familiarity, adaptability, and relevance of fengshui provided people at the bottom with grounds to lawfully protest to their superiors while allowing people at the top to retain exclusive authority over the final decision.

Qing authorities perceived geomantic anxieties across Chinese society as pronounced in the dynasty's last century. From reigning emperors to provincial governors and all the way down to county magistrates, many in the ruling elite made note of concerns related to fengshui. The *Great Qing Code* increasingly came to reflect that unease, with mandated punishments for grave destruction and tree cutting trending stricter and harsher from the late 1700s onward. To be clear, fengshui had always been important in Qing law, but factors in the nineteenth century elevated its presence in legal discourse. These factors included growing tensions over dwindling fuel and resources, the shrinking capacity of the Qing state to control events and shape messaging in the way it desired, and its increasing tendency to relegate governing functions to local elites. Concerns over mining, economic development, and fengshui had longstanding foundations. Still, as it entered the nineteenth century, the dynasty confronted unprecedented challenges, and its legal system adjusted to meet them—within the politics of the possible.

How did the system adjust on the ground? Courts rigorously mapped mountain landscapes, protected resource commons, selectively banned mining, and sought local consent and consensus on many rural development matters. When there was a chance that a felled tree or grove might harm the fengshui of an area, county magistrates tended to order that the axes be put away. When striving scholars facing nearly impossible odds of success could not focus on their studies due to elevated noise pollution, officials redirected raucous industries outside city walls. When a coal or gold mine fomented community conflict, officials tried to keep rocks and ore in place. On the other hand, when a construction or demolition project offering considerable long-term societal benefit needed to be pursued without gentry interference or popular objection, officials issued public notices announcing that claims about fengshui were off the table for the duration of the development. At that point, authorities might summon Daoists to seal the deal with rituals that announced no harm would come from breaking the land.

None of these actions were exclusively related to ancient tradition, although fengshui masters constantly appealed to antiquity to legitimate their work.

Nor could appeals to fengshui by officials or commoners prevent environmental degradation in either Sichuan or other parts of China during the dynasty. But these actions helped Qing officials maintain the stability of the social order in what became the most populated province of the empire by 1850. Further afield, related actions also bought a dynasty in distress needed time for its bureaucrats to regroup and assess the government's next steps. Although vigorous debates over industrialization continually preoccupied late Qing statesmen and local officials, the key point is that those strategic discussions and policy proposals were built on a shared understanding of fengshui and the law.[5] On map and earth alike, the two aligned a boundary that had no border.

Law and Knowledge in Qing China

These findings invite several interpretive questions. One concerns the source base and existing historical record. Chinese local and national archives, which, for the imperial era, are most intact for the nineteenth and early twentieth centuries, indicate that fengshui was a part of the legal landscape at that time. Still, they do not explain how fengshui originally entered that landscape. To be sure, fengshui had very ancient roots, but its conspicuous presence in Chinese legal affairs seems to have been a relatively recent phenomenon. When and why did fengshui become so prevalent in Chinese legal life?

At present, it appears a combination of forces during and following the Song dynasty (960–1279) worked to elevate fengshui as a law within the law. One might speculate as to what those forces were: the movement of the population center to the mountainous south by the end of the medieval era; the development of the civil service examinations as the primary means for selecting officials in the Song alongside fengshui's endorsement by some prominent Neo-Confucians; the government registration of Yin-yang households in the Mongol Yuan (institutionalized in later dynasties as Yin-yang officers); the growth of the Chinese family lineage system and an empire-wide gentry culture in the Ming; the domination of Jiangnan publishing houses over much of the late imperial era; and, finally, the distinctive demographic, economic, and political forces at work under the Manchu Qing.[6] The nineteenth century marked the culmination of a long institutional process that enshrined fengshui's presence on the ground and in the courts.

As a matter of source bias, it is possible that county magistrates and even higher officials occasionally relied on fengshui's principles to justify difficult rulings even while their decisions were shaped by a wider range of factors that they would not be willing to openly admit. These factors included an awareness of the harsh realities of rural power relations, a desire to minimize embarrassing discrepancies in contractual or tax records, or just sheer corruption. For that reason, the Qing legal record as preserved in local archives obviously does not convey the complete picture of administrative calculations. Yet, even so, the records that do survive reveal that fengshui was strong enough to serve as laws of the land. The imperial legal system operated through the strength of that dynamic, not in spite of it.

Avenues also abound for research about family and marriage.[7] If fengshui served as laws of the land, then yin-yang cosmology and "eight-character" horoscope analysis were components of the laws of marriage. Before matrimony, a prospective groom and bride's birth times were written on a *gengtie* (also called *huntie* or *hunshu*) and then exchanged to calculate the potential for a harmonious match. This document often served as a "marriage contract," and its exchange lent agency to the bride's family agency during matchmaking negotiations.[8] Legal disputes occurred at every point in the process: people submitted *gengtie* horoscopes as legal evidence for annulling an engagement or for proving two people were already married.[9] People also "forged" horoscopes to legitimize and disguise illicit wife sales.[10] When considered alongside fengshui, these practices reveal the influence of Yin-yang cosmology and the Qing calendar in legal affairs governing land and marriage. Historians grappling with the contentious question of "civil law" in the late imperial era might look at the interplay of contract, custom, and cosmology to explain the deep resonances found across Chinese legal practice and geographically across China itself.[11] Yin-yang cosmology rooted in the *Classic of Changes* appears to be a missing part of the puzzle.

Fengshui's legal invocations also offer further possible inquiries into the global history of science. Lurking behind the cases of this book was a broader trend. From 1600 to 1800, Europe gradually took the global lead in the mathematical, mechanical, and technical fields that would later lay the groundwork for industrialization. The arrival of the Jesuits in China around that time brought news of Western advances, appealing to late Ming literati interest in what came to be known as concrete studies (*shixue*). During the Qing, the philologically based empirical school of evidentiary studies (*kaozheng*) rose among literati, building on the earlier foundation of concrete studies.[12] The extent to which Western learning directly influenced the rise of evidentiary studies remains debated. Still, one narrative holds that these intellectual trends contributed to a "decline" in imperial cosmology in favor of a greater focus on a more "skeptical and secular classical empiricism."[13] That narrative has some merit, but we can speculate about another.

Historians may consider the possibility that more accurate eclipse predictions and calendars strengthened—rather than undermined—salient features of state ritual, imperial cosmology, and Qing governance during the eighteenth and nineteenth centuries. The Qing state continued to mandate eclipse rescue rituals across China through the end of the dynasty, with the last recorded rescue ritual in Nanbu held in late 1910, less than a year before the outbreak of the Xinhai Revolution.[14] Besides the dynasty's fixation on eclipses, Yin-yang specialists employed Western learning to update astrological knowledge, the imperial calendar, and, by extension, fengshui and the geomantic compass from the seventeenth to the nineteenth centuries. These examples certainly support Benjamin Elman's contention that Chinese scholars dealt with knowledge transmitted by Westerners "on their own terms."[15] They also accord with Catherine Jami's proposal of a dynamic interaction between Western and Chinese learning over the prevailing "Western action," "Chinese reaction" model. Specifically, there was no single reaction to the Jesuit legacy in China.[16] From this insight and others, *Laws of the Land* joins the rising chorus of historians drawing attention to the questions people asked about the natural world and the methods employed to answer them.

For understanding the questions, people, particularly literati, asked about the natural world during the Qing, the law holds untapped explanatory value. As alluded to above, some aspects of imperial cosmology fell into relative disrepute by the mid-Qing period; the *fenye* system's twenty-eight astral correlations with the historical geography of ancient China constitute one example. However, because the twenty-eight constellations of antiquity were intrinsically tied to other knowledge systems (i.e., fengshui, medicine, the calendar, etc.), they never were rendered irrelevant in Chinese society. As we have seen, literati continued to meticulously update astrological and geomantic knowledge, which could be consequential during community debates over development or during trials in court. When the members of Sichuan's gentry wanted to bring a sharply worded petition against an unwanted edifice or activity in town, related information from a local gazetteer or reputable manual could come in handy for making the case. Evidently not a few litigation masters felt compelled to keep their fengshui knowledge up to date—and ready to deploy. In these ways, the Qing legal system and the privileges accorded to the gentry class accentuated the social relevance, practicality, and usefulness of applied cosmology, contributing to its continued dynamism over time.

This final observation helps explain the remarkable resilience of the Qing empire and the broader imperial system it embodied during its 268-year reign over the provinces of China. The statutes of the *Great Qing Code* radiated a commanding potency, as did the magnificent palaces of Beijing and the thoughtful layouts of county yamens. But there was enduring influence of a different kind among the leafy groves of the dead, along the flowing rivers that kept drought far and fortune close, and in the patient wait for the perfect day to celebrate a wedding or offer a final goodbye. Across the life cycle of families and the dynasty that ruled them, much power in Qing China rested with those who knew the laws of the land.

The fall of the imperial state in 1912 officially dismantled fengshui's previous status in Chinese political and legal life, but it remained relevant in broader society.[17] Even after the rise of the New Culture Movement and the anti-superstition campaigns of the the Republican era (1912–49), the inherited cosmology of fengshui remained remarkably tenacious, and divination of all kinds remained popular, although they were occasionally forced underground.

With the founding of the People's Republic of China (PRC) in 1949, fengshui was officially labeled as a "feudal superstition" (*fengjian mixin*). That attitude has since softened into the increasingly more common designation of a "popular custom" (*minsu*). Today, official attitudes toward fengshui on the Chinese mainland remain ambivalent, in no small part because of its obvious political, economic, and legal implications. The complex legacy of the Qing dynasty looms large if quietly in mainland China, where the Communist Party seeks firm control over critical sectors, especially real estate and energy. The past decade alone has seen the removal of high-ranking party officials due to their associations with "superstitions" like fengshui.[18] While it is tempting to assume PRC policies have curtailed fengshui's influence, such scandals reveal its enduring relevance. Other signs do as well. The many packed university courses on the *Classic of Changes*, rising demand for environmentally conscious development, and a resurgence of interest in traditional architecture are suggestive of a vast and continuing audience for fengshui in mainland China.

Despite a difficult twentieth century in its place of origin, fengshui remains a vibrant living tradition, well deserving of recognition as part of the intangible cultural heritage of Sinophone peoples. Hong Kong's skyline shines bright with looming "fengshui towers" (i.e., skyscrapers) that have been woven into new narratives befitting the high concentration of humans and capital within dense urban spaces. Taiwan's National Revolutionary Martyrs' Shrine dignifies the fallen by embracing the Keelung River at its front and the Chingshan Mountains at its back. Singapore's Changi Airport welcomes visitors with a consoling balance of the five agents—including a cascading waterfall encompassed by a terraced forest—earning it pride of place among the world's great travel hubs. But fengshui is not only visible in grand monuments; it is also audible in everyday conversations. Over dim sum in Sheung Wan or espresso in Taipei, people draw on fengshui to discuss the ups and downs of the stock and real estate markets, the latest political intrigues, crises, and scandals, and, more intimately, the highs, lows, joys, and sorrows that come packaged with the people called family.

Although it still inspires disputes of all sorts, fengshui sustains communities by magnifying the environment with two empowering and hopeful messages: it is never too late to make a change for the better, and a rare kind of resilience can be found by recognizing the meaning embedded in the places we live. Throughout the ages, people in China have been driven to engage, debate, and improve the study of fengshui for many reasons, not least of which is a stubborn feeling that there remain places on earth where the land must speak for itself.

APPENDICES

Appendix A: Legal Cases of the Nanbu County Archive

TABLE A.1.

Era Names	Legal Cases Concerning Graves, Burials, Trees, Etc.
Qianlong (1736–96)	2
Jiaqing (1796–1821)	2
Daoguang 1821–51)	69
Xianfeng (1851–62)	49
Tongzhi (1862–75)	55
Guangxu (1875–1909)	878
Xuantong (1909–12)	137
Total	1,192

The table above includes data for legal cases concerning graves, burials, and trees as recorded in the archive's index. Among those 1,192 case files, the nonmapped cases considered in this book include the following references. Nanbu County Qing Archive: (1787) 2.60; (1781) 2.66; (1816) 3.81; (1819) 3.86; (1840) 4.59; (1832–3) 4.79; (1835) 4.83; (1841) 4.103.17; (1833) 4.108; (1835) 4.109; (1835) 4.110.07; (1838) 4.111; (1839) 4.112; (1840) 4.113.02; (1840) 4.115; (1840) 4.116; (1840) 4.117; (1840) 4.118; (1840) 4.119; (1841) 4.120; (1825) 4.122; (1825) 4.123; (1826) 4.124; (1829) 4.126; (1829) 4.127; (1829) 4.128); (1829) 4.129; (1829) 4.130; (1829) 4.131; (1830) 4.132; (1831) 4.135; (1831) 4.137; (1831) 4.138; (1831) 4.139; (1832) 4.143; (1835) 4.144; (1835) 4.146; (1837) 4.147; (1837) 4.150; (1840) 4.155; (1840) 4.159; (1840) 4.160; (1841) 4.162; (1844) 4.163; (1847) 4.164; (1829) 4.265; (1837) 4.284; (1857) 5.44; (1858) 5.47; (1858) 5.48; (1857) 5.62; (1852) 5.68; (1852) 5.69; (1852–3) 5.70 (1853) 5.71; (1853) 5.73; (1855) 5.80; (1855) 5.82; (1856) 5.83; (1857) 5.86; (1857) 5.87; (1857) 5.88; (1857) 5.90; (1857) 5.92; (1857) 5.94; (1857) 5.95; (1857) 5.96; (1859) 5.100; (1861) 5.107; (1861) 5.108; (1861) 5.109; (1861) 5.110; (1863) 5.143; (1854) 5.145; (1867) 6.70; (1870) 6.80; (1870) 6.81; (1873) 6.107; (1865) 6.130; (1867) 6.131; (1870) 6.132; (1872) 6.136; (1872) 6.138; (1872); 6.139; (1872) 6.141; (1873) 6.144; (1873) 6.147; (1874) 6.153; (1874) 6.154; (1874) 6.155; (1874) 6.156; (1869) 6.179; (1870) 6.181; (1862) 6.414; (1862) 6.419; (1862) 6.421; (1865) 6.425; (1866–7) 6.431; (1864) 6.425; (1871) 6.449; (1868) 6.488; (1872) 6.501; (1876) 7.171; (1876) 7.181; (1876) 7.182; (1877) 7.303; (1877) 7.304; (1877) 7.305; (1877) 7.306; (1877) 7.312; (1877) 7.314; (1877) 7.316; (1877) 7.320; (1877) 7.322; (1877) 7.326; (1877) 7.327; (1877–8) 7.336; (1877) 7.338; (1877) 7.445; (1878–8) 7.517; (1878) 7.526; (1878) 7.528; (1878) 7.625; (1878) 7.640; (1878) 7.644; (1879) 7.720; (1879) 7.721; (1879) 7.732; (1880) 7.740; (1879) 7.741; (1879) 7.742; (1878) 7.533; (1879) 7.745; (1878–9) 7.746; (1879) 7.844; (1880) 8.49; (1880) 8.51; (1880) 8.56; (1880) 8.66; (1880) 8.69; (1880) 8.70; (1880) 8.71; (1880) 8.73; (1880–1) 8.74; (1880) 8.194; (1880) 8.215; (1880) 8.216; (1881)

Continued on next page

TABLE A.1. (*continued*)

8.217; (1882) 8.440; (1882) 8.561; (1882) 8.562; (1882) 8.564; (1882) 8.566; (1882); 8.655; (1882) 8.684; (1883) 8.741; (1883–4) 8.743; (1883) 8.744; (1883) 8.746; (1882) 8.747; (1883) 8.748; (1883) 8.749; (1884) 8.906; (1885) 8.907; (1884) 8.932; (1885) 9.50; (1885) 9.51; (1885) 9.52; (1885) 9.53; (1885) 9.54; (1885) 9.55; (1885) 9.57; (1885) 9.58; (1885) 9.60; (1885) 9.64; (1885) 9.65; (1885) 9.66; (1885) 9.67; (1885) 9.68; (1885) 9.78; (1886) 9.86; (1885–6) 9.194; (1885) 9.205; (1885) 9.267 (1886) 9.340; (d.u.) 9.345; (1886) 9.364; (1886) 9.365; (1886) 9.366; (1887) 9.651; (1887) 9.654; (1887) 9.655; (1887) 9.658; (1887) 9.659; (1887) 9.661; (1887) 9.662; (1887) 9.666; (1887–8) 9.673; (1886) 9.676; (1886) 9.681; (1887) 9.684; (1887) 9.691; (1887) 9.726; (1889) 10.304; (1890) 10.624; (1890) 10.631; (1890) 10.646; (1890–1)10.650; (1890) 10.656; (1890) 10.657; (1890) 10.658; (1890) 10.661; (1890) 10.663; (1890) 10.667; (1890) 10.668; (1890–1) 10.669; (1890) 10.670; (1892) 10.713; (1890) 10.959; (1891) 11.209; (1892) 11.210; (1892) 11.480; (1893–4) 11.576; (1893) 11.578; (1893) 11.579; (1893–4) 11.865; (1893) 11.872; (1894) 12.40; (1894) 12.41; (1894) 12.44; (1895) 12.49; (1894) 12.60; (1894) 12.62; (1895) 12.416; (1895) 12.429; (1895) 12.430; (1895) 12.431; (1895) 12.433; (1895) 12.436; (1895) 12.439; (1895) 12.440; (1895) 12.444; (1895) 12.445; (1895) 12.446; (1895–6) 12.447; (1896) 12.451; (1896) 12.452; (1895) 12.457; (1895) 12.458; (1895) 12.459; (1895) 12.460; (1895) 12.461; (1895) 12.462; (1895) 12.463; (1895) 12.464; (1895) 12.465; (1895) 12.466; (1895) 12.475; (1895–6) 12.476; (1896) 12.478; (1896) 12.479; (1894) 12.558; (1898) 14.59; (1898) 14.62; (1898) 14.68; (1898) 14.69; (1898) 14.71; (1898) 14.72; (1898) 14.73; (1899) 14.79; (1888) 14.423; (1899) 14.637; (1900) 15.65; (1900) 15.67; (1900) 15.68; (1900) 15.73; (1902) 15.958; (1903) 16.74; (1904) 17.35; (1905) 17.102; (1905) 17.103; (1904–5) 17.104; (1907) 18.463; (1902) 19.266; (1909) 20.213; (1909) 20.267; (1909–10) 20.306; (1910) 21.156; (1910) 21.220; (1910) 21.243; (1910) 21.244; (1910–11) 21.480; (1910) 21.732; (1911) 22.177; (1910) 22.181; (1911) 22.186; (1911) 22.188; (1911) 22.191; (1911) 22.202; (1911) 22.490.

Appendix B: Maps of the Nanbu County Archive

TABLE B.1. Origin and Types of Maps in the Nanbu Archive 181 Maps (Dated 1829–1911)

Origin and Types of Maps in the Nanbu Archive 181 Maps (Dated 1829–1911)					
	Graves and Houses	Trees	Farmland	Corpse Forensic Diagrams	Misc. (City Walls, Roads, Etc.)
County Yamen (Unattributed Department)	8	3	3	0	4
Dept. of Works	66	33	30	0	1
Dept. of Punishments	0	2	3	10	0
Dept. of Rites	2	0	2	0	1
Dept. of Revenue	0	0	4	0	1
Litigant (Privately Created)	0	0	8	0	0
Total	76 (42%)	38 (21%)	50 (27%)	10 (6%)	7 (4%)

For the case files containing maps, see Nanbu County Qing Archive: (1829) 4.74.3; (1840) 4.98.8; (1852) 5.55.5; (1852) 5.67.9; (1853) 5.76.2; (1872) 6.140.6; (1873) 6.143.3; (1873) 6.145.4; (1873) 6.146.7; (1873) 6.148.4; (1874) 6.151.4; (1874) 6.152.5; (1874) 6.157.5; (1874) 6.159.3; (1871) 6.499.9; (1875) 7.42.5; (1877) 7.308.5; (1877) 7.308.11; (1877) 7.317.4; (1878) 7.321.4; (1877) 7.325.5; (1877) 7.343.13; (1877) 7.344.5; (1877) 7.354.5; (1878) 7.499.4; (1878) 7.508.4; (1878) 7.515.7; (1878) 7.523.4; (1878) 7.527.9; (1878) 7.530.5; (1864) 7.566.1; (1881) 8.41.4; (1880) 8.59.4; (1880) 8.61.6; (1880) 8.72.4; (1881) 8.76.7; (1881) 8.322.4; (1881) 8.323.4; (1881) 8.349.6; (1881) 8.443.5; (1882) 8.543.5; (1882) 8.558.6; (1882) 8.563.5; (1882) 8.585.6; (1882) 8.586.5; (1882) 8.609.1; (1884) 8.900.10; (1884) 8.911.2; (1885) 9.37.6; (1885) 9.77.14; (1885) 9.77.15; (1885) 9.80.2; (1886) 9.87.4; (1886) 9.354.7; (1886) 9.354.9; (1886) 9.355.5; (1886) 9.371.3; (1887) 9.643.7; (1887) 9.645.4; (1887) 9.652.7; (1887) 9.663.5; (1887) 9.664.6; (1886) 9.675.5; (1887) 9.679.4; (1887) 9.680.6; (1887) 9.682.4; (1887) 9.683.6; (1887) 9.685.5; (1887) 9.687.6; (1887) 9.689.5; (1887) 9.690.2; (1887) 9.706.5; (1887) 9.720.3; (1887) 9.725.3; (1887) 9.819.5; (1886) 9.887.15; (1888) 10.16.3; (1888) 10.17.5; (1888) 10.23.3; (1888) 10.25.4; (1888) 10.29.5; (1888) 10.35.10; (1889) 10.276.8; (1889) 10.280.4; (1889) 10.283.4; (1889) 10.291.2; (1889) 10.292.5; (1889) 10.294.7; (1889) 10.302.5; (1889) 10.303.5; (1889) 10.305.5; (1889) 10.314.4; (1890) 10.437.4; (1890) 10.627.4; (1890) 10.638.4; (1890) 10.640.7; (1890) 10.649.6; (1890) 10.654.5; (1890) 10.659.4; (1890) 10.660.4; (1890) 10.662.4; (1890) 10.671.5; (1890) 10.689.5; (1892) 11.251.4; (1892) 11.262.6; (1892) 11.262.9; (1892) 11.263.7; (1892) 11.281.3; (1893) 11.478.8; (1893) 11.623.2; (1894) 12.34.11; (1894) 12.45.3; (1895) 12.50.2; (1894) 12.54.6; (1894) 12.55.2; (1894) 12.58.4; (1894) 12.226.5; (1896) 13.116.2; (1896) 13.119.1; (1897) 13.792.3; (1899) 14.631.4; (1899) 14.861.7; (1900) 15.59.6; (1900) 15.74.5; (1900) 15.82.5; (1901) 15.450.6; (1902) 15.722; (1902) 15.805.7; (1903) 16.54.9; (1903) 16.73.13; (1903) 16.251.1; (1904) 16.640.6; (1904) 16.658.3; (1904) 16.752.2; (1906) 17.579.4; (1907) 18.46.4; (1907) 18.124.3; (1907) 18.355.12; (1907) 18.565.12; (1908) 18.792.6; (1908) 18.802.8; (1908) 18.898.10; (1908) 18.903.1; (1908) 18.1011.6; (1908) 18.1126.1; (1908) 18.1127.3; (1908) 18.1295.5; (1886) 19.27.1; (1909) 20.212.8; (1910) 20.272.6; (1909) 20.418.1; (1909) 20.935.7; (1911) 21.210.8; (1911) 21.230.3; (1911) 21.245.7; (1910) 21.247.2; (1910) 21.654.5; (1910) 22.142.4; (1911) 22.146.8; (1911) 22.146.9; (1911) 22.146.10; (1911) 22.160.6; (1911) 22.173.5; (1911) 22.174.3; (1911) 22.180.8; (1911) 22.180.9; (1911) 22.183.6; (1911) 22.193.5; (1911) 22.195.5; (1911) 22.199.6; (1911) 22.201.4; (1911) 22.203.5; (1911) 22.204.7; (1911) 22.344.5; (1911) 22.350.4; (1911) 22.366.11; (1911) 22.409.2; (1911) 22.524.6; (1911) 22.764.4; (1910) 23.132.1; (1910) 23.222.6.

Appendix C. Mappers of the Nanbu County Yamen

TABLE C.1. Mappers of the Nanbu Yamen Clerks from the Department of Works

This chart profiles prolific mappers who worked at the yamen. The men listed—all of whom were identified as clerks employed at the Department of Works (*gongshu*)—created most of the litigation maps surviving in the archive. The twenty-two names listed together composed ninety-six maps preserved in the article, slightly more than half of the total. Some maps preserved in the archive lack a clear attribution, while other clerks only left a single map. This chart only profiles clerks whose names appear more than once in the collection.

Mappers of the Nanbu Yamen Clerks from the Department of Works

Name of Mapper from the Department of Works	Years Active	Maps Created
Jin Lianfang 進聯芳	c. 1873–74	Farmland (1), Trees (1)
Wang Heping 汪和平	c. 1874–80	Graves (3), Trees (2)
Guo Yongsheng 郭永昇	c. 1875–78	Graves (2)
Mei Kaixian 梅開先	c. 1877–78	Graves (2), Trees (1)
Liu Qingxuan 劉青選	c. 1877–80	Graves (1), Farmland (1)
Zhang Dengyun 張登雲	c. 1877–80	Farmland (2), Graves (1)
Chen Hongxue 陳洪學	c. 1877–94	Graves (6), Farmland (2), Trees (1)
Cheng Ruiyun 程瑞雲	c. 1878–82	Trees (2), Graves (1), Farmland (1)
Wang Changjiang 汪長江	c. 1880–99	Graves (5), Trees (3)
Cao Chunyun 曹春雲	c. 1884–87	Graves (1), Trees (1)
He Qingyan 何清晏	c. 1886–88	Graves (1), Trees (1)
Wang Qingxi 汪清溪	c. 1886–1912	Graves (8), Farmland (5), Trees (3)
Zhang Zhongyou 張仲友	c. 1887–92	Trees (3), Farmland (2), Houses (1)
Wang Fuqing 汪輔清	c. 1887	Farmland (1), Trees (1), Graves (1)

TABLE C.1. (*continued*)

Name of Mapper from the Department of Works	Years Active	Maps Created
Chen Qingtai 陳清泰	c. 1887–1903	Farmland (3), Trees (1), Roads (1), Graves (1)
Xian Fengyu 鮮俸玉	c. 1890	Graves (2)
Yang Xinde 楊新德	c. 1804–1912	Graves (1), Trees (1)
Du Changwen 杜長文	c. 1889–94	Trees (2), Graves (1)
Deng Wenjie 鄧文傑	c. 1912	Trees (2)
Yang Maode 楊懋德	c. 1900–12	Farmland (4), Graves (2)
Wang Dianchen 王殿臣	c. 1912	Farmland (1), Graves (1)
Yang Xinchun 楊新春	c. 1912	Trees (3), Farmland (1)

LIST OF CHINESE TERMS

ai 礙
anshan 案山
aoce 廠冊
aofeng 鼇峰
baihu 白虎
baimai 敗脈
baojia 保甲
bazi 八字
bingyin 丙寅
boshi 博士
bu 步
budi 卜地
changnian jishui 常年積水
chi 尺
chishi 赤石
chongsheng ci 崇聖祠
chuan 串
Chuanbei dao 川北道
Chuanzhu 川主
chukui 初虧
cong jiupin 從九品
daowa 盜挖
daozang 盜葬
dianshi 殿試
dianxian 電線
difang gongshi 地方公事
dili 地理
dimai 地脈
ding 丁
dishi 地師
dixue 地學
dizhan 地占

dongtu 動土
dong xiang zhi wu 動響之物
enke 恩科
famu 伐木
fen 墳
fengbing 風病
fengjian mixin 封建迷信
fengshui 風水
fengshui ta 風水塔
fengshui xin you ran ye 風水信有然也
fenye 分野
fu 福
ge-dang, ge-dang 咯當咯當
gen 艮
geng 庚
gengtie 庚貼
gengzi 庚子
gong 弓
gongshu 工書
gui 鬼
hun 魂
hunshu 婚書
jia 甲
jiangjun jian 將軍箭
jiangren 匠人
jiansheng 監生
jianzhu 建築
jiemai jingzhong 截脈驚冢
jiemai qiao 接脈橋
jiao 醮
jin chunyao 禁春窯

jin qi huanwang huo min 禁其幻妄惑民
jing 井
jingcheng 京城
jinshi 進士
jintie 津貼
jinyu mantang 金玉滿堂
jitian 祭田
jiugong tu 九宮圖
juanshu 捐輸
juren 舉人
kaifang jili 開方計里
kaiqu chuanjing 開渠穿井
kanyu 堪輿
kaozheng 考證
kuang 礦
kun 坤
kuixing ge 魁星閣
kui 奎
lailong 來龍
liangmin 良民
li 理
li 里
lijin 釐金
liqipai 理氣派
liquan 立券
longmai 龍脈
longxue yipo 龍穴已破
luopan 羅盤
lu 祿
luye 爐冶
mao 卯
mingmai 命脈
mingsheng 名勝
mingtang 明堂
minsu 民俗
mixin 迷信
mu 畝
mu ru kangong 木入坎宮
nacai 納彩
nian 碾

nianbiao 年表
panchi 泮池
panlong shi 蟠龍勢
piao 票
pingzhi daotu 平治道涂
po 魄
potu 破土
qi 氣
qian 乾
qianbulang 千步廊
qianfa 錢法
qilong jiemai 騎龍截脈
qing 情, *li* 理, *fa* 法
qinglong 青龍
qingming 清明
renxu 壬戌
ribiao 日表
rupan 入泮
sanhe 三合
sanyuan 三元
sha 煞
shajin 沙金
shanghui 傷毀
shaojiu 燒酒
shen 申
sheng 省
shengkuang 生壙
shengmai 省脈
shengyuan 生員
shenshi 紳士
shi'e 十惡
shigu 屍骨
shixian li 時憲曆
shixian shu 時憲書
shixue 實學
shou 壽
shuifa 水法
shuiju 水局
shuikou 水口
shumin 庶民
shumu 樹木

shuotie 說帖
songshi 訟師
suicha 歲差
Tian'anmen 天安門
tianjing 天井
tiansheng qiao 天生橋
tiaoli 條例
tongsheng 童生
tu 圖
tushi 土師
wang 王
wenhan 文翰
wenming 問名
wu 午
wuguan fengshui 無關風水
wusong qiao 無訟橋
wuxing 五行
xiangshen 鄉紳
xiangzhai 相宅
xiangzhi 相址
xin 辛
xing 姓
xingshipai 形勢派
xinzheng 新政
xiucai 秀才
xiudu 宿度
xiuzao 修造
xiyang lifa 西洋曆法
xue 穴
xun 巽
yamen 衙門
yanglian yin 養廉銀
yangzhai 陽宅

yanwei 鹽委
yimai 一脈
yindi 陰地
yinsi 禋祀
yinyang hu 陰陽戶
yinyang jia yan 陰陽家言
yinyangsheng 陰陽生
yinyangxue guan 陰陽學官
yinzhai 陰宅
yishi 醫士
you 酉
youfang 油坊
yuebiao 月表
zedi 擇地
zhaibing 宅病
zhaizhang 齋長
zhan 棧
zhang 丈
zhangcheng 章程
zhanzhai 占宅
zhaobi 照壁
zhayou jie 榨油街
zheng 正
zhengji 政績
zhenwu shan 真武山
zhongshu zhiyu zhi fa 種樹致雨之法
zhongsong 冢訟
zhongtang 中堂
zhuangyuan 狀元
zhudi 築堤
zhuque 朱雀
zi 子
zui yao zhi duan 最要之端

NOTES

Introduction

1. For a resonant history of a "Ming landscape," see Dardess, "A Ming Landscape."

2. Quoted from Smith, *Fortune-tellers and Philosophers*, 132. Fengshui was first listed in the *Encyclopædia Britannica* in 1797, when it was described as "the name of a ridiculous superstition among the Chinese." See "fong-choui," *Encyclopædia Britannica*, 3rd edition, vol. 7 (Edinburgh, 1797), 323.

3. Said, *Orientalism*. For more on the development of early Orientalism, see Statman, *A Global Enlightenment*.

4. Weber, *The Religion of China*, 199–200. For more on Weber's discussion of fengshui, see Barbalet, *Confucianism and the Chinese Self*, and Brook, "Weber's *Religion of China*," 99.

5. Ernest Eitel (1838–1908) observed, "Feng-shui has a legal status in China . . . [law courts adjudicate cases] on the presumption that Feng-shui is a reality and a truth, not a fiction." See Eitel, *Feng-shui*, 80.

6. Freedman, *Lineage Organization in Southeastern China*; Ahern, *The Cult of the Dead*; Feuchtwang, *An Anthropological Analysis of Chinese Geomancy*; Bruun, *Fengshui in China*; Weller, *Unities and Diversities in Chinese Religion*; Kiong, *Chinese Death Rituals in Singapore*.

7. Segawa Masahisa, *Zokufu*; Chen Jinguo, *Xinyang, yishi yu xiangtu shehui*. See also Zhang Peiguo, *Linquan, fenshan yu miaochan*.

8. For exceptions to this trend from the historical discipline, see Hong Jianrong, "Dang [*fengshui*] chengwei [*huoshui*], 1–46; Smith, *Fortune-tellers and Philosophers*; Smith, *Fathoming the Cosmos and Ordering the World*; Smith, "The Transnational Travels of Geomancy in Premodern East Asia," Parts I and II; Chen Jinguo also has an appendix of geomantic litigation from Fujianese genealogies. See Chen Jinguo, *Xinyang, yishi yu xiangtu shehui*, vol 2.

9. Chen, *Chinese Law in Imperial Eyes*; Ruskola, *Legal Orientalism*.

10. Chen, *Chinese Law in Imperial Eyes*, 3–6.

11. Allee, *Law and Local Society in Late Imperial China*; Huang, *Civil Justice in China*, 1–2; Linxia Liang, *Delivering Justice in Qing China*; Sommer, *Polyandry and Wife-Selling in Qing Dynasty China*.

12. Note that the number of substatutes grew over time: the 1740 edition of the law code had 1,049, while the 1870 edition had 1,892. For a classic work on Qing legal practice, see Shiga Shūzō, *Shindai Chūgoku no hō to saiban*. For Shiga's conceptions of Qing legal practice as informed by *qing* (sentiment); *li* (principle); and *fa* (law), see Shiga Shūzō, "Qingdai susong zhidu zhi minshi fayuan de gaikuoxing kaocha: qing, li, fa," 19–53. For the role of model cases in shaping local adjudication, see Constant, "Thinking with Models," 417–73.

13. Huang, *Civil Justice in China*, 226.

14. Katherine Carlitz, "Genre and Justice in Late Qing China," 255. For a related impression of the "distinct, even mutually exclusive, realms" of religious communities and the state, see Brook, *Praying for Power*, 31; see also Ibid., 29, 93.

15. Bodde and Morris, *Law in Imperial China*, 10.

16. Joseph Needham cited the lack of a divine lawgiver to explain trends in the development of Chinese law and science. See Needham, *Science and Civilisation in China*, vol. 2, 582. For a summary of relevant historiography, see Peerenboom, "Law and Religion in Early China," 99.

17. Marsh, "Weber's Misunderstanding of Traditional Chinese Law," 281–302.

18. Ibid., 288.

19. Brown, "The Deeds of the Dead in the Courts of the Living," 109–55. For more on the resonances between law and religion, see also Brook, Bourgon, and Blue, *Death by a Thousand Cuts*, 143–48.

20. Katz, *Divine Justice*. See also Katz, "Divine Justice in Late Imperial China," 869–902.

21. Jiang, *The Mandate of Heaven and the Great Ming Code*, 4.

22. Denis Twitchett has suggested that China was not a "purely secular state." See Twitchett, "Law and Religion in East Asia," 469–72. For a study on the importance of ritual in the construction of Qing law, see Keliher, *The Board of Rites and the Making of Qing China*.

23. Bodde and Morris identified cosmological resonances in the law. See Bodde and Morris, *Law in Imperial China*, 561–62. See also Bodde, "'Chinese Laws of Nature,'" 139–55.

24. Witte Jr., *The Reformation of Rights*.

25. Zgonjanin, "Quoting the Bible," 41.

26. Latour, *We Have Never Been Modern*; Josephson-Storm, *The Myth of Disenchantment*.

27. Fischel, *The Economics of Zoning Laws*.

28. Schlosberg and Carruthers, "Indigenous Struggles, Environmental Justice, and Community Capabilities," 12–35.

29. Hayes, "Specialists and Written Materials in the Village World," 75–111.

30. Qintianjian louke ke, *Qintianjian Dili xingshi qieyao bianlun*, 68–69. Note that this title applied to the original 1740 edition (preface dated QL5.11); upon its reissuance in 1746, the text's title was changed to *Qintianjian Fengshui zhenglun*.

31. For the pervasiveness of these concepts, see Smith, *Fortune-Tellers and Philosophers*, 49–172; 260–70; Smith, "The Legacy of Daybooks in Late Imperial and Modern China," 336–72. For a few examples of their application in Chinese medicine, see Porkert, *The Theoretical Foundations of Chinese Medicine*; Liu, *The Essential Book of Traditional Chinese Medicine*; Lee, *The Philosophical Foundations of Chinese Medicine*.

32. Smith, *Fathoming the Cosmos and Ordering the World*; Zhu Xi, *The Original Meaning of the Yijing*.

33. Wang, *Yinyang*.

34. Sivin, *Traditional Chinese Medicine in Contemporary China*, 47.

35. For a comprehensive discussion of these and other cosmic operators, see Needham, *Science and Civilisation in China*, vol. 2, 216–395.

36. Bray, *Technology and Gender*, 76–77.

37. Duara, *Culture, Power, and the State*, 210–11.

38. The heavenly stems and earthly branches were also used for timekeeping, astronomy, and other sciences from antiquity through later periods. For details and illustrations of the compass, see Meyer, "'Fengshui' of the Chinese City," 148–55; Feuchtwang, *An Anthropological Analysis of Chinese Geomancy*, 37–49; Smith, *Fortune-tellers and Philosophers*, 131–71.

39. Qintianjian louke ke, *Qintianjian Dili xingshi qieyao bianlun*, 7.

40. The term "fengshui master" (*fengshui xiansheng*) is common today, but did not start appearing in gazetteers until the early twentieth century. I have not seen it in Qing legal cases.

41. For a reference to female diviners, see Crook and Gilmartin, *Prosperity's Predicament*, 192–95. For a discussion of the gendered dimensions of divination as an occupation, see Wang, *Physiognomy in Ming China*, 70–72. See also Kory, "Presence in Variety," 3–48.

42. Earth veins can be traced to the early empire, with the concept appearing in Sima Qian's (145–86 BCE) record of allegations that Qin (221–206 BCE) wall-building injured the "earth veins" of the land. See Paton, *Five Classics of Fengshui*, 81.

43. Qintianjian louke ke, *Qintianjian Dili xingshi qieyao bianlun*, 13–14.

44. Lin, "Boiling Oil to Purify Houses (*zhuyou jingwu*)," 151–69.

45. Whiteman, *Where Dragon Veins Meet*. Geomantic veins were identified across China in the late imperial era. In the Max Planck Insitute of Science's LoGaRT Database, there are some 5,921 Qing-era references to just one of geomancy's many appellations (*kanyu*); 1,833 references to earth veins (*dimai*); and 1,504 references to Arriving Dragons (*lailong*). Some of these references are redundant and derive from a single gazetteer or repeatedly cited texts among several gazetteers, while others refer simply to place names. Nonetheless, most of the references—which appear across north, south, east, and west China—concern fengshui practices or geomantic veins that people identified in their communities.

46. Schurmann, "Traditional Property Concepts in China," 507–16. See also Perdue, *Exhausting the Earth*; Faure and Siu, *Down to the Earth*; Mazumdar, "Rights in People, Rights in Land," 89–107; Isett, *State, Peasant, and Merchant in Qing Manchuria*; Zhang, "Cultural Paradigms in Property Institutions," 347–413.

47. White Jr., "The Historical Roots of Our Ecologic Crisis," 1203–7.

48. Bruun, *Fengshui in China*, 232–39.

49. For Qing definitions of "nature," see Schlesinger, *A World Trimmed with Fur*, 14–15.

50. Cruikshank, "'Are Glaciers Good to Think With?'" 239–50.

51. Coggins, "When the Land Is Excellent," 97–126; Tang, Yang, Ohsawa, Momohara, Mu, and Robertson, "Survival of a Tertiary Relict Species," 2112–19.

52. Elvin, *The Retreat of the Elephants*, 470–71.

53. Weller, *Discovering Nature*, 29–30.

54. For a discussion of fengshui's political significance in the Ming-Qing transition period, see Bello, *Across Forest, Steppe, and Mountain*, 28–40.

55. *Da Qing lüli* (1740 Rpt.; 1998) *juan* 17, *lilü*, *yizhi*, *sangzang*: 296 (Statute 181). In 1746 another substatute condemning people who uncovered the dead in pursuit of geomantic benefit was added to Statute 276 ("Uncovering Graves"). See Xue Yunsheng, *Duli cunyi* (Beijing: Hanmao zhai, 1905), *juan* 31, *xinglü*: 6.

56. Macauley, *Social Power and Legal Culture*, 14.

57. Clunas, *Fruitful Sites*, 183.

58. Zhang, *Circulating the Code*.

59. Sommer, *Polyandry and Wife-Selling in Qing Dynasty China*.

60. Von Glahn, *An Economic History of China*, 504; Zelin, *The Magistrate's Tael*.

61. Li, *The Making of the Modern Chinese State, 1600–1950*, 63; 136. Revenue was raised not by registering new land in Sichuan, but by adding "extra charges" (*jintie*) in the wake of the White Lotus Rebellion (1796–1804), and then "voluntary surcharge taxes" (*juanshu*) during the Taiping Civil War (1850–64). For more, see Gou Deyi, *Qingdai jiceng zuzhi yu xiangcun shehui guanli: Yi Sichuan Nanbu xian wei ge'an de kaocha*, 386–87. For Qing population figures, see Vermeer, "Population and Ecology along the Frontier in Qing China," 271; Deng, "China's Population Expansion and Its Causes during the Qing Period."

62. Yang Guanqiong, *Dangdai Zhongguo xingzheng guanli moshi yange yanjiu*, 166. Note that this statistic does not include subprefectures and departments.

63. Luo Dajing, *Helin yulu* (Preface 1248; Beijing: Zhonghua shuju, 1983), 344.

64. For more on cremation and its critics in Song times, see Ebrey, "Cremation in Sung China," 406–28.

65. Juefei shanren and Chiu Pengsheng, *Juefei shanren Erbi kenqing dian jiao ben*, in *Mingdai yanjiu* 13 (2009): 240–42. This text is one of the earliest extant litigators' manuals, with a composition date in the mid-to-late Ming dynasty (c. after 1582). For more, see Miller, *Fir and Empire*, 91–92.

66. Brokaw, *Commerce in Culture*, 456–66. In the 1700s, Sibao also saw the publication of a novel based on a fengshui lawsuit; ibid., 318.

67. Will, *Handbooks and Anthologies for Officials in Imperial China*, 1092. One is tempted to compare courtroom investigations of fengshui to the forensic knowledge of the *Xiyuanlu*. This Song-era manual formed the basis of homicide investigations through the late imperial era, when it was periodically updated with new editions bearing modified versions of the title. The analogy is limited by the fact that the government did not control the development of geomantic theories, whereas forensic science evolved within its legal-administrative tradition. See Will, "Developing Forensic Knowledge through Cases in the Qing Dynasty," 62–100.

68. Haraway, "Situated Knowledges," 575–99.

69. Katz and Goossaert, *The Fifty Years that Changed Chinese Religion*, 107–9.

70. A helpful analogy is state rain-making during droughts. See Snyder-Reinke, *Dry Spells*.

71. Paton, *Five Classics of Fengshui*, 158, 240, 262, 264, 265, 273, 291, 294, 302, 316, 318, 324–28, 362, and 364.

72. Zeng Guofan, *Zeng Guofan quanji*, 564.

73. Smith, *Fortune-tellers and Philosophers*, 160.

74. Reed, *Talons and Teeth*.

75. The change in title occurred when the the character *li* became taboo with the ascension of Hongli, the Qianlong Emperor. See Chang, *The Chinese Astronomical Bureau, 1620–1850*, 19–20.

76. Yingcong Dai, *The Sichuan Frontier and Tibet*, 8–12.

77. Liang Yong, *Yimin, guojia yu difang quanshi: Yi Qingdai Ba xian weili*.

78. For tiger attacks in Sichuan, see Lan Yong, "Qingchu Sichuan huhuan yu huanjing fuyuan wenti," 203–10. A similar pattern of tiger attacks applied in Lingnan. See Marks, *Tigers, Rice, Silk, and Silt*, 323–27.

79. Wang Di, *Street Culture in Chengdu*, 236–39. See also Ownby, *Brotherhoods and Secret Societies in Early and Mid-Qing China*; Ownby, "Recent Chinese Scholarship on the History of Chinese Secret Societies," 139–58.

80. Zelin, "The Rights of Tenants in Mid-Qing Sichuan," 499–526. Nanbu's annual land tax quota was less than half of Ba's. See Zhou Xun, *Shuhai congtan*, 192 (Ba), 218 (Nanbu).

81. Gou Deyi, *Qingdai jiceng zuzhi yu xiangcun shehui guanli*, 18.

82. Ying Liangeng, *Sichuansheng zudian zhidu*, 23–24.

83. Pomeranz, "Land Markets in Late Imperial and Republican China," 101–50.

84. Xihua shifan daxue quyu wenhua yanjiu zhongxin and Nanbu xian difangzhi bangongshi, eds., *Tongzhi zengxiu Nanbu xian zhi* (1870 Rpt.; Chengdu: Ba-shu shushe, 2014), 389–90.

85. Yingcong Dai, *The White Lotus War*.

86. Wu Peilin, *Qingdai xianyu minshi jiufen yu falü zhixu kaocha*; Cai Dongzhou et al., *Qingdai Nanbu xianya dang'an yanjiu*.

87. Wu Peilin and Wan Haiqiao, "Qingdai zhouxianguan renqi 'sannian yiren' shuo zhiyi: Jiyu Sichuan Nanbu xian zhixian de shizheng fenxi," 63–72.

88. Xihua shifan daxue and Nanchongshi dang'anju, eds. *Qingdai Nanbu xianya dang'an mulu*; Sichuansheng Nanchongshi dang'anguan, ed., *Qingdai Sichuan Nanbu xian yamen dang'an*.

89. This number is a conservative estimate. I have certainly missed some disputes about graves, cemetery trees, and fengshui that are buried deep in the litigation files of other matters, such as irrigation. This figure also does not include petitions concerning Yin-yang officers or legal cases over the funding of projects likely intending to improve fengshui, such as pagoda construction.

90. For a list of the legal issues designated to the county departments, see SUCSBRC: (Nanbu Collection, GX19.10.7/1893) 466.12.3. This observation about land disputes simplifies since Punishments handled acts of violence, many of which arose from land conflicts, and Revenue handled land tax disputes. Note that the number of statutes under "Laws Relating to the Board of Works" in the *Great Qing Code* is the lowest of all Six Boards, with only thirteen specified articles, compared with over 160 relating to the Board of Punishments. Trends in local litigation did not reflect the proportions of statutes in the Qing law code. With frequent rural disputes over land boundaries, bridges, infrastructure, and irrigation, the clerks and runners at Nanbu's Department of Works were busy. In total, 1,599 surviving litigation files from the Qing were passed to the county's Department of Punishments, while 2,460 files were processed by the Department of Works. For a complete statistical breakdown, see Yu Zhengsong and Zheng Jiewen, "Qingdai Nanbu xian dang'an ji qi jiazhi," *Wenxian* 1 (2008): 87.

91. The total included 261 cases about graves out of a thousand land disputes. See Wei Shunguang, "Qingdai zhongqi de 'jiefen zisong' xianxiang yanjiu: jiyu Ba xian dang'an wei zhongxin de kaocha," *Qiusuo* 4 (2014): 160.

92. Wu Peilin, "Jin sanshi nian lai guonei dui Qingdai zhouxian susong dang'an de zhengli yu yanjiu," *Beida falü pinglun* (2011): 258–72. The Ba County yamen was also larger than Nanbu's. In Ba, clerks worked out of ten departments, compared to Nanbu's seven. See Reed, *Talons and Teeth*, 33–34.

93. Zhao Jishi, *Jiyuan ji suoji* (1695 Rpt; Hefei: Huangshan shushe, 2008), 872.

94. *Qianlong Changsha fu zhi* (1747) 23: 31.

95. Li Xiaofang and Wen Xiaoxing, "Ming Qing shiqi Gannan kejia diqu de fengshui difang yu zhengfu kongzhi," *Shehui kexue* 1 (2007): 108–14.

96. Guangyuan Zhou, "Beneath the Law: Chinese Local Legal Culture During the Qing Dynasty," 165.

97. Watson, "Funeral Specialists in Cantonese Society," 109–34.

98. For more on grave mobility, see Snyder-Reinke, "Afterlives of the Dead," 1–20.

Chapter 1: Litigating Graves

1. *Da Qing lüli* (1740 Rpt.; 1998) juan 17, *lilü, yizhi, sangzang*: 296 (Statute 181). While the Qing never eliminated the practice of cremation, imperial law remained consistent in banning it. The final edition of the *Imperially Endorsed Collected Statutes and Precedents of the Great Qing* reiterated its illegality. See, for instance, *Qinding Da Qing huidian shili* (1899) juan 400, *libu*: 19. Decree Issued in TZ7 (1868). De Groot (1854–1921) saw cremation as not visibly present "in the many provinces through which we have traveled." See de Groot, *The Religious System of China*, vol. 3, 1415. Feng Xianliang argues that the destruction wrought by the Taiping Civil War allowed authorities to reemphasize burial requirements on newly created waste land in the late 1800s. See Feng Xianliang, "Fenying yizhong: Ming-Qing Jiangnan de minzhong shenghuo yu huanjing baohu," *Zhongguo shehui lishi pinglun* 7 (2006): 161–84.

2. *Da Qing lüli* (1740 Rpt.; 1998) juan 25, *xinglü, zeidao xia, fazhong*: 408 (Statute 276). A nominal sentence of a hundred blows typically meant only forty in practice. See Mühlhahn, *Criminal Justice in China*, 32.

3. For the relevant statute on grave trees, see *Da Qing lüli* (1740 Rpt.; 1998) juan 23, *xinglü, zeidao shang, dao yuanling shumu*: 372 (Statute 263).

4. A substatute first adopted in 1780 and later amended in 1801, 1809, and 1814 increased the penalty for stealing grave trees to one hundred blows and one month in the cangue for a first-time offender and harsher terms for repeat offenders. Xue Yunsheng, *Duli cunyi* (Beijing: Hanmao zhai, 1905), juan 25, *xinglü*: 47. Another substatute added in 1817 mandated additional punishment if a person destroyed more than one grave for farming. Ibid., *juan 31, xinglü*: 18.

5. For an overview of property law in China's history, see Long Denggao, *Zhongguo chuantong diquan zhidu jiqi bianqian* (Beijing: Zhongguo shehui kexue chubanshe, 2018); for conditional sales, see Zhang, *The Laws and Economics of Confucianism*.

6. *Da Qing lüli* (1870) juan 25, *xinglü, fazhong*: 13 (Statute 176.22). The relevant line reads: "Those descendants who because of poverty sell land [around graves] but retain the graves for ritual sacrifice and do not level them for farmland or sell them to others—they are not included in this substatute." For a case that hinged on the illicit sale of grave land dating from prior to the legal amendment, see Nanbu County Qing Archive: (QL51.11.23/1787) 2.60.01.

7. Nanjing guomin zhengfu sifa xingzheng bu, *Minshi xiguan diaocha baogao lu*, vol. 1, Hu Xusheng, Xia Xinhua, and Li Jiaofa, eds. (Beijing: Zhongguo zhengfa daxue chubanshe, 2000), 301, 321, and 352. For the Ba County custom, see Sichuansheng dang'anguan and Sichuan daxue lishixi, eds., *Qingdai Qian Jia Dao Ba xian dang'an xuanbian*, vol. 1 (Chengdu: Sichuan daxue chubanshe, 1989), 84, 101, and 202.

8. The notice (dated JQ25.12.17) reads: "Sichuan people are naturally fond of lawsuits. The writers of lawsuits stir up the two sides: if the contest was over mountains, they would claim in

their plaints that tombs were opened, and skeletons were tossed.... The people of Sichuan are obsessed with fengshui and listen to the trickery of geomancers; instigated by the grave-managing tenant farmers, when they see that another person's gravesite is auspicious, they immediately become covetous." See Sichuansheng dang'anguan and Sichuan daxue lishixi, eds., *Qingdai Qian Jia Dao Ba xian dang'an xuanbian*, vol. 2 (Chengdu: Sichuan daxue chubanshe, 1996), 349–50.

9. Guo, "Social Practice and Judicial Politics in 'Grave Destruction' Cases in Qing Taiwan, 1683–1895," 114.

10. Katz, *Divine Justice*.

11. This observation supports Joanna Waley-Cohen's suggestion that "the judicial recognition of supernatural occurrences ... was relatively frequent." See Waley-Cohen, "Politics and the Supernatural in Mid-Qing Legal Culture," 330–53.

12. Liu Bingxue, "Qingdai fengshui zhengsong yanjiu: Yi fenzang jiufen weili," *Zhengfa luntan* 4 (2012): 18–29.

13. *Da Qing lüli* (1740 Rpt.; 1998) juan 17, lilü, yizhi, shushi wangyan huofu: 293 (Statute 178).

14. *Da Qing lüli* (1740 Rpt.; 1998) juan 17, lilü, yizhi, sangzang: 296 (Statute 181).

15. *Qinding Da Qing huidian* (1764) juan 86: 5.

16. Maurice Freedman emphasized what he saw as fengshui's amoral character against the ethical principles associated with ancestor worship. See Freedman, *The Study of Chinese Society*, 189–211. Although the morality of geomantic principles is a complex question, what is clear from the legal record is that people mobilized fengshui with a range of intentions.

17. Faure, *Emperor and Ancestor*, 222.

18. See Cohen, "Souls and Salvation," 180–202. See also Pregadio, *The Encyclopedia of Taoism*, vol. 1, 521–23.

19. The notion of gravesites causing harm can be traced back to the early medieval era. See Bokenkamp, *Ancestors and Anxiety*.

20. Of the ninety-seven persons listed in the county's gazetteer who served as magistrate from 1650 to 1870, fifty-two (54 percent) hailed from southern provinces. Nine arrived from Hunan or Hubei, five from Fujian, five from Jiangxi, four from Guangdong or Guangxi, six from Yunnan or Guizhou, and twenty-three from Jiangsu, Anhui, or Zhejiang. Of the remaining forty-five magistrates, eleven (11 percent) were bannermen and thirty-three (34 percent) came from northern provinces. Note that this list did not include all men who served as county magistrate, and for one given name a place of origin was not included. *Tongzhi zengxiu Nanbu xian zhi* (1870), 195–201.

21. Magistrates relied on legal specialists hired for their knowledge of the law. See Li Chen, "Legal Specialists and Judicial Administration in Late Imperial China, 1651–1911," 1–54.

22. Snyder-Reinke, "Cradle to Grave," https://chinesedeathscape.supdigital.org/read/cradle-to-grave.

23. Jianpeng Deng, Chen, and Wang, "Classifications of Litigation and Implications for Qing Judicial Practice," 27.

24. Nanbu County Qing Archive: (GX5.6.19/1879) 7.844.01.

25. Nanbu County Qing Archive: (GX5.7.6/1879) 7.844.03.

26. Nanbu County Qing Archive: (GX5.6.24/1879) 7.844.02. Five days after Zhang's first plaint, the court issued a *piao* (typically translated as an "order" or sometimes "warrant"

depending on the context) to a yamen runner instructing him to investigate the matter. The runner's report does not survive in the case file, but the magistrate drew upon a description of the site in his verdict.

27. Nanbu County Qing Archive: (GX5.7.9/1879) 7.844.05.

28. Malpractice allegations (pulse misreading, incorrect medicines, etc.) were also levied against doctors in court. See for instance Nanbu County Qing Archive: (GX17.6.26/1891) 11.208.

29. Brokaw, *Commerce in Culture*, 466.

30. Huang Zongsheng, *Xincan houxu baizhong jing* (Chanshan: Fuwentang, d.u./c.1840), 20. This text lacks a date of publication, but its horoscopes extend through the *gengzi* year of 1840. Note that *gengzi* years in the Chinese calendar have been closely associated with unpropitious turns of events, national tragedies, and natural disasters. In the past two centuries, *gengzi* years have corresponded to the First Opium War of 1840, the Boxer Rebellion of 1900, the Great Leap Forward Famine of 1960, and the COVID-19 pandemic of 2020. Many notorious *gengzi* years exist further back in China's history, including the year 220 CE, which marked the end of the Han Dynasty (202 BCE–220 CE) with the abdication of Emperor Xian, the death of the warlord Cao Cao, and the official commencement of the Three Kingdoms Era (220–280 CE).

31. For another reference, see Anonymous, *Xiao'er guansha tujie* (1889 Rpt; Baise: Baise keyin faxingsuo, Undated Reprint), 1, 18.

32. Mugerwa, Nyangito, John, and Bakuneta, "Farmers' Ethno-Ecological Knowledge of the Termite Problem in Semi-Arid Nakasongola," 3183–91.

33. See, for instance, Zhao's preface: "In 1750, after my grandfather passed away, within three years my father died in the prime of his life and I encountered numerous difficulties. Some people said that my grandfather's grave had been obstructed." See Zhao Jiufeng (Zhao Tingdong), *Dili wujue* (1786 Rpt.; Chongwentang, 1841), 1.

34. *Da Qing lüli* (1740 Rpt.; 1998) *juan* 17, *lilü, yizhi, sangzang*: 296 (Statute 181). In one instance, members of the Ye family accused their relative of not burying a corpse for "over three years" since the person's death. Nanbu County Qing Archive: (XT1.6.15/1909) 20.265. See also Sutton, "Death Rites and Chinese Culture," 125–53.

35. This case is cited from Wei Shunguang, "Qingdai zhongqi fenchan zhengsong wenti yanjiu: Jiyu Ba xian dang'an wei zhongxin de kaocha," 185.

36. For similar cases about graves and water control in Nanbu, see Nanbu County Qing Archive: (DG15.5.26/1835) 4.83, (TZ11.7.6/1872) 6.139, (GX9.1.4/1883) 8.746, (GX11.12.8/1886) 9.86, (GX12.8.22/1886) 9.364, and (GX13.10.20/1887) 9.673.

37. *Da Qing lüli* (1740 Rpt.; 1998) *juan* 25, *xinglü, zeidao xia, fazhong*: 410 (Statute 276).

38. *Da Qing lüli* (1740 Rpt.; 1998) *juan* 25, *xinglü, zeidao xia, fazhong*: 411 (Substatute 276.4).

39. After the 1788 revision, the relevant substatute was the third (276.3) under the article "Uncovering Graves." Xue Yunsheng, *Duli cunyi* (Beijing: Hanmao zhai, 1905), *juan* 31, *xinglü*: 34.

40. Nanbu County Qing Archive: (GX26.3.12/1900) 15.74.07.

41. See, for instance, Ba County Qing Archive: (JQ5/1800) 6.03.1489 and (JQ7/1802) 6.03.1560. Wei Shunguang's work brought these cases to my attention. For a discussion of the cases, see Wei Shunguang, "Qingdai zhongqi fenchan zhengsong wenti yanjiu," 177–78.

42. Ibid., 91; 186.

43. Even in Ba County, for cases where illicit graves existed for over a year before adjudication, magistrates might permit them to remain in place. See Ibid., 28.

44. While legal appeals over "family, marriage, and land" cases were meant to be resolved below the prefectural level, appeals to the circuit level nonetheless happened. See Qiang Fang, "Hot Potatoes," 1105–35.

45. Wang Dingzhu, *Shusong pi'an, xiang*: 20–21.

46. For more on the constructed category of "false accusations," see Javers, "The Logic of Lies," 27–55.

47. Nanbu County Qing Archive: (GX21.3.29/1895) 12.429.03.

48. "The court rules that Zhang Zifeng should assist Zhang Jinyu et al. with the labor costs of 6,000 cash, and after [Jinyu] receives this money, all return home to rebuild the grave's forbidden area and alter the road so that it runs beneath it, thereby ceasing litigation." Nanbu County Qing Archive: (GX11.2.15/1885) 9.64.08.

49. Wei Shunguang, "Qingdai zhongqi de 'jiefen zisong' xianxiang yanjiu," 160.

50. David Graham (1884–1961) describes the fengshui groves of Sichuan. See Graham, *Folk Religion in Southwest China*, 113–14.

51. The mechanisms by which trees were divided depended on the region and the family. Myron Cohen describes one arrangement in the following terms: "The common arrangement was to require that the person given rights to whatever the cemetery might yield be responsible at least for providing the food for the lineage Qingming feast." See Cohen, "Lineage Organization in North China," 522.

52. This practice is known from the details of lawsuits and court verdicts. In one case in which a magistrate ordered the protection of fengshui trees, he specified that "only collection [of firewood] and fostering the trees is permissible, but the further cutting of the trees is not permissible." Nanbu County Qing Archive: (GX13.4*.15/1887) 9.655.08.

53. Qinggang Oak grew throughout Northern Sichuan. They were celebrated for hosting colonies of Wood Ear (*mu'er*), a popular ingredient in Chinese dishes. Some Qinggang were also identified as fengshui trees. See, for instance, Nanbu County Qing Archive: (GX11/1885) 9.77.13. The Li millstone dispute of chapter 2 offers another example.

54. One petition to the court begins, "Our ancestor came to Sichuan and forbade the felling of the trees upon his arrival, so as to cultivate fengshui in perpetuity." Nanbu County Qing Archive: (GX3.10.13/1877) 7.316.02.

55. Cohen, "Lineage Organization in North China," 522.

56. *Jiaqing Sichuan tong zhi* (1816) 71:10.

57. Qi Shouhua and Zhong Xiaozhong, *Zhongguo difang zhi meitan shiliao xuanji* (Beijing: Meitan gongye chubanshe, 1990), 427.

58. For a small sample of cases wherein a felled tree allegedly severed an earth vein, see Nanbu County Qing Archive: (DG20.4.1/1840) 4.155, (TZ6.7/1867) 6.131, (GX3.5.16/1877) 7.308, (GX3.9.21/1877) 7.314, (GX3.8.17/1877) 7.445, and (GX5.1.28/1879) 7.720.

59. See Pomeranz, *The Making of a Hinterland*, 125. For a reference to "drastic fuel scarcity" in relation to mining and timber in nineteenth-century Sichuan, see Jin, "Mint Metal Mining and Minting in Sichuan, 1700–1900," 34–35; 96–97.

60. This local idiom may also reflect more elite understandings of woodlands, which were cosmologically associated "with patterns of growth and branching, a model of successful reproduction." See Bray, *Technology and Gender*, 77.

61. Buck, *Chinese Farm Economy*, 32.

62. Duara, *Culture, Power, and the State*, 99. Tombstones were installed and oriented by geomancers, and fengshui manuals contained explicit instructions on how to do so.

63. The opening lines of one contract, composed after members of the Chen family were detained for cutting trees, read: "We, Chen Xiyuan and Chen Ankang, establish a contract to [promise] never again to cut trees held in common and beg to be released and prevent future misfortune. Members of the three branches of the lineage maintain several areas of cypress trees for fostering fengshui; these trees are forever forbidden from being cut." Nanbu County Qing Archive: (DG15.2.12/1835) 4.144.02. For cases about the theft of fengshui tree proceeds, see Nanbu County Qing Archive: (XF6.9.3/1856) 5.83.05, (GX8.1.22/1882) 8.655, (GX12.2.7/1886) 9.340.02, (GX13.4*.15/1887) 9.655.06, (GX13.9.5/1887) 9.666, and (GX19.3.25/1893) 11.872. See also SUCSBRC: (Nanbu Collection, MG7/1918) 466.1176.17.

64. Nanbu County Qing Archive: (GX20.2.25/1894) 12.44.09. For the reference in the law code, see *Da Qing lüli* (1740 Rpt.; 1998) juan 17, lilü, yizhi, sangzang: 296 (Statute 181.2). Since the issue at stake in this case was not a burial, but rather harm to fengshui, the verdict may be understood as not departing from the law code.

65. See the following two trial verdicts. Nanbu County Qing Archive: (GX11.6.10/1885) 9.51.05 and (GX20.3.27/1894) 12.44.07.

66. For trials resulting in orders of fengshui protection, see Nanbu County Qing Archive: (GX5.3.1/1879) 7.721.06 and Ba County Qing Archive (JQ15/1810): 6.03.01871. This source was brought to my attention by Wei Shunguang, "Qingdai zhongqi fenchan zhengsong wenti yanjiu," 179.

67. Nanbu County Qing Archive: (DG12.3/1832) 4.141.01. For another official order of protection, see Nanbu County Qing Archive: (DG15.6*.3/1835) 4.146.01.

68. In one instance from 1825, a group of monks explicitly asked the court's permission to cut trees for needed temple repairs. Ten months later, the same monks requested a notice of protection for the temple's forest. Nanbu County Qing Archive: (DG4.12.29/1825) 4.122.01 and (DG5.10.20/1825) 4.123.03.

69. Nanbu County Qing Archive: (GX16.1.22/1890) 10.638.01.

70. Nanbu County Qing Archive: (GX16.2.17/1889) 10.638.04.

71. Nanbu County Qing Archive: (GX13.2.19/1887) 9.651.04.

72. For reference to the drought of 1877, see Gou Deyi, *Qingdai jiceng zuzhi yu xiangcun shehui guanli*, 71.

73. Nanbu County Qing Archive: (GX13.9.3/1877) 7.312.01.

74. Robert Marks's observation that Buddhist monasteries were the most important forest preserves in late imperial China may be applied to some of Sichuan's temples and gravesites. See Marks, *Tigers, Rice, Silk, and Silt*, 37–39. I thank Gilbert Chen for bringing this connection to my attention. See also Marks, *China: An Environmental History*, 386.

75. Nanbu County Qing Archive: (GX30.3.16/1904) 16.640.06.

76. Nanbu County Qing Archive: (GX30.3.16/1904) 16.640.05. Officials also flagged the sizes of fengshui trees contested during appeal cases. See Wang Dingzhu, *Shusong pi'an, xiang*: 5.

77. The relevant line reads, "... [I]t is up to He Binglin to stockpile wood and grass in this area; these activities should not concern Xu Shaojin's side." Nanbu County Qing Archive: (GX30.3.27/1904) 16.640.08.

78. From the Song (960–1279) onward, a phenomenon called "graves for the living" appeared. These empty graves were constructed on auspicious plots and were thought to extend the lifespan of the future grave occupant. See Bai, "Daoism in Graves," 548–600.

79. Nanbu County Qing Archive: (GX12.12.3/1886) 9.675.01. For a discussion of wind diseases, see Zhang and Unschuld, *Dictionary of the Ben cao gang mu*, vol. 1, 159.

80. For the map drawn by the clerk (not shown here), see Nanbu County Qing Archive: (GX12.11.24/1886) 9.675.05.

81. Nanbu County Qing Archive: (GX12.12.13/1887) 9.675.06.

82. Nanbu County Qing Archive: (GX12.12.13/1887) 9.675.06.

83. Nanbu County Qing Archive: (GX13.1.21/1887) 9.675.12.

84. For an instance where contracts and tax records were sufficient for resolving a fake grave dispute, see the following case. It began when a man surnamed Zhang accused another, surnamed Chen, of leveling his ancestral graves. Chen traveled to court and presented a land deed that did not mention the existence of any graves. The magistrate was convinced and ordered Zhang and his collaborator beaten for burying bones in Chen's land. The magistrate did not speculate what kind of bones were involved, but they were likely not human bones. If human bones had been buried, the grave would have been "real," and the case would have been adjudicated along the lines of "stealing a burial" (*daozang*). For the trial and verdict, see Nanbu Qing County Archive: (GX3.12.17/1878) 7.336.09.

85. Sichuansheng dang'anju, ed., *Qingdai Sichuan Ba xian yamen Xianfengchao dang'an xuanbian*, vol. 9 (Shanghai: Shanghai guji chubanshe, 2011), 50–51.

86. For another case alleging the creation of a fake grave that saw the court map the site, see Nanbu County Qing Archive: (GX7.3.10/1881) 8.323.04.

87. For the nineteen cases, see Nanbu County Qing Archive: (TZ5.12.13/1867) 6.70.01, (TZ9.7.5/1870) 6.80.01, (TZ9.7.16/1870) 6.81, (TZ12.4.4/1873) 6.107.03, (TZ9.10.23/1870) 6.132.01, (TZ11.3.3/1872) 6.136.01, (TZ11.7.6/1872) 6.139.01, (TZ11.6.27/1872) 6.141.01, (TZ12.6*.9/1873) 6.144.01, (TZ12.6*.16/1873) 6.145.01, (TZ12.7.23/1873) 6.146.01, (TZ12.8.11/1873) 6.147.01, (TZ13.3.11/1874) 6.153, (TZ13.3.12/1874) 6.154.01, (TZ13.3.26/1874) 6.155.01, (TZ13.5.3/1874) 6.156.01, (TZ13.5.21/1874) 6.157.01, (TZ8.8.11/1869) 6.179.10, and (TZ9.1.27/1870) 6.181.

88. Zhu Qingqi and You Shaohua (ed.), *Xing'an huilan quanbian*, vol. 13, 12: 518–19.

89. Nanbu County Qing Archive: (GX3.10.9/1877) 7.336.01, (GX7.2.30/1881) 8.323, (GX11.3.14/1885) 8.977.20, (GX11.2.28/1885) 9.69, (GX12.12.3/1886) 9.675.01, and (GX12.12.1/1886) 9.676.02.

90. Zhang, *Circulating the Code*, 138–40.

91. See for instance Nanbu County Qing Archive: (GX11.11.14/1885) 9.194.03 and (GX13.8.6/1887) 9.686.07.

92. In one case contesting ownership of an ancient grave, a magistrate claimed that the hundred-year-old contracts presented by the warring sides were "the same as wastepaper." Nanbu County Qing Archive: (GX18.6.8/1892) 11.262.01.

93. Huang, *Code, Custom, and Legal Practice in China*, 81.

94. Sommer, *Polyandry and Wife-Selling in Qing Dynasty China*, 117–20.

95. For one example of a trial between two parties contesting the identity of an ancient grave, see Nanbu County Qing Archive: (GX11.11.30/1886) 9.194.04.

96. Nanbu County Qing Archive: (XT3.6.6/1911): 22.204.01 and (XT3.6.18/1911) 22.204.02.

97. Nanbu County Qing Archive: (XT3.6*.18/1911) 22.204.06.

98. Nanbu County Qing Archive: (XT3.10.10/1911) 22.204.14.

99. Szonyi, *Practicing Kinship*, 26–55.

100. Cohen, "Family Management and Family Division in Contemporary Rural China," 357–77.

101. *Qinding Da Qing lüli* (1870) juan 9, hulü, tianzhai: 5 (100.15).

102. Late Qing surveyors recognized forged contracts as a significant problem in Sichuan. See Wu, *Qingdai xianyu minshi jiufen yu falü zhixu kaocha*, 414. For a case concerning an allegedly altered contract and fengshui, see Nanbu County Qing Archive: (GX24.2.3/1898) 14.59.01.

103. For a case where a genealogy was submitted to the court, see Nanbu County Qing Archive: (GX5.10.28/1879) 7.845.03. For a case where the absence of a genealogy was deemed important by the court, see Nanbu County Qing Archive: (GX18.6.30/1892) 11.262. For cases where genealogies were composed in court, see Nanbu County Qing Archive: (DG6.3.10/1826 and XF8.1.16/1858) 5.48.05, and Nanbu County Qing Archive: (XF6.9.3/1856) 5.83.05.

104. Hansen, *Negotiating Daily Life in Traditional China*.

105. Around 20 percent of the three hundred land deeds preserved in Sichuan's Longquanyi Archive concern grave lands. Other documents, such as household division registers, reference the communally maintained fengshui trees of local lineages. See Hu Kaiquan and Su Donglai, eds., *Chengdu Longquanyi bainian qiyue wenshu, 1754–1949* (Chengdu: Ba-shu shushe, 2012).

106. "With the two lineages as witnesses, we now state clearly, that from today hereafter, the base land of Du Zhongyao et al.'s house marks the upper boundary. The Wei lineage is not allowed to add a grave to the left of this house and break the stability of the Du lineage's rear earth vein, but the Wei lineage can create one grave on the left border of the property. It is permissible for the Wei lineage to establish an inscription and offer sacrifices at the graves, but they are not permitted to add a grave and break the earth vein. If a grave is added and the earth vein is broken, the two lineages will stand as witnesses." SUCSBRC: (Nanbu Collection, GX32.3.12/1906) 466.77.6.

107. People "sold" the lands around graves and cemeteries since properties proven to be auspicious could fetch a high price. See, for instance, "[A report on] investigating the matter of Wu Yangke's accusation that Wu Yangquan sold the central vein of the Wu ancestral cemetery to Hou Dashi." Nanbu County Qing Archive: (GX11.12.21/1886) 9.87.03. For the judicial map of the vein in contention, see Nanbu County Qing Archive: (GX11.12.21/1886) 9.87.04.

108. Nanbu County Qing Archive: (GX—.4.13) 17.394.08. Note that this contract lacks a written year.

109. For more on the right of first refusal, see Taisu Zhang, "Moral Economies in Early Modern Land Markets," 113.

110. *Da Qing lüli* (1740 Rpt.; 1998) juan 9, hulü, tianzhai, dianmai tianzhai: 199 (Statute 95.3). See also Zhang, *The Laws and Economics of Confucianism*, 47–48.

111. Buoye, *Manslaughter, Markets, and Moral Economy*, 96.

112. For the related substatute, see: *Qinding Da Qing lüli* (1870) juan 9, hulü, tianzhai: 3 (100.4). The substatute derived from a 1756 memorial from the then-governor of Jiangsu Zhuang Yougong (1713–67). For a copy of Zhuang's original memorial, which mentioned ancestral trees connected to fengshui, see National Palace Museum Palace Memorial and Grand Council Archive: (QL21.05.08/1756) 403011858.

113. Here, ancestral sacrifices (*yinsi*) is likely a reference to "ritual fields," or land held in the common trust of the lineage for the upkeep of the necessary rituals performed for ancestors.

114. Wang Dingzhu, *Shusong pi'an, xiang*: 64.

115. Ba County Qing Archive: (DG14/1834) 6.08.03321. This case was brought to my attention through Wei Shunguang's scholarship. Wei Shunguang, "Qingdai zhongqi fenchan zhengsong wenti yanjiu," 91; 189.

116. Zhou Xun, *Shuhai congtan*, 26–27.

117. Grave lands, cemeteries, and charitable estates were employed as tax havens in Jiangnan in the 1800s. See Meyer-Fong, *What Remains*, 129–30.

118. The ruling read: "When the court investigated the contract for the acquisition of grave land presented by Shi Yingliang, we found it bears the date of the Jiaqing era (1796–1821), yet when examining its stamp, it is clear you have intentionally evaded taxation." See Nanbu County Qing Archive: (GX26.4.5/1900) 15.74.07. See also Nanbu County Qing Archive: (XF7.9.29/1857) 5.44 and (XT3.6.30/1911) 22.199.08.

119. Crook and Crook, *Revolution in a Chinese Village*, 115.

120. Langzhong Municipal Archive: (GX13/1887) 343.7.04-05. See Ruf, *Cadres and Kin*, 196.

121. Duara, *Culture, Power, and the State*, 91.

122. "Granary books" (*aoce*) were local versions of tax records that were kept in the rural townships and community granaries.

123. Sommer, *Polyandry and Wife-Selling in Qing Dynasty China*, 155, 185, 268, 343.

124. The verdict read: ". . . [The court orders that] Li Zuohong et al. not change their surname, forget their roots, and extinguish the kindness of their ancestors. Deng Maohu's name and his originally registered grain tax obligation must be re-recorded in the granary books, so Zuohong and Yuanchun fulfill their annual tax obligations. The lineage should inspect Deng Banlong, Deng Zun, and Deng Maohu's tomb steles and worship their ancestral tablets. Now, we will destroy the fake ('illicit') contract for the sale of the trees witnessed by Deng Deng'er and [Deng] Dengwang." SUCSBRC: (Nanbu Collection, DG27.10.3/1847) 466.1829.128. For a similar case involving the felling of fengshui trees and the "extinguishing" of the ancestral sacrifices, see Nanbu County Qing Archive: (GX24.6.1/1898) 14.68.01.

125. See the 1729 case of Zhang Juan and his illicit registration of another's tomb property in Shifang County. *Jiaqing Shifang xian zhi* (1813) 46: 1. For another example, see the case of Li Xingshan in Pengxi County. *Kangxi Pengxi xian zhi* (1713) *shang juan*: 48.

126. See Nanbu County Qing Archive: (GX3.4.22/1877) 7.306.01.

127. Bodde and Morris, *Law in Imperial China*, 123; Idema, *Judge Bao and the Rule of Law*, xxx; Cao Qiangxin, "Qingdai xingbu jianyu yuanliu kaoxi," *Fanzui yu gaizao yanjiu* 10 (2018): 71–74.

128. Geomantic manuals contained yamen dimensions, advice for constructing granaries, formulae for the slope of roofs, and a range of practical information about when and where to construct important buildings. For more information, see Ruitenbeek, *Carpentry and Building in Late Imperial China*, 36–39; 207.

129. See the layout of the Shuntian Prefectural Yamen in Susan Naquin's study of Beijing. See Naquin, *Peking: Temples and City Life, 1400–1900*, 369. See also the maps of yamens under the administration of Shaoxing Prefecture in Zhejiang. *Kangxi Shaoxing fu zhi* (1683) 3: 5–7.

130. For a discussion of this process, see Sommer, *Polyandry and Wife-Selling in Qing Dynasty China*, 392–93.

254   NOTES TO CHAPTER 1

131. First Historical Archives of China: (JQ16.3.28/1811) 02-01-07-09500-015. For the copy of a routine memorial written in Manchu and Chinese concerning a violent crime arising from a disputed earth vein in Guizhou, see Academia Sinica Archives of the Grand Secretariat: (DG8.2.1/1828) 015659.

132. Zhu Qingqi and You Shaohua, *Xing'an huilan quanbian*, vol. 3 (1834 Rpt.; Beijing: Falü chubanshe, 2008), 21: 1182. Note that this case is discussed within the context of another incident.

133. The text reads: "Pan Mei'an, out of resentment arising from the fact Xu Yanyi did not give him newly harvested grain, nailed a peach tree branch to Xu's ancestral grave, seeking to destroy his fengshui and leave Xu plagued by evil spirits, disease, and suffering; his action is comparable to using spells and incantations to curse someone with misfortune, Pan's punishment thus should be reduced from plotting murder on account of the fact that some action was taken but there was no injury, with eighty strokes of the heavy stick and two years penal servitude." Zhu Qingqi and You Shaohua (ed.), *Xing'an huilan quanbian*, vol. 4, 28: 1543.

134. Ocko, "I'll Take It All the Way to Beijing: Capital Appeals in the Qing," 291–315.

135. Hung, *Protest with Chinese Characteristics*, 137.

136. National Palace Museum Palace Memorial and Grand Council Archive: (DG8.6.20/1828) 061107.

137. Paton, *Five Classics of Fengshui*, 249, 251, 254, 284, 287, 291, 305–6, 310, 376, and 396. Also cited in Ouyang Chun, *Fengshui ershu xingqi leize*, 2:143.

138. Cao Xueqin and Gao E, *Honglou meng* (Beijing: Renmin wenxue chubanshe, 2005), 170. Wrongdoings that fell under the "Ten Abominations" (*shi'e*) of imperial law, such as plotting rebellion, could see the graves of condemned evildoers destroyed by the state. For more on this topic, see Smith, *Fortune-tellers and Philosophers*, 156–57.

139. For a related discussion, see Ellickson, "The Costs of Complex Land Titles," 281–302.

140. Zhang, "Moral Economies in Early Modern Land Markets"; Huang, *The Peasant Economy and Social Change in North China*, 74; Isett, *State, Peasant, and Merchant in Qing Manchuria, 1644–1862*. For a discussion of *zhaotie* payments, see Huang, *Code, Custom, and Legal Practice in China*, 73–76.

141. Taussig, *The Devil and Commodity Fetishism in South America*, 15.

142. The phrase "obsessed with fengshui" also appears in Ming sources. Zhang Aihua, *Wenhua ruan quanli shiye xia de jiapu yanjiu: Yi Ming-Qing Anhui Jing xian Zhushi xilie jiapu wei yangben* (Tianjin: Tianjin renmin chubanshe, 2020), 199.

143. For references to other provinces, see: Zhang Chuanyong, "Qingdai 'tingsang bude shijin' lun tanxi: Jianji Qingdai guojia zhili 'tingsang buzang' wenti de duice," *Zhongguo shehui lishi pinglun* 1 (2009): 289. For references to "fengshui obsession" in northern China, see the digitized Gazetteer Database of the Max Planck Institute for the History of Science. Such mentions are present in the local gazetteers from Gansu, Shanxi, Shaanxi, Shandong, Henan, and Zhili.

144. The informal bureaucracy of yamen staff such as clerks and runnners increased in size, along with the collection of informal revenue to pay them. For more on this subject, Reed, *Talons and Teeth*, 147.

145. Marks, *China: An Environmental History*, 221.

146. Cited from Peng Yuxin, *Qingdai tudi kaiken shi* (Beijing: Nongye chubanshe, 1990), 113.

147. National Palace Museum Palace Memorial and Grand Council Archive: (QL13.6.9/1748) 002612.

148. *Guangxu Pengxi xian xu zhi* (1899), 14: 14.

149. *Tongzhi Pi xian zhi* (1870), 18: 2. The 1813 edition of the gazetteer contains the same passage.

150. *Guangxu Yulin zhou zhi* (1894), 4:12.

151. Osborne, "The Local Politics of Land Reclamation in the Lower Yangzi Highlands," 1.

152. Xiaoye Zhang sees this trend stretching across the Ming-Qing period. See Zhang, "Legitimate, but Illegal," 73–94.

153. Perry, *Rebels and Revolutionaries in North China, 1845–1945*, 156.

154. Scott, *Weapons of the Weak*.

Chapter 2: Mapping Fengshui

1. For a breakdown of judicial maps in Nanbu's archive, see Appendix B.

2. For instance, a judicial map was commissioned for the following case, but it does not survive in the archive. Nanbu County Qing Archive: (TZ13.4.14/1874) 6.156.02.

3. Schäfer, *The Crafting of the 10,000 Things*; Nappi, *The Monkey and the Inkpot*; Bray, *Technology and Gender*.

4. Elman, *On Their Own Terms*, 5. See also Sivin "Science and Medicine in Imperial China," 41–90.

5. Luo Guanzhong, *Sanguo yanyi*, vol. 1 (Beijing: Renmin wenxue chubanshe, 2005), 383.

6. Macauley, *Social Power and Legal Culture*.

7. Bray, "Introduction: Science and Confucian Statecraft in East Asia," 13.

8. Taussig, "Visicerality, Faith, and Skepticism," 455.

9. Clunas, *Pictures and Visuality in Early Modern China*, 104–8.

10. Schäfer, *The Crafting of the 10,000 Things*, 141–42. For more on the functions of *tu* in the late imperial era, see Bray, *Technology and Gender*, 132–33, 200.

11. One exception is Lucille Chia's discussion of grand genealogical images produced in woodblock printed books. See Chia, "Text and *Tu* in Context Reading the Illustrated Page in Chinese Blockprinted Books," 241–76.

12. Lu Yao and Yao Lide, *Shandong yunhe beilan*, in *Gugong zhenben congkan*, vol. 234, ed. (1776 Rpt.; Haikou: Hainan chubanshe 2000), 235.

13. Ibid., 246

14. There were also *tu* that represented the fengshui of much larger spaces, including countrywide spaces such as the *San da ganlong zonglan zhi tu*. See Smith, "The Transnational Travels of Geomancy in Premodern East Asia, c. 1600—c. 1900: PART I." Mapmakers were not above adding features like hills or mountains to their productions to give the appearance of more favorable fengshui. See Forêt, *Mapping Chengde*, 100–38.

15. Zhu Huimin, "Ming-Qing Huizhou jiapu xiangzhuan chutan," *Ningxia daxue xuebao (Renwen shehui kexueban)*, 39 (2017): 45–50. For an example of a Huizhou tomb map from 1853, see Zhang, *Linquan, fenshan yu miaochan*, 104.

16. Yu, "Publishing at the Grassroots."

17. *Yongzheng shangyu*, vol. 19 (YZ7.3), 9.

18. The edict broadly corresponded with an imperial directive to record and cultivate land—the last significant campaign to officially record land prior to the twentieth century. See Rowe, *Saving the World*, 58–59.

19. Counties' departments of works were required to abide by the following regulations in mapping ancient tombs: "Whether there is any damage to the ancient tombs and mausoleums or any new additions of shrines—either constructed by the government or by individuals—all need to be mapped accurately and presented along with a detailed record of their sizes, styles, and dimensions; this is to be submitted within three months of receiving this order." Nanbu County Qing Archive: (GX1.3.3/1875) 7.104.06.

20. Huang Liuhong, *Fuhui quanshu*, 22: 1–3. Republished in Guanzhenshu jicheng bianzuan weiyuanhui, ed., *Guanzhenshu jicheng*, vol. 3, 465–66.

21. Tian Wenjing and Li Wei, *Qinban zhouxian shiyi*. Republished in Guanzhenshu jicheng bianzuan weiyuanhui, ed., *Guanzhenshu jicheng*, vol. 3 (Hefei: Huangshan shushe, 1997), 659–60.

22. Will, *Handbooks and Anthologies for Officials in Imperial China*, 293.

23. Cited in He Changling and Wei Yuan, eds., *Huangchao jingshi wenbian* (Shanghai: Jiang zuo shu lin cang ban, 1873), 22: 57.

24. Ibid., 22: 57.

25. Wang Zhi (1681–1766), who served as a magistrate in Guangdong, recommended examining maps of grave land in an essay that was republished in Xu Dong's (1793–1865) handbook, *Book for Magistrates*. "For lawsuits about grave mountains . . . investigate the mountain's topography, using the land to verify the map; where a lineage is buried, people of other lineages should not be buried there; when people of different surnames are buried in the same area, one must establish boundary markers between the tombs." Xu Dong, *Muling shu*, 18: 6. Republished in Guanzhenshu jicheng bianzuan weiyuanhui, ed., *Guanzhenshu jicheng*, vol. 7 (Hefei: Huangshan shushe, 1997), 399.

26. Chen Quanlun, Bi Kejuan, and Lü Xiaodong, eds., *Xu gong yanci: Qingdai mingli Xu Shilin pan'an shouji* (Jinan: Qi-Lu shushe, 2001), 193–95.

27. Ibid., 581–83.

28. Ibid., 349–50.

29. Wang Dingzhu, *Shusong pi'an, yuan*: 22; *xiang*: 79.

30. Wang Dingzhu, *Shusong pi'an, xiang*: 28–29.

31. Wang Dingzhu, *Shusong pi'an, xiang*: 31.

32. Wang Dingzhu, *Shusong pi'an, xiang*: 4.

33. Taiwan Dan-Xin Qing Archive: (GX5.10.26/1879) 22507.36. For the second analysis of the gravesite, see (GX5.11.19/1879): 22507.39. For other cases preserved in the Dan-Xin Archive that reference or contain grave maps as legal evidence, see Taiwan Dan-Xin Qing Archive: (Undated, c. 1817) 17301.07, (Undated, c. 1852) 35201.04, (Undated, c. 1879) 35207.03, (GX5.5.28/1879) 22508.06, (GX12.11.17/1886) 22515.05, and (GX15.5.8/1889) 22425.52.

34. Sichuansheng dang'anju, ed., *Qingdai Sichuan Ba xian yamen Xianfengchao dang'an xuanbian*, vol. 9, 38. See also Ibid., 37, 48.

35. For a reference to a Ba County judicial map of graves, see Sichuansheng dang'anguan and Sichuan daxue lishixi, eds., *Qingdai Qian Jia Dao Ba xian dang'an xuanbian*, vol. 1, 180. The following record preserved in Wang's legal collection is another example: "A person from Ba County, Ren Tianlin, presents a map and plaint with the intention of expelling harm caused by

Zeng Guoquan's wicked destruction of a grave. Comment: I approve a summons for interrogation." Wang Dingzhu, *Shusong pi'an, yuan*: 22.

36. Nanbu County Qing Archive: (GX12.11.29/1886) 9.354.7 and (GX12.11.29/1886) 9.354.9. Note that the two clerks created the maps on the same day.

37. See Appendix C for more information on Nanbu's yamen mappers.

38. Faure, "Between House and Home," 281–94.

39. Carroll, *Between Heaven and Modernity*, 160.

40. Steinhardt, *Chinese Architecture*, 2. Related terms for craftsmen who specialized in building included *jiangren* and *tushi*.

41. Carroll, *Between Heaven and Modernity*, 160–61.

42. Ruitenbeek, *Carpentry and Building in Late Imperial China*.

43. Feng, *Chinese Architecture and Metaphor*, 90.

44. Knapp, *Chinese Houses*.

45. Knapp, "Siting and Situating a Dwelling," 99–138.

46. For discussions of these diverse audiences, see Zhang Fang, "Yi Yueyang Zhangguyingcun weili tantao Xiangbei minju fengshui," *Shanxi jianzhu* 12 (2006): 19–29; Jia Lingli, "Shaanxi Guanzhong diqu nongcun juzhu jianzhu wenhua tantao," *Sichuan jianzhu kexue yanjiu* 1 (2007): 160–163.

47. Xiong Dian (Zaofu sanren), Xiong Weiyao (Preface), and Yuan Shizhen (Epilogue), *Xinzeng Buju misui quanji* (Nanjing: Jinshantang, 1595), *shang juan*: 12b. Note that this reference refers to the *shang juan* of the "Newly Updated" (*xinzeng*) edition of *Buju misui*.

48. Building and geomantic manuals paid considerable attention to the ideal slopes of roofs, which was usually around 1:3.3. There were auspicious days for determining the slope of a roof. See Ruitenbeek, *Carpentry and Building in Late Imperial China*, 37, 161.

49. Sweeten, *China's Old Churches*, 63–64.

50. *Da Qing lüli* (1740 Rpt.; 1998) juan 34, *xinglü, zafan, shihuo*: 536–37 (Statute 382 and Statute 383).

51. *Jiaqing Shifang xian zhi* (1813), 34: 22. See Brokaw, *Commerce in Culture*, 459.

52. Zhou Nan and Lü Lin, *Anju jinjing* (Hangzhou: Shounantang, 1780), 1: 5–6.

53. Ibid., 6: 12.

54. Ibid., 6: 14.

55. Ibid., 3:12–13.

56. Bray, *Technology and Gender*, 91–122.

57. *Anju jinjing* 7: 29.

58. Ibid., 7: 29.

59. For the map, see Nanbu County Qing Archive: (GX16.3.9/1890) 10.689.05.

60. See Appendix C.

61. For the verdict, see Nanbu County Qing Archive: (GX16.5.11/1890) 10.689.07.

62. *Anju jinjing* 6: 38.

63. The coffin was typically stored in the house before burial. See De Groot, *The Religious System of China*, vol. 2, 371–72.

64. One relevant quotation is: "What could be more absurd than their imagining that the safety of a family, honors, and their entire existence must depend upon such trifles as . . . one roof being higher than another?" See Ricci, *China in the Sixteenth Century*, 85.

65. Wang Junrong, *Xinke huitu yangzhai shishu jicheng* (1590 Rpt.; Saoye shanfang, 1882), 1:7–28. For the text's advice on securing fortune for government *yamens*, see Ibid., 4: 40–41.

66. Nanbu County Qing Archive: (GX13.7.18/1887) 9.689.5, (GX14.7.25/1888) 10.17.5, (GX15.8.20/1889) 10.276.8, (GX16.2.17/1890) 10.638.4, and (GX18.3.3/1892) 11.251.4.

67. Nanbu County Qing Archive: (XF7.9.8/1857) 5.95.01, (GX5.1.28/1879) 7.720.01, and (GX10.10.3/1884) 8.906.01. The case (GX13.11.12/1887) 9.673.04 involved a grave and a house. See also SUCSBRC: (Nanbu Collection, MG19.4/1930) 466.6096, and Sichuansheng dang'anju, ed., *Qingdai Sichuan Ba xian yamen Xianfengchao dang'an xuanbian*, vol. 9, 229. For a case involving the negotiation of the use of fengshui trees in relation to building a house, see Nanbu County Qing Archive: (DG3.3.28/1823 and XT3.3.20/1911) 22.177.07. Note that the two dates refer to, first, the date of composition of the original document, and second, the date when that piece of evidence was leveraged in a legal case.

68. Nanbu County Qing Archive: (GX3.6.16/1877) 7.326.03. For another example of a fengshui dispute over a manure pit, see Nanbu County Qing Archive: (GX16.8.20/1890) 10.631.

69. SUCSBRC: (Nanbu Collection, MG11.7.11/1922) 466.5376.

70. Langzhong Municipal Archive: (Undated, c. mid-1930s) 343.25.05.

71. *Pan* was a general term for local government-run schools in imperial China.

72. Li Jie, "Qingdai wanqi Sichuan beibu diqu xiangtu jianzhu jiqi wenhua neihan: Yi Nanbu xian Songjia dayuan weili," *Xihua daxue xuebao* 1 (2016): 1–6.

73. *Anju jinjing* 1: 6.

74. Knapp, "Siting and Situating a Dwelling," 47.

75. Nanbu County Qing Archive: (GX16.11.27/1891) 10.649.05. For the litigation map, see 10.649.06.

76. Fan Zengxiang, *Fanshan pipan* in *Lidai panli pandu*, vol. 11, eds. Yang Yifan and Xu Lizhi (Beijing: Zhongguo shehui kexue chubanshe, 2005), 604.

77. Ibid., 282.

78. The twenty-four directions are marked by the twelve branches and only eight of the ten stems in addition to the four trigrams.

79. For a discussion of Buddhist monks engaging in the geomantic arts, see Pan Sheng, *Zhishi, lisu yu zhengzhi: Songdai dilishu de zhishi shehui shitan* (Nanjing: Jiangsu renmin chubanshe, 2018), 276–80.

80. Fan Zengxiang, *Fanshan pipan*, 281.

81. Ibid., 281. For more on the question of belief, see Hymes, *Way and Byway*, 12–13.

82. Fan Zengxiang, *Fanshan pipan*, 114.

83. *Tongzhi zengxiu Nanbu xian zhi* (1870), 78–79.

84. Wang Daolü, *Nanbu xian xiangtu zhi* (1906 Rpt.), in Yao Leye, ed., *Sichuan daxue tushuguan guancang zhenxi Sichuan difangzhi congkan*, vol. 3 (Chengdu: Ba-shu shushe, 2009), 388.

85. Nanbu County Qing Archive: (GX13.7.23/1887) 9.819.01.

86. For other millstone disputes involving graves and fengshui, see Nanbu County Qing Archive: (TZ9.2.18/1870) 6.181.05 and SUCSBRC: (Nanbu Collection, MG7/1918) 466.1176.17.

87. Skinner, "Marketing and Social Structure in Rural China, Part I," 6.

88. In a 1926 document, a member of the Li lineage was referenced as the sole militia head for Pig's Trough Pass. Another Li-surnamed individual was named as the head of forestry in the settlement. SUCSBRC: (Nanbu Collection, MG15/1926) 466.4360.

89. Nanbu County Qing Archive (GX13.7/1887): 9.819.05.

90. *Guangxu Jingyan zhi* (1900), 3:3.

91. Another case about a waterwheel and fengshui sharing many of the same details was litigated at Bishan's county court in the 1930s. There, the county court ruled in favor of the fengshui claim, but the Superior Court of Chongqing overturned that verdict. The river and waterwheel were mapped for the court's consideration. SUCSBRC: (Bishan Collection: MG20.2/1931) 12.3-4.4810.

92. For a classic overview of local governance and corruption, see T'ung-tsu Ch'ü, *Local Government in China Under the Ch'ing*. For corruption and law, see Park, "Corruption in Eighteenth-century China," 967–1005.

93. Li Rong, *Shisanfeng shuwu quanji* (Preface 1892), in *Li Shenfu xiansheng quanji*, ed. Jiang Dejun (Shanghai: Youhaishan fangshi, 1899), 7:6.

94. *Tongzhi zengxiu Nanbu xian zhi* (1870), 27–32. Gou, *Qingdai jiceng zuzhi yu xiangcun guanli*, 277; 28–288.

95. Geographic and geomantic writings profilerated in the "age of disunion" (220–589 CE). For more on the emergence of geographical writing at this time, see Felt, *Structures of the Earth*.

96. Wu, *Qingdai xianyu minshi jiufen yu falü zhixu kaocha*, 96–99.

97. Li Chenghua, ed. *Sichuansheng Cangxi xian hexi fenghuanggong hedong yangyueshan Lishi zongpu* (Cangxi: Cangxixian xingzi diannao wenyinbu, 2002), 11–12.

98. Chen Tingxian, *Sichuan Nanbu xian Chenshi zupu* (Preface Dated 1858). Langzhong Municipal Library, Local Rare Books Collection, 3:5–6. A microfilm copy of the genealogy is also held by the Family History Library in Salt Lake City, Utah.

99. *Tongzhi zengxiu Nanbu xian zhi* (1870), 78.

100. Chen Tingxian, *Sichuan Nanbu xian Chenshi zupu* (1858), 2:13–14.

101. Wei, "Village Fengshui Principles," 41.

102. Gong Yilong, *Zuqun ronghe yu shehui zhenghe: Qingdai Chongqing yimin jiazu yanjiu* (Beijing: Zhongguo wenshi chubanshe, 2015), 165–71. See Szonyi, *Practicing Kinship*, 36.

103. Chen Tingxian, *Sichuan Nanbu xian Chenshi zupu* (1858), 2:13–14.

104. This choice of rooting a lineage dragon vein back to a central market may speak to broader trends in geomantic mapping in Sichuan, as the map of the Li dragon vein (figure 2.9) also placed an annotation for the area's market center in the upper right of its page.

105. Wang Daolü, *Nanbu xian xiangtu zhi* (1906), 357–58.

106. Huang Liuhong, "Baojia sanlun," in Wei Yuan, *Wei Yuan quanji*, vol. 17, Wei Yuan quanji bianji weiyuanhui, ed. (Changsha: Yuelu shushe, 2007), 133.

107. For a discussion of rural militization in the last century of Qing rule, see Kuhn, *Rebellion and Its Enemies in Late Imperial China*.

108. For the Henan site, see *Da Ming Yitong zhi* (1461) 26:19; for the Sichuan site, see *Da Ming Yitong zhi* (1461) 68: 11.

109. Cai Dongzhou, "Langzhong Chenshi zupu kaolun," *Wenxian* 3, 1997: 134–51. Cai Dongzhou and Zhang Liang, "Nanbu dang'an zhong youguan Songdai Langzhou Chenshi jiazu mu dang'an yanjiu," *Zhonghua wenhua luntan* (2014): 98–104.

110. Chen Tingxian, *Sichuan Nanbu xian Chenshi zupu* (1858), 2:12. Some of the cited local writings of the genealogy date from the eighteenth century.

111. Historians have highlighted the links between genealogies and gazetteers, with the editorial compilers of some Ming and Qing gazetteers drawing on genealogies for much content. See Dennis, *Writing, Publishing, and Reading Local Gazetteers in Imperial China, 1100–1700*, 64–116.

112. *Tongzhi zengxiu Nanbu xian zhi* (1870), 61–66.

113. Ibid., 70–73; 443–44.

114. *Tongzhi zengxiu Nanbu xian zhi* (1870), 330–56. A total of 225 exemplary women were listed, but a few elite surnames dominated. The surnames of Chen (seventeen), Li (eighteen), and Xu (fifteen) collectively constituted around a quarter of the total. Official recognitions of virtuous women were liberally bestowed in the nineteenth century. See Elvin, "Female Virtue and the State in China," 111–52; Theiss, *Disgraceful Matters*.

115. The medieval treatise, *On the Origins and Symptoms of Various Illnesses* (*Zhubing yuanhou lun*), and the Song dynasty medical text, *Complete Collection of Effective Prescriptions for Women* (*Furen daquan liangfang*), claim that a failure to conceive may be related to improper sacrificial offerings to graves. Chao Yuanfang and Lu Zhaolin (ed.), *Zhubing yuanhou lun* (c. 610 CE rpt.; Shenyang: Liaoning kexue jishu chubanshe, 1997), 185–86. For more on the subject of women's health and religion, see Wu, *Reproducing Women*, 79–81.

116. *Dili yuanxu*, 6.1a. I obtained *Preface to the Origins of Earthly Principles* during my fieldwork in Sichuan. The text is a handwritten manuscript ascribed to the geomantic lineage of the great Tang master, Yang Yunsong (c. 834–900). Possible dating of the text ranges from the late Qing into the early Republic. An argument for a late Qing date of composition relates to the fact that the manual is replete with references to the civil and military examinations. The manuscript does not identify a place of composition, but it was owned by a family in Langzhong, Sichuan through the twentieth century.

117. Nanbu County Qing Archive: (GX16.10.13/1890) 10.890.03.

118. Cai Dongzhou and Zhang Liang, "Nanbu dang'an zhong youguan Songdai Langzhou Chenshi jiazu mu dang'an yanjiu," 103.

119. Long Xianzhao and Huang Haide, eds., *Ba-shu daojiao beiwen jicheng* (Chengdu: Sichuan daxue chubanshe, 1997), 328; 336–37; 351.

120. Nanbu County Qing Archive: (XF7.9.5/1857) 5.92.05

121. For the 1854 official order protecting fengshui, see Nanbu County Qing Archive: (XF4.5.18/1854): 5.92.07.

122. Nanbu County Qing Archive: (XF7/1857) 5.92.03.

123. See the follow county order from the nineteenth century: "In places where two counties meet, residents usually create a tax identity with both counties. Whenever a mishap occurs, a plaintiff has already submitted a lawsuit in this county, while the defendant will run to the neighboring county to present a separate accusation." Cited from Wu, *Qingdai xianyu minshi jiufen yu falü zhixu kaocha*, 286.

124. Nanbu County Qing Archive: (XF7.9.24/1857) 5.92.07.

125. SUCSBRC: (Nanbu Collection; MG31/1942) 462.681.81. The report read: "Gaoguan Temple of Nanbu County: this place juts into Langzhong County, it should be drawn into the jurisdiction of Langzhong."

126. This insight provides a new angle with which to consider Matthew Mosca's astute observation that "before the late Qing, foreign geography was studied almost entirely through word rather than image." See Mosca, *From Frontier Policy to Foreign Policy*, 26.

127. Henderson. *The Development and Decline of Chinese Cosmology*.

128. Yee, "Traditional Chinese Cartography and the Myth of Westernization," 170–202; Smith, *Mapping China and Managing the World*; Pegg, *Cartographic Traditions in East Asian Maps*.

129. For further discussion of popular fortune-telling books of the Qing, see Suleski, *Daily Life for the Common People of China, 1850–1950*, 173–98.

130. Bian, *Know Your Remedies*, 120.

131. Hanson, "The Golden Mirror in the Imperial Court of the Qianlong Emperor, 1739–1742," 112.

Chapter 3: Examining Fortune

1. *Qianlong Suining xian zhi* (1747) 11: 26.

2. This summary simplifies. For an elaborate chart of all civil service examination levels and degrees in the Qing, see Elman, *A Cultural History of Civil Examinations in Late Imperial China*, 659. For other useful works on the examination system, see Miyazaki, *China's Examination Hell*, and Elman, *Civil Examinations and Meritocracy in Late Imperial China*.

3. Elman, *A Cultural History of Civil Examinations in Late Imperial China*, 311.

4. See, for instance, Qintianjian louke ke, ed., *Qintianjian Dili xingshi qieyao bianlun*, 12–14; 65. The Astronomical Bureau looked to powerful geomantic case studies from the southeast (e.g., "the Maos of Qiantang... the Lins of Fujian") for bolstering the imperial government's own knowledge of fengshui. Ibid., 13.

5. For discourses and practices surrounding examination success, see Burton-Rose, "Terrestrial Rewards as Divine Recompense." For the cult of Wenchang, see Kleeman, *A God's Own Tale*. For limits to the discourse of "talent" in Qing China, see Zhang, "The Legacy of Success," 259–97.

6. Schneewind, *A Tale of Two Melons*, 55.

7. For more on this theme, see Hymes, "Truth, Falsity, and Pretense in Song China," 1–26.

8. Richard J. Smith has observed, "We should remember that the vast majority of individuals who practiced the medical arts in China were low-status individuals, 'artisans' who were viewed by Chinese society as mere technicians. To the degree that they were well educated, they might enjoy considerable status, but their occupation itself was not socially esteemed." See Smith, *The Qing Dynasty and Traditional Chinese Culture*, 364

9. Maurice Freedman made a pertinent observation: "Underlying fengshui is a fundamental notion in and about Chinese society: all men (that is, all Chinese fully accepted within society, not boat people or other marginal elements) are in principle equal and may legitimately strive to improve their station in life. The peasant in his cottage has as much right to hope for advancement as the mandarin in his yamen." See Freedman, *Chinese Lineage and Society*, 125.

10. Wenchang, the god of the examinations, had been worshipped for centuries, but was only formally promoted into the imperial state cult in 1801, during the White Lotus War. *Tongzhi zengxiu Nanbu xian zhi* (1870), 164.

11. Yu Li, "Social Change During the Ming-Qing Transition and the Decline of Sichuan Classical Learning in the Early Qing," *Late Imperial China* 19.1 (1998): 26–55.

12. Huang, *Reshaping the Frontier Landscape*, 132–51.

13. See, for instance, the essays in Faure and Ho Ts'ui-p'ing, eds., *Chieftains into Ancestors*.

14. Even more numerous were *jiansheng* ("Titular Imperial Academy Student") who had purchased degrees to obtain baseline gentry status.

15. Bird, *The Yangtze Valley and Beyond*, 274.

16. Elman, *Civil Examinations and Meritocracy in Late Imperial China*, 318.

17. Ibid., 318.

18. Li Runqiang, "Qingdai jinshi de shikong fenbu yanjiu," *Xibei shida xuebao* 1 (2005): 65.

19. Li Chaozheng, ed., *Qingdai Sichuan jinshi zhenglüe* (Chengdu: Sichuan daxue chubanshe, 1986), 273. Li Runqiang, "Qingdai jinshi de shikong fenbu yanjiu": 66.

20. Zhou Xun, *Shuhai congtan*, 115–16.

21. Esherick, *The Origins of the Boxer Uprising*, 28–37; Pomeranz, *The Making of a Hinterland*.

22. Wang Daolü, *Nanbu xian xiangtu zhi* (1906), 384.

23. Nanbu's gazetteer lists thirteen provincial degrees awarded between the examination years of 1800 and 1867, with a success rate of one provincial degree holder from the county every two exam cycles. *Tongzhi zengxiu Nanbu xian zhi* (1870), 245–46.

24. Consider by comparison Tongxiang County, which produced seventy provincial degree holders between 1796 and 1889. See Shih, *Chinese Rural Society in Transition*, 86.

25. Liang Yong, *Yimin, guojia yu difang quanshi*, 177.

26. *Tongzhi zengxiu Nanbu xian zhi* (1870), 235. Provincial sources, like the one cited below, list fifty-seven provincial graduates from Nanbu over the dynasty, while the county's gazetteer generously counts eighty-four through 1870. Li Chaozheng provides four *jinshi* graduates for Nanbu and twenty-one for Langzhong. Li Chaozheng, ed., *Qingdai Sichuan jinshi zhenglüe*, 276. *Tongzhi zengxiu Nanbu xian zhi* (1870), 223–46.

27. Esherick, *The Origins of the Boxer Uprising*, 1–38.

28. Brown, "A Mountain of Saints and Sages," 437–91.

29. Some late imperial geographies of foreign lands, including of Western countries, contained descriptions of their fengshui. See Hostetler, *Qing Colonial Enterprise*, 58.

30. *Xianfeng Langzhong xian zhi* (1851), 3:34.

31. Yu Zhengui and Lei Xiaojing, eds., *Zhongguo Huizu jinshilu* (Yinchuan: Ningxia renmin chubanshe, 2001), 491–92.

32. Liu Xiancheng, *Langyuan Bianlian jijin* (Yinchuan: Ningxia renmin chubanshe, 2010), 216.

33. The Khoja's shrine has more official dedications than any other structure in the area. See Liu, *Langyuan: Bianlian jijin*, 187–230. Liu Xiancheng observes: "The place in Langyuan [Langzhong] with the richest collection of inscribed wooden boards is the famous Pavilion of Lingering Illumination of the Shrine of the Wiseman (the Khoja's shrine)." Ibid., 3.

34. Jin Tianzhu and Hai Zhengzhong, eds., *Qingzhen shiyi* (Yinchuan: Ningxia renmin chubanshe, 2002), 118–20; Murata, *The First Islamic Classic in Chinese*, 220.

35. *Guangxu Ninghe xian zhi* (1880), 3: 5. In another instance in Huili Department of southern Sichuan, a mosque's property was confiscated during a mid-century rebellion. *Tongzhi Huili zhou zhi* (1874) *xuzhi* section, *xia*: 7–8.

36. Gladney, *Muslim Chinese*, 162.

37. For discussions of these conflicts, see Lipman, *Familiar Strangers*; Atwill, *The Chinese Sultanate*.

38. Baoning's main mosque, for instance, was created in 1669 on the grounds of the old Ming-era Circuit Yamen. Langzhongshi lishi wenhua mingcheng yanjiuhui, ed., *Mingcheng yanjiu*, vol. 14. (Langzhong: Gazetteer Office of Langzhong Municipality, 2012), 80.

39. Brown, "The Muslims of 'All Under Heaven,'" 79–106.

40. Yu Zhengui and Lei Xiaojing, eds., *Zhongguo Huizu jinshilu*, 304–6.

41. For more information about the importance of imperial calendars for determining auspicious days and times, see Smith, "A Note on Qing Dynasty Calendars," 123–45.

42. See the following record from 1850 concerning the ownership of a mountain by Yi (Lolo) people in Mianning County: "We again received favor by a court's ruling that the 'stolen jade' [the mountain] be returned to its former owner, with its branches and leaves forbidden from disturbance so that it may grow and prosper to cultivate fengshui. If Han commoners from near or far attempt to steal its trees, we will present a lawsuit at the county court according to the facts." Liangshan Yizu zizhizhou bowuguan and Liangshan Yizu zizhizhou wenwu guanlisuo, eds., *Liangshan lishi beike zhuping* (Beijing: Wenwu chubanshe, 2011), 146. For another dispute involving the invocation of fengshui by non-Han peoples in Sichuan, see Wang Dingzhu, *Shusong pi'an, li*: 19–21.

43. National Palace Museum Palace Memorial and Grand Council Archive: (JQ20.2.11/1815) 404017816.

44. Will and Wong, *Nourish the People*, 84; Dai, *The White Lotus War*, 25; Entenmann, "Migration and Settlement in Sichuan, 1644–1796," 211–42.

45. First Historical Archives of China: (JQ21/1816) 04-01-01-0563-007. Note that this document is not a palace memorial but rather a descriptive survey of the region.

46. "Nourishing honesty silver" (*yanglian yin*) was a supplementary allowance paid to government officials in the Qing in addition to their official salary. It often amounted to ten or more times the official's regular compensation. See Will, "Officials and Money in Late Imperial China," esp. 30–37. See also Zelin, *The Magistrate's Tael*, 100.

47. *Daoguang Baoning fu zhi* (1821) 11:2.

48. Celebrations of imperial officials by an area's residents often held strategic political messaging to future officials. See Schneewind, *Shrines to Living Men in the Ming Political Cosmos*.

49. *Tongzhi zengxiu Nanbu xian zhi* (1870), 389–390.

50. Prior to the late Qing campaign to build new-style schools in rural temples, there was one other effort, in 1869, to establish charity schools in Nanbu, though it appears that the effort focused on shoring up the schools that had been established thirty years earlier. *Tongzhi zengxiu Nanbu xian zhi* (1870), 180.

51. Nanbu County Qing Archive: (GX22.10.8/1896) 13.402.02. See also Woodside, "Some Mid-Qing Theorists of Popular Schools," 3–35.

52. Kleeman, *A God's Own Tale*, 46

53. Burton-Rose, "Wenchang Buildings in Late Imperial China."

54. The character *ao* denotes an ancient marine turtle from Chinese mythology. The term *aofeng* was one of the titles of the Hanlin Academy; many academies were named with the character.

55. *Tongzhi zengxiu Nanbu xian zhi* (1870), 215–17, 429–30, and 450–51.

56. Gao Jiannan, *Xiangzhai jingzuan* (Fuzhou: Weigen caotang, 1844), juan 2.

57. See, for instance, Long and Huang, eds., *Ba-shu daojiao beiwen jicheng*, 354, 357–58, 426–27, 472–73, 474–75, and 490–91. For a discussion of fengshui pagoda construction in Fujian, see Dean, *Lord of the Three in One*, 87.

58. Jiang Haitao, *Qingdai Sichuan diqu fengshui ta shikong fenbu yanjiu*, 34–38.

59. For more reading on the transformation of the southwestern borderlands in relation to "Confucian" ritual, see Giersch, *Asian Borderlands*, 190–92.

60. Little made this observation while traveling through Badong in the Upper Yangzi. See Little, *Through the Yang-tse Gorges*, 130.

61. A magistrate selected the orientation of a Kui Star Pavilion in Cangxi County during the late Qianlong reign. *Qianlong Cangxi xian* (1783), 3: 233.

62. *Minguo Mianyang xian zhi* (1932), 1 *guji* section: 17.

63. Nanbu County Qing Archive: (GX29.10.1/1903) 16.448.01. The petition begins, "We humbly submit a budget for constructing a tower and soliciting donations for the fostering of the county's 'literary wind' [i.e., fengshui]."

64. See, for instance, the following two cases concerning funds for a tower's repair. Nanbu County Qing Archive: (GX19.8.30/1893) 11.688.07 and (GX19.12.18/1894): 11.912.03.

65. Zhu Qingqi and You Shaohua, *Xing'an huilan quanbian*, vol. 7, 50: 2625–26.

66. For walled fortifications, see McMahon, "Geomancy and Walled Fortifications in Late Eighteenth Century China," 373–93.

67. *Daoguang Nanjiang xian zhi* (1827), *zhong juan*: 7

68. *Jiaqing Peng xian zhi* (1813), 10: 17.

69. Long Xianzhao, ed., *Ba-shu fojiao beiwen jicheng* (Chengdu: Ba-shu shushe, 2004), 732.

70. Rowe, *Saving the World*, 436–37.

71. Kieschnick, *The Impact of Buddhism on Chinese Material Culture*, 38–44. See also Long Xianzhao, *Ba-shu fojiao beiwen jicheng*, 626; 730.

72. Wenqu, Kui, and Wenchang were all invoked as deities and as astrological bodies. Wenqu was the ancient name for *Delta Ursae Majoris* within the Big Dipper. Wenqu was one of the four stars of *Kui*, which derived its name from two sources. *Kui* (written with the *da* radical) referred to one of the twenty-eight ancient lodges; *Kui* (written with the *gui* radical) by contrast referred to a grouping of four stars in *Ursa Major*, which was part of the Dipper lodge—not the Kui lodge. Over time, these two similar-sounding *kui* stars were conflated in the popular imagination. Wenchang referred to a grouping of six stars within *Ursa Major* that shined in a crescent formation above the Big Dipper in the night sky. By the Qing period, they were all worshiped for their associations with literary talent and examination success.

73. Li Wenfu and Li Yongqi, eds., *Langzhong guji* (Beijing: Zhongyang wenxian chubanshe, 2009), 24–25.

74. "[By improving fengshui], we will see excellent results on the examinations and carry forward the scholarly tradition of past ages. Members of the gentry and commoners from the town are enthusiastically donating for this venture." Ibid., 24.

75. For a discussion of fengshui and examinations in the Song, see Pan Sheng, *Zhishi, lisu yu zhengzhi*, 365–89.

76. Elman, "Political, Social, and Cultural Reproduction via Civil Service Examinations in Late Imperial China," 14.

77. Elman, *A Cultural History of Civil Examinations in Late Imperial China*, 287–93.

78. Ibid., 287–93.

79. Zhang, "Legacy of Success."

80. For one of many examples, consider the following record from 1851. When Magistrate Xu recommended moving the site of the temple school, he celebrated the money that had poured in with a statement saying, "From the Tang dynasty (618–907 CE) onward, with the establishment and rise of the examination system, geomancy became ablaze with popularity. It is said that the scholarly wind depends on the rising and falling of fengshui. Therefore, the temple school must be in an appropriate location." Li Wenfu and Li Yongqi, eds., *Langzhong guji*, 25–26. Readers should be aware that the examination system existed in Tang times as conveyed by this quote, but that the use of the examinations as the primary method for selecting officials is associated with the Song (960–1279 CE).

81. *Qinding Da Qing huidian* (1764), 55: 2.

82. Ibid.

83. Johnson, *Spectacle and Sacrifice*, 180–82; 136–37; Overmyer, *Local Religion in North China in the Twentieth Century*, 79–80; Jones, *In Search of the Folk Daoists in North China*, 47–49.

84. Nanbu County Qing Archive: (DG20.6.5/1840) 4.216.01.

85. Liu, "The World of Rituals," 103–8.

86. For reference to local Yin-yang officers in the *Collected Statutes of the Great Ming*, see *Da Ming huidian* (1585) 4: 5. For reference to the promotion and examination of local Yin-yang officers to the Ming Astronomical Bureau, see *Da Ming huidian* (1585) 104: 2.

87. Ba County Qing Archive: (DG21/1841) 6.07.00311.

88. *Da Qing huidian* (1764) 55: 2.

89. *Jiaqing An xian zhi* (1812), 24: 25.

90. Ba County Qing Archive: (DG21/1841-XF8/1858) 6.07.00311.

91. Wang Daoheng and Wang Fangzhi, *Luojing toujie* (Taiyuan: Sihetang, 1824).

92. These understandings of solar and lunar eclipses can be traced back to the writings of Dong Zhongshu (179–104 BCE). Note that in Yin-yang cosmology, the *yin* force comprises "punishments," hence the association of lunar eclipses with the legal order. Loewe, *Dong Zhongshu, A "Confucian" Heritage and the Chunqiu Fanlu*, 90.

93. Chang, "Chinese Hereditary Mathematician Families of the Astronomical Bureau, 1620–1850," 145. See also the author's newly published book, Chang, *The Chinese Astronomical Bureau, 1620–1850*.

94. Lingfeng, "Eclipses and the Victory of European Astronomy in China," 127–45. For much of the dynasty's history, officials at the Astronomical Bureau simply copied calculated predictions for eclipse observations, thus concealing inaccuracies. This laxity derived from success: "[A]fter the Western method had built up its credibility through repeated success in observational tests, emperors and officials of the Qing dynasty became so trusting of it that they thought it would not make any essential difference to take 'observational data' either from the heavens itself or only from predictions." Ibid., 139–40.

95. Chang, "Chinese Hereditary Mathematician Families of the Astronomical Bureau, 1620–1850," 25–26.

96. See for instance, Nanbu County Qing Archive: (GX21.11/1895) 13.393.01.

97. Nanbu County Qing Archive: (GX13.11.9/1888) 9.867.04.

98. Nanbu County Qing Archive: (GX13.11.9/1888) 9.867.04. Bibliographic information about this specific lunar eclipse is found in the following article Walravens, "Vorhersagen von Sonnen- und Mondfinsternissen in Mandjurischer und Chinesischer Sprache," 446.

99. For a description of an eclipse ritual at a yamen, see Doolittle, *Social Life of the Chinese*, 247–50. Doolittle's description of the ceremony at Fuzhou, Fujian Province, bears similarities to the written descriptions available in Nanbu's archive.

100. Pankenier, "Characteristics of Field Allocation (*fenye*) Astrology in Early China," 499–513.

101. Elman observes that the first major attacks on the *fenye* system as "outdated" were in the seventeenth century. See Elman, *On Their Own Terms*, 195–96.

102. Qiu Jingjia, *Tiandi zhijian: Tianwen fenye de lishixue yanjiu* (Beijing: Zhonghua shuju, 2020), 232–33.

103. *Tongzhi Chengdu xian zhi* (1873), tianwen zhi: 6–7. The specific line concerns the positions of the Triaster and Turtle Beak lodges, which was a fierce topic of astronomical debate in the early Qing. See Huang Yinong, *Shehui tianwenxue shi shijiang* (Shanghai: Fudan daxue chubanshe, 2004), 239–268. See also Jami, *The Emperor's New Mathematics*, 121–25.

104. Wu, "Guns, Maize, and Europeans," forthcoming.

105. For a discussion of historical changes to Chinese astrology, see Kotyk, "Buddhist Astrology and Astral Magic in the Tang Dynasty," 75–76.

106. There are three kinds of geomantic compasses: *sanyuan*, *sanhe*, and *zonghe*, the latter of which combines elements from the first two. The compass under discussion in *A Thorough Explanation of the Geomantic Compass* is the *sanhe* compass, which is divided into three plates for heaven, earth, and people. The problématique posed by the Western calendric method was how to arrange the newly calculated values of the twenty-eight lodges against the twenty-four earthly directions of the compass.

107. For a discussion of precession in Qing astronomy, see Jami, *The Emperor's New Mathematics*, 123–24.

108. Wang Daoheng and Wang Fangzhi, *Luojing toujie* (Taiyuan: Sihetang, 1824), xia juan: 38. This quotation is ascribed to Zhang Jiuyi's late seventeenth-century geomantic manual, *Dili qian danzi sha shui yaojue* (Preface 1681).

109. This exact passage was cited in other geomantic manuals as well. See, for instance, Gūwalgiya Lianrui, *Qingnang dili jiyao*, xia juan (Beijing: Liulichang Wenbaotang, 1907), *Shafa yaojue*, chapter 6. Note that this geomantic manual was composed by a Manchu in the late Qing, suggesting that some bannerman had become well read on fengshui by that time.

110. Yuan Shushan, *Zhongguo lidai buren zhuan* (Shanghai: Runde shuju, 1948), 19:38.

111. Zeng Xiaomei and Wu Mingran, eds., *Qiangzu shike wenxian jicheng*, vol. 3 (Chengdu: Ba-shu shushe, 2017), 1341–43. The annotations on this transcribed version of the inscription mistakenly place these events in the early Ming. In this Qing-era inscription, *gengwu* (KX29) refers to 1690, *guiyou* (KX32) to 1693, and *bingzi* (KX35) to 1696. *Guiyou* and *bingzi* were provincial examination years.

112. Nanbu County Qing Archive: (XF8.5.13/1858) 5.201.05 and (TZ3.4.24/1864) 6.533.01. According to *Imperially Endorsed Collected Statutes and Regulations of the Great Qing*, from 1645 onward, one medical officer (*yishi*) could be summoned for the provincial examinations. As Sichuan employed Yin-yang officers as well as medical personnel during the exams, there were

variations between provinces. *Qinding Da Qing huidian zeli* (1764), 66: 22. Note that these page numbers correspond to the 1764 edition held in Harvard University's Chinese Rare Books Collection and not to the *Siku quanshu* reprinted edition.

113. Miyazaki, *China's Examination Hell*, 46–47.

114. Nanbu County Qing Archive: (XF9.4.26/1859) 5.201.02.

115. Naquin, "The Transmission of White Lotus Sectarianism in Late Imperial China," 255–91. See also the discussion in Ter Haar, *The White Lotus Teachings in Chinese Religious History*, 109–10.

116. Nanbu County Qing Archive: (DG20.6.5/1840) 4.216.01, (GX4.1.16/1878) 7.644.01, (GX14.7.16/1888) 10.182.01, and (GX15.6.23/1889) 10.481.01. For more on the complex relationship between Daoists and spirit mediums, see Davis, *Society and the Supernatural in Song China*.

117. Graham, *Folk Religion in Southwest China*, 112.

118. Jingyan County counted around two hundred local geomancers that its Yin-yang officer oversaw in 1900. *Guangxu Jingyan zhi* (1900), 2.2:18.

119. In 1841, Nanbu's Yin-yang officer was instructed to coordinate with local elites to arrest a French Catholic missionary, "Mu Daoyuan" (François Alexis Raméaux, 1802–45). Nanbu County Qing Archive: (DG21.2.19/1841) 4.217.01.

120. Spirit mediums were reported to authorities for spreading heresy during the nineteenth century. See, for instance, Nanbu County Qing Archive: (GX5.3*.30/1879) 7.810.01.

121. The Yin-yang officer "Master Xu" calculates the time of death of Ximen Qing's wife in *The Plum in the Golden Vase*. Bai Weiguo and Bu Jian, annot., *Jin Ping Mei cihua jiaozhu* (Changsha: Yuelu shushe, 1995), 1746–48.

122. Asen, *Death in Beijing*, 38–39.

123. This observation extends to Yin-yang officers' roles in selecting the sites of charitable cemeteries. *Guangxu Zhecheng xian zhi* (1896) 6: 22.

124. Ba County Qing Archive: (XT2/1910) 6.54.00660.

125. Bai Weiguo and Bu Jian, annot., *Jin Ping Mei cihua jiaozhu*, 305

126. The tradition of Buddhist monks performing geomantic services can be traced back to the Song. Pan Sheng, *Zhishi, lisu yu zhengzhi*, 276–80.

127. *Illustrated Earthly Principles for Guiding the Original Truth* was likely modeled on earlier works such as *Illustrated Survey of Optimi during the Ming Dynasty* (*Ming zhuangyuan tukao*; 1607), which recorded the "eight characters," fate calculations, and astrological portents behind the examinations successes of Ming first-ranked palace scholars (*zhuangyuan* or "optimi"), the highest ranked candidate for an examination year's *jinshi* class. Elman, *Civil Examinations and Meritocracy in Late Imperial China*, 180.

128. Kong Wenxing, *Huitu dili zhizhi yuanzhen*, 1: shangyuan 5a.

129. Kong Wenxing, *Huitu dili zhizhi yuanzhen*, 1: mulu 1b. *Kangxi Zhejiang tong zhi* (1684) 50: 23. Note that in the gazetteer, Kong is referred to by his Buddhist name, Che Ying. Note also that the late Qing reprinting of Kong's manual—the version consulted here—added "illustrated" (*huitu*) to the text's title.

130. *Dili yuanxu*, 29b.

131. For a critique of cultural standardization, see Szonyi, "The Illusion of Standardizing the Gods," 113–35.

132. For a discussion of banned books in the Qing, see Guy, *The Emperor's Four Treasuries*, 157–96. For a discussion of the geomantic texts included in the Four Treasuries, see Li Dingxin, *Siku quanshu: kanyulei dianji yanjiu* (Shanghai: Shanghai guji chubanshe, 2007).

133. For the image of the prefecture school, see Xiong Dian (Zaofu sanren), Xiong Weiyao (Preface), and Yuan Shizhen (Epilogue), *Buju misui* (Nanjing: Jinshantang, 1595), *xia juan*: 13.

134. Li, "The Evolution of the *zhaobi*," 53.

135. *Daoguang Yunmeng xian zhi lüe* (1840), *juan shou*: 1–2.

136. *Qianlong Hezhou zhi* (1748), *juan shou*: 3.

137. *Tongzhi zengxiu Nanbu xian zhi* (1870), 427–28.

138. Eckfeld, *Imperial Tombs in Tang China, 618–907*, 101–5.

139. For more Qing-era examples, see Steinhardt, *Chinese Architecture*, 234.

140. Presumably not all elements of the school were included in the image; for instance, Anqing's school almost surely had a *pan* Pond and bridge in the first courtyard by the Qing period if not earlier. See Shryock, *The Temples of Anking and Their Cults*, 58.

141. Great Perfection Palaces (*dachengdian*), also called Central Palaces (*zhengdian*), were named after an imperially bestowed title of Confucius and marked the central place in the complex where he was worshipped. See Huang, *Confucianism and Sacred Space*, 45.

142. In Baoning, for instance, references to five-phase layouts are explicit in Ming-era school and temple inscriptions but largely absent in Qing ones. See, for instance, the Ming-era construction of a "water" agent Hall of the Perfected Warrior to quell an excess of "fire" *qi* in Baoning's prefectural town. Long and Huang, eds., *Ba-shu daojiao beiwen jicheng*, 216.

143. The subject of legal privileges for the gentry is discussed at length for a related case in chapter 5.

144. The sesame oil presses of early twentieth-century Qingdao occupied one of the most economically precarious positions of all craftsmen in the city. See Moll-Murata, *State and Crafts in the Qing Dynasty (1644–1911)*, 282.

145. *Minguo Sichuan xin dizhi* (1946), 80–149.

146. Bu Yingtian and Xie Zhidao (ed.), *Chongding jiaozheng kui ban jujie xiaosha jing jietu dili jueyao xuexin fu*," in *Yuwai hanji zhenben wenku, di 5 ji*, vol. 19 (Chongqing: Xihua shifan daxue chubanshe, 2015), 52. I thank Ye Hua of Fudan University for directing me to this source.

147. Rowe, *Hankow*, 161–62.

148. Zhou Nan and Lü Lin, *Anju jinjing*, 1:4.

149. Cited from Zhu Cishou, *Zhongguo gongye jishu shi* (Chongqing: Xinhua shudian jingxiao, 1995), 1453.

150. Rowe, *Hankow*, 160.

151. *Qingdai Qian Jia Dao Ba xian dang'an xuanbian*, vol. 1, 335.

152. Ba County Qing Archive: (DG16/1836) 6.08.03482. He was ordered punished for presenting a false accusation. See also Wei Shunguang, "Qingdai zhongqi fenchan zhengsong wenti yanjiu," 190.

153. For a case involving a dispute over an oil press in Nanbu, see Nanbu County Qing Archive: (GX4.4.21/1878) 7.526.01.

154. Fan Zengxiang, *Fanshan pipan*, 147.

155. *Qingdai Qian Jia Dao Ba xian dang'an xuanbian*, vol. 1, 336.

156. Ibid., 336.

157. Cao Xueqin and Gao E, *Honglou meng*, 97.

158. Knowledge of the province's astrologically-affiliated lodges remained prevalent among the scholarly class through the end of the imperial era. At the opening of a 1913 article that was published as a book two years later, James Hutson Edgar (1872–1936) remarked, "I have heard it said that the Province of Szechwan was controlled astrologically, as to its intellectuals, by the demon and well stars," with demon here referring to the Ghost (*gui*) lodge. Hutson was correct. The continued salience of this astrological knowledge over time related to the fact that the province's celestial governance by select heavenly bodies provided a forum in which local elites could engage and petition provincial authorities, hence Magistrate Shen's invocation of them during the legal conflict over flour-sieving benches. Hutson, *Mythical and Practical in Szechwan*, 1–2.

159. *Guangxu Qingfu xian zhi* (1876), 49: 383.

160. Zhu Cishou, *Zhongguo gongye jishu shi*, 1453.

161. For one such example, see Wu Jianxin and Zhong Haiyan, "Ming-Qing Guangdongren de fengshuiguan: Difang liyi yu shehui jiufen," *Xueshu yanjiu* 2 (2007): 98–106.

162. Elvin, *The Retreat of the Elephants*, 187–90; 199.

163. People cited the appearance of banditry as legal evidence for harmed fengshui. For instance: "Calamities have been piling up in our community: not only have we suffered from hot diseases, the disasters of bandits, but also the inexplicable harming of life." SUCSBRC: (Nanbu Collection, MG16.12.1/1927) 466.3079a. Note that in that case, the county court ordered the cemetery in contention mapped. For the map, see 466.30791.51 in the same case file.

164. Wang Di, *Kuachu fengbi de shijie: Changjiang shangyou quyu shehui yanjiu, 1644–1911* (Beijing: Zhonghua shuju, 2001), 35

165. Yuan Yongbin, *Nanbu xian yutu kao* (1853; 1896 Rpt.), in Yao Leye, ed., *Sichuan daxue tushuguan guancang zhenxi Sichuan difangzhi congkan*, vol. 3 (Chengdu: Ba-shu shushe, 2009), 255.

166. Jingmenshi shuili zhi bianzuan weiyuanhui, ed., *Jingmenshi shuili zhi* (Wuhan: Hubei Jiaoyu chubanshe, 1989), 282. For Sichuan-based references to "Connecting Vein" (*jiemai qiao*) bridges, see *Daoguang Chongqing fu zhi* (1843), 1: 17 and *Guangxu Xuzhou fu zhi* (1895), 13:4.

167. This comment simplifies, as the law code mandated that official inspect the conditions of existing roads and bridges. *Da Qing lüli* (1740 Rpt.; 1998) juan 39, *gonglü, hefang, xiuli qiaoliang daolu*: 618 (Statute 436).

168. *Qianlong Danling xian zhi* (1761), 2: 35.

169. *Qingdai Qian Jia Dao Ba xian dang'an xuanbian*, vol. 1, 336.

170. *Jiaqing Shifang xian zhi* (1813), 12: 4–5.

171. See the description of the "Dragon Bridge" (*longqiao*) in Tongliang County, Chongqing Prefecture. *Guangxu Tongliang xian zhi* (1875), 2: 23.

172. For fengshui-related litigation in Longchang County involving the damming of a waterway around Returning Dragon Mountain, see *Qianlong Longchang xian zhi* (1764) 1:3. At the culmination of litigation, a magistrate ordered the waterway's original flow restored.

173. "For five *li* around the temple in all directions, it is not permissible to construct raised beam bridges." Shi Hansheng, ed. *Shishi zongpu* (Guangyuan: Shishi qingminghui: 2005), 19. Bandits also stole the components of wooden bridges. See Knapp, *Chinese Bridges*, 241. Note that shrines and temples were often linked to bridges Ibid., 71.

174. *Qianlong Rong xian zhi* (1756): 38: 5.

175. For Zhang's petitions, see Nanbu County Qing Archive: (GX3.6.9/1877) 7.401.18 and (GX16.2.25/1890) 10.657.01.

176. The Qing state formulated sumptuary laws for tombstone dimensions. Officials of the first rank were permitted tombstones of sixteen *chi* (512 cm, 16.8 ft.) upon their deaths, while commoners could only erect tombstones of up to four *chi* (128 cm, 4.2 ft.). Li Zhe, *Zhongguo chuantong shehui fenshan de falü kaocha: Yi Qingdai wei zhongxin* (Beijing: Zhongguo zhengfa daxue chubanshe, 2017), 35.

177. Nanbu County Qing Archive: (GX16.3.6/1890) 10.657.05.

178. Wang Daolü, *Nanbu xian xiangtu zhi* (1906), 403.

179. Nanbu County Qing Archive: (GX16.2*.17/1890) 10.657.04.

180. Some commoners did present plaints that touched on the imperial examinations as they related to fengshui. For one such case, see Nanbu County Qing Archive: (GX3.10.13/1877) 7.316.02.

181. Wesley-Smith, "Identity, Land, Feng Shui and the Law in Traditional Hong Kong," 238.

182. Wheatley, *The Pivot of the Four Quarters*.

183. Wright, "The Cosmology of the Chinese City," 33–73.

184. Meyer, "'Fengshui' of the Chinese City," 138–55.

185. Loewe, "The Pivot of the Four Quarters," 290.

186. Schneewind, *Shrines to Living Men in the Ming Political Cosmos*.

187. Ibid., 270.

188. Yang, *Religion in Chinese Society*, 263.

189. Verellen, "'Evidential Miracles in Support of Taoism,'" 217–63.

Chapter 4: Mining Sichuan

1. The edict reads, in part, "For areas with rich mineral outcrops, direct a capable official to closely inspect them, making detailed notes of all local circumstances, and then memorialize the throne with proposals. Do not fall back into the old habits of letting unworthy officials and clerks place a perfunctory ban on mining and then secretly engage in schemes. If you respond to this edict with 'there is an obstruction to fengshui' or 'a brawl has ensued,' then you are not doing your job conscientiously." Lu Zijian, *Qingdai Sichuan caizheng shiliao*, vol. 2 (Chengdu: Sichuansheng shehui kexueyuan chubanshe, 1988), 372.

2. Zelin, *The Merchants of Zigong*.

3. Zelin, "Eastern Sichuan Coal Mines in the Late Qing," 105–22; Jin, "Mint Metal Mining and Minting in Sichuan, 1700–1900."

4. Zelin, *The Merchants of Zigong*, xv.

5. See, for instance, many references to the province in Xia Xiangrong, Li Zhongjun, and Wang Genyuan, *Zhongguo gudai kuangye kaifashi* (Beijing: Dizhi chubanshe, 1980).

6. Golas, *Science and Civilisation in China*, vol. 5, Part 13, 411.

7. See the "four harms" of mining as identified by Shandong native Lu Jianzeng (1690–1768) during his tenure as magistrate in Hongya County, Sichuan. According to Lu, the "four harms" included higher grain prices, the loss of land and livelihoods, rising crime, and conflicts between mining operators and locals of the sort that broke out in the late Ming. Lu Zijian, *Qingdai Sichuan caizheng shiliao*, vol. 2, 370. Lu Zijian, *Qingdai Sichuan caizheng shiliao*, vol. 2, 370.

8. Bello, *Across Forest, Steppe, and Mountain*; Miller, *Fir and Empire*; Zhang, *Timber and Forestry in Qing China*.

9. Around 13 percent of the government's annual revenue came from salt in the mid-1800s. See Li, *The Making of the Modern Chinese State, 1600–1950*, 91.

10. Ibid., 91.

11. I drew this observation from Peter Golas's work. See Golas, *Science and Civilisation in China*, 425.

12. For more on Chen Hongmou's advocacy for mining, see Rowe, *Saving the World*, 243–45.

13. Jiang Huang, Wu Shengzu, and Wang Kaizhuo, eds., *Jiuyi shanzhi (liangzhong) Yanling zhi* (Changsha: Yuelu shushe, 2008), 126.

14. Governor Chen wrote that "the arriving dragon vein and node of the mausoleum of Yu Shun meanders amongst these mountains, which protect [the tomb] from each side; for these things, the evidence is clear." Ibid, 125–26.

15. Collins, *Mineral Enterprise in China*, 38–44.

16. Ibid., 41.

17. Elvin, "On Water Control and Management during the Ming and Ch'ing Periods," 405.

18. Golas, *Science and Civilisation in China*, 409. For a discussion of fengshui in relation to mining from the Ming period, see Dardess, "A Ming Landscape": 295–364.

19. He Wei, "Jindai kanyu tankuang lilun chutan," *Zhongguo kuangye daxue xuebao* 5 (2015): 69–73.

20. Chiu Pengsheng identifies a similar situation in Yunnan, where fengshui and public security were also identified by officials as concerns. Chiu Pengsheng, *Dang jingji yushang falü: Ming-Qing Zhongguo de shichang bianhua* (Taipei: Lianjing chuban shiye gufen youxian gongsi, 2018), 230.

21. For a discussion of silver mining, see Yang Yuda, "Silver Mines in Frontier Zones," 87–114. For a discussion of gold acquisition by the imperial court, see Su and Lai, "Resplendent Innovations," 157–86.

22. Chen, *Zinc for Coin and Brass*, 172.

23. *Da Qing lüli* (1740 Rpt.; 1998) *juan* 24, *xinglü, zeidao zhong, dao tianye gumai*: 397 (Statute 271.1-2).

24. *Da Qing lüli* (1740 Rpt.; 1998) *juan* 11, *hulü, cangku shang, qianfa*: 216–17 (Statute 118.1). For the history of the substatute on mining and graves, see Xue Yunsheng, *Duli cunyi* (Beijing: Hanmao zhai, 1905), *juan* 13, *hulü*: 2.

25. *Qinding Da Qing huidian zeli* (1764) 137: 44–45.

26. *Da Qing lüli* (1740 Rpt.; 1998) *juan* 9, *hulü, tianzhai, daomai tianzhai*: 197 (Statute 93.3).

27. Naquin, *Peking*, 11–12; see also Ibid., 258–65.

28. A substatute (Statute 393.1) was added to the law code in 1742 and revised in 1822 specifying that coal miners working in the Western Hills were required to register their names and places of origin with the local government. Any mining around the Western Hills that was not registered with the government in advance was to be punished, as were acts of gambling, the exploitation of (unpaid) mining labor, or failures to report a miner's illness or death. Xue Yunsheng, *Duli cunyi* (Beijing: Hanmao zhai, 1905), *juan* 47, *xinglü*: 24. Toward the end of his reign, in 1787, the Qianlong Emperor rebuffed officials' arguments to restrict registered mining in the Western Hills area because of fengshui. For more, see Golas, *Science and Civilisation in China*, 422.

29. Zhao Erxun, ed., *Qingshigao* (Beijing: Qingshiguan, 1928), *juan* 124, *zhi* 99 (*shihuo zhi*, No. 5): 15.

30. Chen, *Zinc for Coin and Brass*, 199, 204.

31. For other examples, see Zhongguo renmin daxue qingshi yanjiusuo, ed., *Qingdai de kuangye*, vol. 1 (Beijing: Zhonghua shuju, 1983), 331, 391, 412, 441–42, 452, 456–57, 461–63, 487, 489, 490. See also Chen, *Zinc for Coin and Brass*, 142.

32. Adapted translation and citation from Chen, *Zinc for Coin and Brass*, 142.

33. Guy, *Qing Governors and Their Provinces*, 348–49.

34. Chen, *Zinc for Coin and Brass*, 359–60.

35. National Palace Museum Palace Memorial and Grand Council Archive: (YZ3.2.25/1725) 402018348.

36. Zhongguo renmin daxue qingshi yanjiusuo, ed., *Qingdai de kuangye*, vol. 1, 8. For the relevant passage in the *Imperially Endorsed Collected Statutes and Precedents of the Great Qing*, see *Da Qing huidian shili* (1818) 716: 9.

37. Ibid., 15.

38. Skinner, "Sichuan's Population in the Nineteenth Century," 38–39.

39. Li, *The Making of the Modern Chinese State, 1600–1950*, 90.

40. For an example of one such memorial from Sichuan, see Lu Zijian, *Qingdai Sichuan caizheng shiliao*, vol. 2, 389.

41. National Palace Museum Palace Memorial and Grand Council Archive: (JQ22.05.11/1819) 053861.

42. Yunlu, ed., *Qinding xieji bianfang shu*. Composed in 1739 (Beijing: Wuyingdian, 1741), 20: 4.

43. Some kinds of extraction were strictly controlled, such as saltpeter mining. See Greatrex, "The Illegal Trade in Saltpetre in Southern China in the Eighteenth and Early Nineteenth Centuries," 349–78.

44. Zelin, *The Merchants of Zigong*, xiii; 18–21.

45. Wu Peilin and Deng Yong, "Qingdai Sichuan Nanbu xian jingyanye gailun: Yi 'Qingdai Nanbu xian yamen dang'an' wei zhongxin de kaocha," *Yanyeshi yanjiu* 1 (2008): 41.

46. The salt industry was the largest employer outside of agriculture in Sichuan. See Zelin, *The Merchants of Zigong*, 3; 116–22.

47. Zelin, *The Merchants of Zigong*, 15.

48. "The salt wells of Northern Sichuan are drilled and extend several hundred *zhang* into the ground. Geomancers point out the spots for drilling, and thousands of taels are contributed to undertake the task, with well diameters around three *chi* across. If one doesn't obtain brine from this investment, bankruptcy soon follows." Liu Xianting, *Guangyang zaji* (c. 1648–95) in *Jifu congshu* (v. 363), ed. Wang Hao (Dingzhou: Qiagdetang, 1879), 1: 58.

49. "The salt industry causes some to get very rich and some to become poor; it's all about luck. Thus, I heard that when opening wells, people rely upon the words of geomancers; I don't know what kind of art can be used to obtain such information [regarding the location of salt brine]." *Yongzheng Sichuan tong zhi* (1733) 45: 1. For a resonant story, see Zelin, *The Merchants of Zigong*, 14–15.

50. Vogel, "'That Which Soaks and Descends Becomes Salty,'" 469–515.

51. For the business career of Xiong Zuozhou, see Zelin, *The Merchants of Zigong*, 215–16. For Xiong's career in fengshui consulting, see Li Pingyi, "Zigong chuantong juluo jingguan zhong de fengshui wenhua jiexi," *Zhonghua wenhua luntan* 7 (2014): 122–23.

52. For one example among many of this usage in the salt business, see Nanbu County Qing Archive: (GX18.7.20/1892) 11.519.01.

53. Liu Yunsheng, *Zigong yanye qiyue yuhui jishi* (Beijing: Falü chubanshe, 2014),182–183.

54. Zigongshi dang'anguan, Beijing jingji xueyuan, and Sichuan daxue, ed., *Zigong yanye qiyue dang'an xuanji (1732–1949)* (Beijing: Zhongguo shehui kexue chubanshe, 1985), 1026–28.

55. Ibid., 259–60.

56. Hu Kaiquan and Su Donglai, eds., *Chengdu Longquanyi bainian qiyue wenshu*, 248.

57. Nanbu County Qing Archive: (GX13.5.20/1887) 9.658.01.

58. Liu Yunsheng, *Zigong yanye qiyue yuhui jishi*, 260. See also Zigongshi dang'anguan, Beijing jingji xueyuan, and Sichuan daxue, eds., *Zigong yanye qiyue dang'an xuanji*, 398, 481.

59. *Minguo Mianzhu xian zhi* (1920), 9: 10.

60. Tu Qinghong, "Lüe lun Qingdai Sichuan yanshang dui difang jiaoyu de zhichi," *Chengdu shifan daxue xuebao* 12 (2019): 7–12.

61. For a discussion of the geography of coal deposits in China, see Pomeranz, *The Great Divergence*, 59–68.

62. Von Richthofen, *Baron Richthofen's Letters, 1870–1872*, 122.

63. In the early twentieth century, surveyors estimated that Sichuan's coal reserves were the eighth largest among the Chinese provinces. Richthofen's high appraisal of the province's reserves was related to the fact that they stretched over such a wide area of eastern Sichuan—comparable in range to the Shanxi coal belt. Golas, *Science and Civilisation in China*, 187–88.

64. For coal in modern Chinese history, see Wu, *Empires of Coal*; Seow, *Carbon Technocracy*.

65. Zelin, "Eastern Sichuan Coal Mines in the Late Qing," 118.

66. One record from Guang'an County from the late Qing claimed that "hundreds" of people had been buried in collapsed coal mines dug deep into the county's mountains. Qi Shouhua and Zhong Xiaozhong, eds., *Zhongguo difang zhi meitan shiliao xuanji*, 428–29.

67. Golas, *Science and Civilisation in China*, 328–29.

68. Schäfer, *The Crafting of the 10,000 Things*, 170–73.

69. Smith, *Chinese Characteristics*, 260.

70. Qi Shouhua and Zhong Xiaozhong, eds., *Zhongguo difang zhi meitan shiliao xuanji*, 379, 410, 421.

71. Zelin, "Eastern Sichuan Coal Mines in the Late Qing," 106.

72. Qi Shouhua, "Chuanyan yu meitan," *Yanyeshi yanjiu* 3 (1988): 52.

73. Qi Shouhua and Zhong Xiaozhong, eds., *Zhongguo difang zhi meitan shiliao xuanji*, 410.

74. Von Richthofen, *Baron Richthofen's Letters, 1870–1872*, 116.

75. Du Guiling, who assumed office as magistrate of Nanbu in 1824, later served as magistrate of Jianzhou, also in Baoning Prefecture. There, the price of coal was said to be expensive due to transportation costs. Magistrate Du constructed a road, thus bringing the cost of coal down. Notably, Jianzhou faced high coal prices despite its location immediately next to Guangyuan

County, the biggest producer of coal in the prefecture. See *Daoguang Fugou xian zhi* (1833) 10: 23.

76. *Qingdai Qian Jia Dao Ba xian dang'an xuanbian*, vol. 1, 283.
77. Zelin, "Eastern Sichuan Coal Mines in the Late Qing," 120.
78. For items that came to be taxed under Sichuan's *lijin*, see Lu Zijian, *Qingdai Sichuan caizheng shiliao*, vol. 2, 629.
79. Ibid., 403. For a discussion of coal taxation in the Qing, see Wu, *Empires of Coal*, 23.
80. Zelin, *The Merchants of Zigong*, 158–59
81. Qi Shouhua and Zhong Xiaozhong, eds., *Zhongguo difang zhi meitan shiliao xuanji*, 402.
82. Ibid., 403.
83. Ba County Qing Archive: (TZ9/1870) 6.26.06018. I thank Gilbert Chen for sharing his transcription of this case file with me.
84. *Qingdai Qian Jia Dao Ba xian dang'an xuanbian*, vol. 1, 285. For the original case file, see Ba County Qing Archive: (DG8/1838) 6.15.17133.
85. *Qingdai Qian Jia Dao Ba xian dang'an xuanbian*, vol. 1, 280.
86. Zelin, "Eastern Sichuan Coal Mines in the Late Qing," 108–9.
87. *Qingdai Qian Jia Dao Ba xian dang'an xuanbian*, vol. 1, 258. See also Ibid., 259, 262, 263, 264, 266, 267, 270, 276.
88. Ibid., 258, 260, 264, 267, 268, 269, 270, 271, 276.
89. Ibid., 271.
90. *Qingdai Qian Jia Dao Ba xian dang'an xuanbian*, vol. 1, 282.
91. Ba County Qing Archive: (TZ9.1.24/1870) 6.25.05832. I thank Gilbert Chen for sharing his transcription of the original case file with me. For a discussion of this case, see Chen, "Living in This World," 160.
92. *Qingdai Qian Jia Dao Ba xian dang'an xuanbian*, vol. 1, 279, 287.
93. Ba County Qing Archive: (QL46/1781) 6.01.00787. This case, presented by members of the Zhou family against the monks of Dragon Chariot Temple, alleged that a previous magistrate had investigated and banned mining around their family's ancestral graves.
94. Qi Shouhua and Zhong Xiaozhong, eds., *Zhongguo difang zhi meitan shiliao xuanji*, 387.
95. Around 1800, the Qianwei salt yard was using "several hundred thousand *jin* of coal daily." One *jin* was equal to approximately 1.3 English pounds. Zhang Xuejun and Ran Guangrong, *Ming-Qing Sichuan jingyan shigao* (Chengdu: Sichuan renmin chubanshe, 1984), 71.
96. Qi Shouhua and Zhong Xiaozhong, eds., *Zhongguo difang zhi meitan shiliao xuanji*, 413–17. Due to its location on a flat plain, with few nearby coal mines, Chengdu remained dependent on firewood for its major fuel source through the early 1900s. See Di Wang, *The Teahouse*, 18–19.
97. Hua Guoying, ed., *Sichuan guanyun yan'an leibian* (Luzhou: Luzhou zongju, 1902), 27: 3.
98. Ibid., 61 :12.
99. Te-cheng Su and Hui-min Lai, "Resplendent Innovations," 157–86.
100. Golas, *Science and Civilisation in China*, 117.
101. *Tongzhi zengxiu Nanbu xian zhi* (1870), 123.
102. Gold placer deposits were also found in Ba County. One contract, dated 1841, tells of a site where gold panning was banned in the late 1700s. However, the local property-owning Dengs appealed to overturn the ban in 1820, when flooding destroyed any potential for

cultivating the surrounding fields. Once the ban was lifted, new disputes over gold mining boundaries ensued. *Qingdai Qian Jia Dao Ba xian dang'an xuanbian*, vol. 1, 317. For a discussion of similar dynamics in Jiangxi Province, see Hornibrook, *A Great Undertaking*, 51.

103. See chapter 5 for a discussion of the 1900s.

104. Nanbu County Qing Archive: (XF7.1.8/1857) 5.86.03.

105. Nanbu County Qing Archive: (XF7.10.1/1857) 5.86.05.

106. *Tongzhi zengxiu Nanbu xian zhi* (1870), 123.

107. For more on inkstones, see Ko, *The Social Life of Inkstones*, 60.

108. Zhong Haiyan, "'Kuangmai' yu 'longmai' zhi zheng: Qingdai guanyu Duanyan kaicai de 'fengshui' lunshuo," *Huanan nongye daxue xuebao* 4 (2007): 101–5.

109. Nanbu County Qing Archive: (GX16.2*.20/1890): 10.654.01, (GX16.2*.21/1890): 10.656.06, and (GX16.2*.22/1890) 10.662.

110. In 1889, a staff member at the local academy of Qijiang County (a "tutor," *zhaizhang*) presented a petition alleging that stones were being quarried across from the local Wenchang Palace, thus injuring the area's fengshui. The magistrate banned further quarrying. *Minguo Sichuan Qijiang xian xu zhi* (1938), 2:15.

111. Nanbu County Qing Archive: (GX2.5*.23/1876) 7.171.06.

112. Nanbu County Qing Archive: (GX19.2.13/1894) 11.576.04.

113. Two similar cases concerning stone quarrying and fengshui were presented by two different kinship groups within a day of each other in 1890. Nanbu County Qing Archive: (GX16.2*.20) 10.654. For the second accusation that "quarrying stones severed our earth vein," see Nanbu County Qing Archive: (GX16.2*.21) 10.656.

114. Nanbu County Qing Archive: (GX16.6.24/1890) 10.662.07.

115. For an overview of Li Benzhong's life and works, see Kim, *Mountain Rivers, Mountain Roads*, 225–69; Kim, "River Control, Merchant Philanthropy, and Environmental Change in Nineteenth-Century China," 660–94.

116. Ibid., 662.

117. Golas, *Science and Civilisation in China*, 300–6.

118. Xiaowei Zhang, *The Politics of Rights and the 1911 Revolution*, 118.

119. Feng Tianyu et al., *Wuhan shizhi: renwuzhi* (Wuhan: Wuhan daxue chubanshe, 1999), 205.

120. Hua Guoying, ed., *Sichuan guanyun yan'an leibian* (1902) 64:2.

121. For a sampling of the official notices invoking fengshui issued in the 1820s and 1830s, see Li Benzhong, *Pingtan jilüe* (Location Unknown: Qingliantang, 1840), 2:1 (DG3.11.30/1823), 2:4 (DG4.3.22/1824), 2:5 (DG4.4.1/1824), 2:5 (DG4.6.9/1824), 2:6 (DG4.5.13/1824), 2:7 (DG4.6.2/1824), 2:8 (DG4.10.15/1824), 2:9 (DG5.10.2/1825), 4.40 (DG5.5.17/1825), 4:46 (DG6.9.4/1826), 4:51 (DG9.10.18/1829), 4:56 (DG10.9.11/1830), 4:56 (DG11.4.16/1831), 4:57 (DG12.9.13/1832), 5:28 (DG11.3.29/1831), 5:30 (DG11.4.18/1831), 5:32 (DG11.12.20/1832), 5:35 (DG11.12.28/1832), and 5:39 (DG12.9.26/1832).

122. *Pingtan jilüe* 2:8.

123. Golas, *Science and Civilisation in China*, 405.

124. Ibid., 212–13.

125. Zelin, "Eastern Sichuan Coal Mines in the Late Qing," 107.

126. Qi Shouhua and Zhong Xiaozhong, eds., *Zhongguo difang zhi meitan shiliao xuanji*, 107; 316; 355; 418; 449; 501–2; 582.

127. Olles, *Ritual Words*.

128. Olles views the rise of the compiler's Liumen School as a scholarly and religious tradition that emerged in Sichuan during the nineteenth century. An offshoot of the Confucian Liumen movement was the Daoist *Fayan tan* ritual lineage, whose priests worked under the patronage of the Liumen patriarchs. See Olles, *Ritual Words*.

129. Hu Daojing, ed., *Zangwai daoshu* (Chengdu: Ba-shu shushe, 1992), 599.

130. Ibid., 598.

131. Recall the growing awareness among Qing elites of the "silver outflow problem" in the early 1800s. For more information, see Man-houng Lin, *China Upside Down*, 112–13.

132. Hu Daojing, ed., *Zangwai daoshu*, 717.

133. Lagerwey, *Paradigm Shifts in Early and Modern Chinese Religion*, 100–1; Raz, "Daoist Sacred Geography," 1404–5.

134. Chinese text from Miao Xiyong, ed., *Nanjie ershisi pian*, in Mao Jin, *Jindai mishu, di si ji* (c. 1630 Rpt.; Shanghai: Boguzhai, 1922), 6–7. This page number refers to the count beginning from the first page of *Nanjie ershisi pian* in volume four of *Jindai mishu*. For this English translation, I have slightly amended Michael Paton's excellent rendering of the text. For a complete translation of the passage, see Paton, *Five Classics of Fengshui*, 167–68.

135. Less than half of Qing governors had obtained the palace degree. Imperial appointment to a governorship was highly dependent on earlier performance. See Guy, *Qing Governors and Their Provinces*, 15.

136. Will and Wong, *Nourish the People*, 302.

137. Zhang Yanmei, "Qingdai Sichuan hanzai shikong fenbu yanjiu," 66–67.

138. Ibid., 46.

139. Von Richthofen, *Baron Richthofen's Letters*, 125–26.

140. First Historical Archives of China: (TZ11.1.22) 03-4656-118.

141. Kathryn Edgerton-Tarpley, *Tears from Iron: Cultural Responses to Famine in Nineteenth Century China*, (Berkeley: University of California Press, 2008), 76–77.

142. Richthofen, *Baron Richthofen's Letters*, 125.

143. Snyder-Reinke, *Dry Spells*, 89–90.

144. *Tongzhi Chengdu xian zhi* (1873), dili zhi: 1–2.

145. Brook, *The Confusions of Pleasure*, 19–21.

146. Wong, Cheng, Boon, Ee, and Chong, *Chinese Auspicious Culture*, 193. For a discussion of the timber market and Jingdezhen, see Miller, *Fir and Empire*, 71–72. For the cost of timber for the porcelain industry, see Gerritsen, *The City of White and Blue*, 99–100; 146.

147. Zhang Yanmei, *Qingdai Sichuan hanzai shikong fenbu yanjiu*, 47.

148. Order issued on TZ11.7.14 (August 7, 1872). *Tongzhi Chengdu xian zhi* (1873), dili zhi 3–4.

149. Edgerton-Tarpley, *Tears from Iron*, 59–60; Will, *Bureaucracy and Famine in Eighteenth Century China*.

150. Guangyuan County, the center of coal production in Qing Baoning, yielded 8.78 million catties (*jin*) of coal annually by 1940. One catty was equivalent to approximately 1.33 pounds, or 0.6 kilograms. Although that number reflects Republican-era production, one suspects the number in Qing times was also significant. *Minguo Chongxiu Guangyuan xian zhi gao* (1940): 3.10:2. Golas, *Science and Civilisation in China*, 424.

151. Faure, *China and Capitalism*, 4.

152. Von Glahn, *An Economic History of China*, 380–82.

153. Pomeranz, *The Great Divergence*; Vries, *State, Economy and the Great Divergence*.

Chapter 5: Breaking the Land

1. See Smith, "Li Hung-chang's Use of Foreign Military Talent," 119–44.

2. Mi Rucheng, *Zhongguo jindai tielu shi ziliao (1863–1911)*, vol. 1 (1963 Rpt.; Beijing: kexue chubanshe, 2016), 25. Li Hongzhang's memorial is dated TZ6.12.6, or December 31, 1867, so it was delivered and read in 1868. Note that Li practiced various forms of divination throughout his life. See, for instance, a letter composed to his son in 1892 containing detailed geomantic instructions for locating a burial spot for his late wife. He Wei, "Jindai Jiangnan kanyuye yanjiu," 124.

3. For a discussion of Li's changing position on technology from the West, see Wu, "Superstition and Statecraft in Late Qing China," 297. For Li Hongzhang's terming the crisis facing China one that had "never been seen in thousands of years," see Li, *The Making of the Modern Chinese State, 1600–1950*, 116–17.

4. Wright, *The Last Stand of Chinese Conservatism*, 9–10.

5. Li Baoping, Deng Ziping, Han Xiaobai, eds., *Kailuan meikuang: dang'an shiliaoji*. vol. 1. (Shijiazhuang: Hebei jiaoyu chubanshe, 2012), 111.

6. For the case of telegraphy, see Zhou, *Historicizing Online Politics*. For railways, see Barbalet, *Confucianism and the Chinese Self*, 177.

7. Elman, "Naval Warfare and the Refraction of China's Self-Strengthening Reforms into Scientific and Technological Failure, 1865–1895," 326. See also Halsey, "Sovereignty, Self-Strengthening, and Steamships in Late Imperial China," 81–111.

8. Lavelle, *The Profits of Nature*. For further discussion of late Qing attempts to exploit natural resources in frontier zones, see Kinzley, *Natural Resources and the New Frontier*.

9. Halsey, *Quest for Power*.

10. Elleman and Paine, *Modern China*, 225.

11. For an introduction to Song Gengping, see Chen, "Creating Intellectual Space for West-East and East-East Knowledge Transfer," 61–62.

12. Song Gengping, *Kuangxue xinyao xinbian* (Location Unknown: Shuxi Guangshi shan fang, 1902), xia juan: 1.

13. Gou Deyi, *Qingdai jiceng zuzhi yu xiangcun shehui guanli*, 70–71.

14. Wang, *Kuachu fengbi de shijie*, 686. Cited in Zhang, *The Politics of Rights and the 1911 Revolution*, 40. See also Young, *Ecclesiastical Colony*, 94.

15. Daigle, "Challenging the Imperial Order," 12.

16. Wyman, "The Ambiguities of Chinese Antiforeignism," 94.

17. For an incident in Chengdu involving fengshui from 1893, see Stapleton, *Civilizing Chengdu*, 43–44.

18. Driscoll, *The Whites Are the Enemies of Heaven*, 118–19.

19. Nanbu County Qing Archive: (TZ12.3.8/1873) 6.628.02.

20. Jin Shengyang, Xie Jiayuan, and Liu Yanwei, "Qingdai Nanbu xian yan li shoushi chuyi," *Yanye shi yanjiu* 3 (2017): 3–9.

21. Nanbu County Qing Archive: (GX12.10.28/1886):9.604.01.
22. Nanbu County Qing Archive: (GX14.7/1888) 14.423.04.
23. Nanbu County Qing Archive: (GX13.4.20/1887) 9.863.02.
24. See chapter 3 for a detailed discussion of astral deities and their connections to examination success.
25. Nanbu County Qing Archive: (GX14.5.8/1888) 14.423.02.
26. Nanbu County Qing Archive: (GX14.5.8/1888) 14.423.02.
27. Jeff Hornibrook's discussion of Ao Island Academy's purchase for a railway's construction in Pingxiang County, Jiangxi Province, echoes this theme. See Hornibrook, *A Great Undertaking*, 105–7.
28. For a discussion of the legal privileges of the gentry in Qing society, see Chung-li Chang, *The Chinese Gentry*, 35–37.
29. Upon its levying, the salt *lijin* brought in more silver than the county's base land-tax and continued to be raised alongside other new taxes. Salt commissioners became important local political actors from the 1880s onward, with several recorded in the county's late Qing gazetteer alongside the magistrates under which they served. Yuan Hui and Jin Shengyang, "Qingdai Nanbu xian yanliju de shezhi," *Zhongguo yan wenhua* 11 (2018): 36. The *lijin* was not the only extra tax placed on the salt trade; consider, for instance, the "added price" tax on salt. See Zelin, *The Merchants of Zigong*, 153–54.
30. Nanbu County Qing Archive: (GX13.1.14/1887) 9.716.01. A similar notice of the regulatory change is found in the form of a copied memorial from the Board of Punishments in the Dan-Xin collection. Taiwan Dan-Xin Qing Archive: (GX12.7.14/1886) 31105.01. For the relevant 1886 amendment to the law code, see *Qinding Da Qing huidian shili* (1899) *juan* 796, *xingbu*: 7. As Weiting Guo has argued, local officials reported relatively few cases of grave destruction to capital authorities, preferring to resolve most disputes at the county level. I suspect capital authorities were aware of this arrangement. The updated punishments for grave destruction signalled to local officials that they needed to emphasize grave protection and swiftly put a stop to troubling trends in their districts. See Guo, "Social Practice and Judicial Politics in 'Grave Destruction' Cases in Qing Taiwan, 1683–1895," 96–97.
31. Nanbu County Qing Archive: (GX21.2.17/1895) 12.1042.01.
32. Nanbu County Qing Archive: (GX19.6.28/1893) 12.266.04.
33. Wang Daolü, *Nanbu xian xiangtu zhi* (1906), 393.
34. Driscoll, *The Whites Are the Enemies of Heaven*, 117–24.
35. Wooldridge, *City of Virtues*, 113–15.
36. Wright, *The Last Stand of Chinese Conservatism*, 274–75.
37. Ibid., 174–78; 274–75.
38. Zhou, *Historicizing Online Politics*, 23–35.
39. Wright, *The Last Stand of Chinese Conservatism*, 175.
40. Ibid., 274.
41. Mi Rucheng, *Zhongguo jindai tielu shi ziliao (1863–1911)*, vol. 1, 23–24.
42. For more on the "disastrous results" of the 1874 attempt to construct telegraph lines, see Baark, *Lightning Wires*, 107–57.
43. Jialuo Yang, ed. *Yangwu yundong wenxian huibian*, vol. 6 (Taipei: Shijie shuju, 1963), 331.
44. Zhou, *Historicizing Online Politics*, 28.

45. Ibid., 33.
46. Halsey, *Quest for Power*, 220.
47. Ibid., 213–37.
48. In the 1898 incident, 150 Chinese miles (around fifty-four English miles) of telegraph lines were destroyed. See Anon., "The Szechuan Telegraph Lines: Chungking, Nov. 7," 516. For the 1892 incidents, see "A memorial concerning the residents of Linfen and other counties who were desperately seeking rain and easily believed false words directing people to destroy telegraph wires and towers," First Historical Archives of China: (GX18.4.15/1892) 03-6723-017, National Palace Museum Palace Memorial and Grand Council Archive: (GX18.3.16/1892) 408011376; Peng Nansheng, "Lun yangwu huodong zhong 'fengshui' guan de yingxiang." *Gansu shehui kexue* 6 (2004): 93–94; Bijie xian difang zhi bianzuan weiyuan hui, ed., *Bijie xian zhi* (Guiyang: Guizhou renmin chubanshe, 1996), 685; Fairbank and Liu, eds., *The Cambridge History of China*, part 2, vol. 2, 302. For a discussion of geomantic beliefs in Hongtong County, where some of the attacks on telegraphs occurred, see You, *Folk Literati, Contested Tradition, and Heritage in Contemporary China*, 179–80.
49. Wang Hongbin, *Wan Qing haifang: Sixiang yu zhidu yanjiu* (Beijing: Shangwu yinshuguan, 2005), 273. For a discussion of the 1898 Yan Manzi incident in Sichuan, see Driscoll, *The Whites Are the Enemies of Heaven*, 122–29.
50. Shuge Wei, "Circuits of Power," 116–17.
51. "Sending and receiving telegrams truly began in 1915. . . . In the early period, there were gentry and merchants who opposed the quick sending of messages and there were those who opposed the wires entering the county seat on account of harm to our town's fengshui." *Minguo Xinghua xian xiao tong zhi* (1934), dianbao pian, 76–77.
52. Thirteen percent of the maintenance costs on the Tianjin-Shanghai line were reserved for the salaries of these personnel. See Yoon, "Dash Expectations," 837. For a discussion of the line's construction, including the engineering effort to avoid "touching areas of important to fengshui harmonies," see Baark, *Lightning Wires*, 166.
53. Yoon, "Dash Expectations," 839. "Recently across many provinces there have been many cases of ignorant people who have repeatedly pointed to abnormal phenomena and improperly caused confusion, leading crowds to destroy telegraph lines." Zhong Qingxi, ed., *Xingbu zouding xinzhang* (Chengdu: Sichuan yanju, 1901), 4:15.
54. China Ministry of Communications, *List of Post Offices, Eleventh Issue*, 146, 147, 154, 258, 275.
55. Nanbu County Qing Archive: (XT2.11.15/1910) 21.160.01.
56. Yoon, "Dash Expectations," 841.
57. For discussions of Liu Xihong's career overseas, see Day, *Qing Travelers to the Far West*, 130–32.
58. For a discussion of twentieth-century treatments of Liu Xihong, see Wu, "Superstition and the State in Late Qing China," 20.
59. Mi Rucheng, *Zhongguo jindai tielu shi ziliao (1863–1911)*, vol. 1, 98.
60. For those crises, see Marks, *China: An Environmental History*, 257–80.
61. Elvin, "How Did the Cracks Open?" 9.
62. Fairbank and Liu, *The Cambridge History of China*, 175.
63. Mi Rucheng, *Zhongguo jindai tielu shi ziliao (1863–1911)*, vol. 1, 104–5.

64. Dong, *Republican Beijing*, 182.

65. For relevant context, see Lillian M. Li, *Fighting Famine in North China*, 18, 309.

66. Cohen, "Lineage Organization in North China," 522.

67. Quoted from Qi Qizhang, *Zhongguo jindai shehui sichaoshi*, 294.

68. For a discussion of the debates surrounding the Tianjin-Tongzhou line, see Kan Li, "On the Road to a Modern City," 93–97.

69. National Palace Museum Palace Memorial and Grand Council Archive: (GX23.9.26/1897) 141905.

70. Chen Qizhang went on to propose an alternative route for the Hangzhou railway that was adopted by developers in the following decade. Yue Qintao, *Yi Shanghai wei zhongxin: Hu Ning, Hu Hang Yong tielu yu jindai Changjiang sanjiaozhou diqu shehui bianqian* (Beijing: Zhongguo shehui kexue chubanshe, 2016), 106–7.

71. Song-Chuan Chen, "The Power of Ancestors: Tombs and Death Practices in Late Qing China's Foreign Relations, 1845–1914," *Past and Present* 239.1 (2018): 133.

72. Köll, *Railroads and the Transformation of China*, 294.

73. Mi Rucheng, *Zhongguo jindai tielu shi ziliao (1863–1911)*, vol. 1, 162.

74. Ibid., 152.

75. Teng and Fairbank, *China's Response to the West*, 101–2.

76. While Zheng Guanying was critical of some geomantic practices, he was nonetheless a devout Daoist who defended the ideal of Daoist cultivation as a resource of national salvation. See Li Zhitian (Lai Chi Tim), "Zheng Guanying 'xiandao' yu 'jiushi' de sixiang he shijian: jianping qi dui Qingmo Minchu daojiao fazhan de yingxiang ji yiyi," *Zhongguo wenhua yanjiusuo xuebao* 67 (2018): 151–202.

77. Chen, "The Power of Ancestors," 123.

78. Archives of Academia Sinica's Modern Historical Institute: (GX6.4.30/1880) 01.12.162.01.031.

79. Ibid.

80. Mi Rucheng, *Zhongguo jindai tielu shi ziliao (1863–1911)*, vol. 1, 96.

81. Ibid., 158.

82. Chen, "The Power of Ancestors," 132–37.

83. Yue Qintao discusses some of the issues that arose in the expropriation of land for railways in the Greater Shanghai region. See Yue Qintao, *Yi Shanghai wei zhongxin*, 156–58.

84. *Qinding Da Qing huidian shili* (1899) *juan* 796, *xingbu*: 7. Imperial authorities amended the statute on "Uncovering Graves" throughout the 1800s. Of the twenty-three substatutes under "Uncovering Graves" that are found in the last published edition of the code (1870), eighteen of them were adopted or amended (from eighteenth-century texts) between 1801 and 1870. Xue Yunsheng's (1820–1901) legal commentary, "Lingering Doubts after Reading the Statutes," addresses these amendments in pointing out where Qing authorities added conditions of leniency and—more often—where they added terms of severity. Xue also adds important historical context. For instance, he suggests that the terms of substatute thirteen be understood as a specific regulation of Zhili Province, since reports of grave destruction had been most numerous there; the memorial introduced below composed by Yuan Shikai during his tenure as governor-general of Zhili is one such example of that phenomenon. See Xue Yunsheng,

Duli cunyi (Beijing: Hanmao zhai, 1905), *juan* 31, *xinglü*: 13; for Xue's discussion of all twenty-three substatutes, see Ibid., *juan* 31, *xinglü*: 3–19.

85. As governor-general of Zhili, Yuan Shikai reported that grave destruction by bandits had increased on the North China Plain and requested permission to execute tomb disturbers on the spot. National Palace Museum Palace Memorial and Grand Council Archive: (GX29.8.22/1903) 408001061. Similar rumors circulated in the area during the buildup to the Boxer Uprising. For a discussion of the Boxers, fengshui, and telegraphs, see Sun Li, *Wan Qing dianbao jiqi chuanbo guannian* (Shanghai: Shanghai shiji chuban jituan, 2007), 104–5.

86. Wu, "Mining the Way to Wealth and Power," 581–99.

87. Wu Yicheng and Shen Weiwei, "Shilun Qingdai yilai de kuangye huanjing baohu," *Lanzhou xuekan* 1 (2011): 123.

88. Ibid., 123.

89. Zhongyang yanjiuyuan jindaishi yanjiusuo, ed., *Zhongguo jindaishi ziliao huibian: Kuangwu dang, Sichuan, Fujian, Guangdong, Guangxi* (Taipei: Zhongyang yanjiuyuan jindai shi yanjiusuo, 1960), 2586.

90. For a reference to Li Zhengyong's assumption of this post, see Nanbu County Qing Archive: (GX25.11.4/1899) 14.600.01.

91. Zhongyang yanjiuyuan jindaishi yanjiusuo, *Zhongguo jindaishi ziliao huibian: Kuangwu dang, Sichuan, Fujian, Guangdong, Guangxi*, 2573.

92. Ibid., 2573.

93. *Minguo Ba xian zhi* (1939) 16: 29.

94. Ch'en, *The Highlanders of Central China*.

95. Joseph Lawson, writing of the Liangshan region, has contended that "state-supported mining meant invasion" for the peoples of southeastern Sichuan during this decade. See Lawson, *A Frontier Made Lawless*, 52.

96. Nanbu County Qing Archive: (GX24.12.23/1899) 14.52.01 and (GX30.2.20/1904) 16.638.

97. For the gold mining cases between 1909 and 1910 involving involving graves, see Nanbu County Qing Archive: (XT1.9.14/1909) 20.267.02, (XT2.9.12/1910) 21.156.01, (XT2.8.11/1910) 21.244, and (XT2.7.22/1910) 21.732.

98. Nanbu County Qing Archive: (XT2/1910) 21.153.

99. Nanbu County Qing Archive: (XT2/1910) 21.153.

100. Nanbu County Qing Archive: (XT2.7.29/1910) 21.732.02.

101. Nanbu County Archive: (MG5/1916) 26.1700.6442.133.

102. Katz, *Religion in China and Its Modern Fate*, 36–57; Goossaert, "1898," 1–29.

103. For a discussion of the increasingly destable conditions across Sichuan after 1898, see Adshead, *Province and Politics in Late Imperial China*.

104. Republican-era gazetteers reference many cases from this campaign. One concerns a community leader in Mianzhu County who attempted to sell the trees of a large cypress forest around a historic gravesite "under the guise of raising money for schools." The leader, Zhang Yigui, was stopped after a member of the county's gentry protested. *Minguo Mianzhu xian zhi* (1920): 16: 5.

105. Xu Yue, "Sichuan's Promotion of Education and Activities of Felling Temple Trees in the Late Qing Dynasty," 428.

106. For the official notice, see Nanbu County Qing Archive: (GX28.3.8/1902) 15.958.01. For a legal dispute concerning the felling of temple trees from around this time, see Nanbu County Qing Archive: (GX29.7/1903) 16.393.02.

107. Xu Yue, "Sichuan's Promotion of Education and Activities of Felling Temple Trees in the Late Qing Dynasty," 427.

108. Nanbu County Qing Archive: (GX33.3.15/1907) 18.463.01.

109. Nanbu County Qing Archive: (XT3.5.15/1911) 22.186.01.

110. Qi Shouhua and Zhong Xiaozhong, eds., *Zhongguo difang zhi meitan shiliao xuanji*, 428.

111. Shaw, *Chinese Forest Trees and Timber Supply*, 146.

112. For discussions of futures markets in timber, see McDermott, *The Making of a New Rural Order in South China*, 369–429; Miller, *Fir and Empire*, 8, 34–36; Zhang, *Sustaining the Market*, 48–79.

113. Shaw, *Chinese Forest Trees and Timber Supply*, 146.

114. For a related discussion of these legal changes in the early twentieth century, see Ch'en, *The Highlanders of Central China*, 18–20.

115. *Minguo Yuanshi xian zhi* (1933), *jiangyu juan*, 39.

116. Pomeranz, *The Making of a Hinterland State*, 127–28.

117. For more on the history of the term "superstition," borrowed from Japan in post-1898 China, see Goossaert, "1898."

118. *Minguo Chongqing xian zhi* (1926) *shiji* 3: 17–18.

119. Duara, *Culture, Power, and The State*, 15–41.

120. SUCSBRC: (Nanbu Collection; MG6.11.9/1917) 466.499.24.

121. "Court Verdict: The lower court ruling is voided. As for the twenty-four trees purchased by Shu Jingzhai, they should be immediately returned to Lady Jing. The price of thirty-six *chuan* that Shu Jingzhai already paid should be recovered from Ren Shun. The fees for the lawsuits should be shared by Ren Shun and Shu Jingzhai." SUCSBRC: (Nanbu Collection; MG7.1.7/1918) 466.500.04.

122. See, for instance, the following example, recorded by land surveyors as an example of an old-style contract: "Sun Guilin establishes a contract for the sale of a house, a foundation, farmland, bamboo and trees, stone tools, old instruments, Yin-yang and fengshui, the foundation of a forest grove, and firewood and thatched grass." Li Zhenghong, *Sichuan nongye jinrong yu diquan yidong zhi guanxi*, in *Minguo ershi niandai Zhongguo dalu wenti ziliao*, vol. 89 (Taipei: Chengwen chubanshe, 1977), 47316. Presumably, the *yin* and *yang* in this contract referred to graves and houses, while the fengshui referred to the land and trees around them. For a discussion of another fengshui case related to the Anyang excavations during the 1930s, see Lam, *A Passion for Fact*, 91–92.

123. Heightened anxieties around fengshui during the decades following the Taiping Civil War may relate to the broader "revival" of religious organizations and redemptive societies. See Nedostup, *Superstitious Regimes*, 12–13; Meyer-Fong, *What Remains*, 63.

124. Yue Qintao, *Yi Shanghai wei zhongxin*, 113.

125. See, for example, Li and Lackner, "Contradictory Forms of Knowledge?" 461–85. See also Harrison, *The Man Awakened from Dreams*.

126. The phrase "changes in the land" is borrowed from William Cronon's classic work on landed property in early New England. Cronon, *Changes in the Land*.

Concluding Remarks

1. Peter Harrison, "'Science' and 'Religion,'" 81–106.

2. A long and celebrated Sinological tradition has seen the use of local materials to explore broader issues in state-society relations. See, for instance, Perdue, *Exhausting the Earth*; Duara, *Culture, Power, and the State*; Dardess, "A Ming Landscape"; Rowe, *Crimson Rain*; Naquin, *Peking: Temples and City Life, 1400–1900*; Sommer, *Polyandry and Wife-Selling in Qing Dynasty China*.

3. Yeoh, *Contesting Space in Colonial Singapore*.

4. Watson, "Standardizing the Gods," 292–324.

5. Though facing vastly different political circumstances and power structures, the Qing officials and scholars who objected to industrial development from the 1860s to the 1890s bear some resemblance to the conservationists involved in the battles over England's Lake District in the 1870s. Although the efforts to stop industrial development in the Lake District ultimately failed, those efforts launched the environmental movement as we have come to know it today. See Ritvo, *The Dawn of Green*.

6. For the relevant history of the Song period, see Ebrey, "Sung Neo-Confucian Views on Geomancy," 75–107. For the Qing period, a multitude of factors likely contributed to the legal trends of the nineteenth century, some of which have not been explored in this book. For example, in 1751, the Qing government lifted the centuries-long ban on the private publication of the imperial calendar. As Huang Yinong has discussed, this liberalization made the calendar more accessible and resulted in the flourishing of private production to meet market demand. Historians may ask what this regulatory change entailed for the dissemination of divinatory, astrological, and geomantic knowledge in the subsequent 161 years of the dynasty's rule. For a Qing-era legal commentary on the relevant statutory amendment, see Wu Tan, *Da Qing lüli tongkao* (c. 1779; 1886 Edition), *juan* 32, *xinglü zhawei*: 9–11. See also Huang Yinong, "Tongshu: Zhongguo chuantong tianwen yu shehui de jiaorong": 169–170.

7. For marital litigation, see Zhao Weini, *Shenduan yu jinxu: Yi wan Qing Nanbu xian hunyin lei anjian wei zhongxin* (Beijing: Falü chubanshe, 2013).

8. Wu Peilin, "Qingmo xinzheng shiqi guanzhi hunshu zhi tuixing: Yi Sichuan wei li," *Lishi yanjiu* 5 (2011): 78–96.

9. Nanbu County Qing Archive: (GX4.6.26/1878) 7.625.02, (GX17/1891) 11.203.02, (GX18.3.28/1892) 11.449.04, and (GX29.8.24/1903) 16.405.01.

10. Ransmeier, *Sold People*, 158–59.

11. Bourgon, "Uncivil Dialogue," 50–90.

12. Henderson, *The Development and Decline of Chinese Cosmology*.

13. Elman, "Early Modern or Late Imperial?" 226.

14. Nanbu County Qing Archive (XT2.9.8/1910): 21.969.01.

15. See for instance: "[T]he Manchu throne proclaimed its legitimacy by resolving the Ming calendar crisis. For self-legitimation, the Qing court presented a new calendar that like its Mongol precedessor would link empire and time far into the future." Elman, *On Their Own Terms*, 133. See also Kurtz, *The Discovery of Chinese Logic*.

16. Jami, *The Emperor's New Mathematics*, 385–92. Reaching a related conclusion, Ori Sela points out that "divination often required astronomical and mathematical knowledge." Sela, *China's Philological Turn*, 137.

17. Josephson-Storm, *The Myth of Disenchantment*.

18. Perhaps most notorious among these officials was the deputy party secretary of Sichuan, Li Chuncheng, who was removed from office in 2012 and sentenced to thirteen years in prison for abuses of power and bribery. Li's involvement with geomantic consultants formed a part of the investigation against him.

BIBLIOGRAPHY

Archives and Libraries

Bayerische Staatsbibliothek
Columbia University C. V. Starr East Asian Library
First Historical Archives of China 中國第一歷史檔案館
Harvard-Yenching Library
Langzhong Municipal Archive 閬中市檔案局
Langzhong Municipal Library 閬中市圖書館
Langzhong Gazetteer Office 閬中市人民政府地方志辦公室
Library of Congress
Nanbu County Archive 南部縣檔案局
Nanchong Municipal Archive 南充市檔案館
National Library of China 中國國家圖書館
Sichuan Provincial Archives 四川省檔案館
Sichuan University's China Southwest Bibliography Research Center (SUCSBRC) 四川大學中國西南地區文獻研究中心, Digitized Archival Collections for Nanbu County 南部縣, Bazhong County 巴中縣, Nanjiang County 南江縣, Santai County 三台縣, and Bishan County 璧山縣
Shanghai Library's Genealogical Database 上海圖書館家譜數據庫
Shanghai Library's Rare Books Depository 上海圖書館古籍館
Staatsbibliothek zu Berlin
(Taiwan) Academia Sinica's Archives of the Grand Secretariat 臺灣內閣大庫檔案
(Taiwan) Archives of Academia Sinica's Modern Historical Institute 臺灣近史所檔案館
(Taiwan) Digital Archives Project of Taiwan National University 臺灣大學典藏數位化計畫
(Taiwan) National Palace Museum Palace Memorial and Grand Council Archive 台灣清代宮中檔奏摺及軍機處檔摺件
Tōyō Bunko (Oriental Library) 東洋文庫
University of California at Berkeley C. V. Starr East Asian Library
University of Michigan Hatcher Library

Law Codes, Judicial Casebooks and Case Collections, and Political Writings

Chen Quanlun 陳全倫, Bi Kejuan 畢可娟, and Lü Xiaodong 呂曉東, eds. *Xu gong yanci: Qingdai mingli Xu Shilin pan'an shouji* 徐公讞詞: 清代名吏徐士林判案手記. Jinan: Qi-Lu shushe, 2001.

Da Ming huidian 大明會典. Beijing: Neifu kanben, 1587. Harvard-Yenching Library.

Da Qing lüli 大清律例. 1740 Rpt. Tian Tao 田濤 and Zheng Qin 鄭秦, eds. Beijing: Falü chubanshe, 1999.

Fan Zengxiang 樊增祥. *Fanshan pipan* 樊山批判. In *Lidai panli pandu* 歷代判例判牘, vol. 11, eds. Yang Yifan 楊一凡 and Xu Lizhi 徐立志. Beijing: Zhongguo shehui kexue chubanshe, 2005.

Guanzhenshu jicheng bianzuan weiyuanhui 官箴書集成編纂委員會, ed. *Guanzhenshu jicheng* 官箴書集成, vols. 3 and 7. Hefei: Huangshan shushe, 1997.

He Changling 賀長齡 and Wei Yuan 魏源, eds. *Huangchao jingshi wenbian* 皇朝經世文編. 1827 Rpt. Shanghai: Jiangzuo shulin cangban, 1873. University of Michigan Library.

Huang Liuhong 黃六鴻. "Baojia sanlun" 保甲三論. In *Wei Yuan quanji* 魏源全集, vol. 17, Wei Yuan quanji bianji weiyuanhui 魏源全集編輯委員會, ed. Changsha: Yuelu shushe, 2007, 131–34.

Juefei shanren 覺非山人 and Chiu Pengsheng 邱澎生, eds. *Juefei shanren Erbi kenqing dian jiao ben* 覺非山人珥筆肯綮點校本. In *Mingdai yanjiu* 明代研究 13 (2009): 233–90.

Li Benzhong 李本忠. *Pingtan jilüe* 平灘紀略. Location Unknown: Qingliantang, 1840.

Li Rong 李榕. *Shisanfeng shuwu quanji* 十三峰書屋全集 (Preface 1892). In *Li Shenfu xiansheng quanji* 李申夫先生全集, ed. Jiang Dejun 蔣德鈞. Shanghai: Haishan fang, 1899.

Liu Xianting 劉獻廷. *Guangyang zaji* 廣陽雜記 (c. 1648–95). In *Jifu congshu* 畿輔叢書, vol. 363, ed. Wang Hao 王灝. Dingzhou: Qiandetang, 1879. Harvard-Yenching Library.

Lu Yao 陸燿 and Yao Lide 姚立德. *Shandong yunhe beilan* 山東運河備覽. In *Gugong zhenben congkan* 故宮珍本叢刊, vol. 234, ed. 1776 Rpt.; Haikou: Hainan chubanshe 2000, 234–493.

Mi Rucheng 宓汝成, ed. *Zhongguo jindai tielu shi ziliao (1863–1911)* 中國近代鐵路史資料 (1863–1911), vol. 1, 1963 Rpt. Beijing: Kexue chubanshe, 2016.

Qinding Da Qing huidian 欽定大清會典. Beijing: Wuying dian, 1764. Columbia University Libraries.

Qinding Da Qing huidian zeli 欽定大清會典則例. Beijing: Wuying dian, 1764. Harvard-Yenching Library.

Qinding Da Qing huidian shili 欽定大清會典事例. Beijing: Huidian guan, 1899. Staatsbibliothek zu Berlin.

Qinding Da Qing lüli 欽定大清律例. Location and publisher not identified, 1870. Harvard-Yenching Library.

Sichuansheng dang'anguan 四川省檔案館, ed. *Qingdai Ba xian dang'an huibian, Qianlong juan* 清代巴縣檔案彙編, 乾隆卷. Beijing: Dang'an chubanshe, 1991.

Sichuansheng dang'anguan 四川省檔案館 and Sichuan daxue lishixi 四川大學歷史系, eds. *Qingdai Qian Jia Dao Ba xian dang'an xuanbian* 清代乾嘉道巴縣檔案選編 (Two Volumes). Chengdu: Sichuan daxue chubanshe, 1989; 1996.

Sichuansheng dang'anju 四川省檔案局, ed. *Qingdai Sichuan Ba xian yamen Xianfengchao dang'an xuanbian* 清代四川巴縣衙門咸豐朝檔案選編. Shanghai: Shanghai guji chubanshe, 2011.

Sichuansheng Nanchongshi dang'anguan 四川省南充市檔案館, ed. *Qingdai Sichuan Nanbu xian yamen dang'an* 清代四川南部縣衙門檔案. Hefei: Huangshan shushe, 2016.

Tang Jiong 唐炯 and Hua Guoying 華國英, eds. *Sichuan guanyun yan'an leibian* 四川官運鹽案類編. Luzhou: Luzhou zongju, 1902. Columbia University.

Wang Dingzhu 王定柱. *Shusong pi'an* 蜀訟批案. Chongqing, Sichuan. c. 1821–27. Tōyō Bunko.

Wu Tan 吳壇. *Da Qing lüli tongkao* 大清律例通考. Location and publisher not identified, c. 1779; 1886 edition. Harvard-Yenching Library.

Xihua shifan daxue 西華師範大學 and Nanchongshi dang'anju 南充市檔案局, eds. *Qingdai Nanbu xianya dang'an mulu* 清代南部縣衙檔案目錄. Beijing: Zhonghua shuju, 2010.

Xue Yunsheng 薛允升. *Duli cunyi* 讀例存疑. Beijing: Hanmao zhai: 1905.

Yongzheng shangyu 雍正上諭. Beijing: Wuying dian, 1741. Harvard-Yenching Library.

Zeng Guofan 曾國藩. *Zeng Guofan quanji* 曾國藩全集, vol. 20. Changsha: Yuelu shushe, 2011.

Zhao Erxun 趙爾巽, ed. *Qingshigao* 清史稿. Beijing: Qingshiguan, 1928.

Zhong Qingxi 鍾慶熙, ed. *Xingbu zouding xinzhang* 刑部奏定新章. Chengdu: Sichuan yanju, 1901. Harvard-Yenching Library.

Zhongyang yanjiuyuan jindaishi yanjiusuo 中央研究院近代史研究所, ed. *Zhongguo jindaishi ziliao huibian: Kuangwu dang, Sichuan, Fujian, Guangdong, Guangxi* 中國近代史資料彙編: 礦務檔四川福建廣東廣西. Taipei: Zhongyang yanjiuyuan jindai shi yanjiusuo, 1960.

Zhou Xun 周詢. *Shuhai congtan* 蜀海叢談. 1948 Rpt. Taipei: Wenhai chubanshe, 1966.

Zhu Qingqi 祝慶祺 and You Shaohua 尤韶華, eds. *Xing'an huilan quanbian*: 刑案匯覽全編. 1834/1840/1886/1887 Rpt. Beijing: Falü chubanshe, 2008.

Note: This fifteen-volume collection contains *Xing'an huilan* 刑案匯覽 (1834; vols. 1–8), *Xuzeng xing'an huilan* 續增刑案匯覽 (1840; vols. 9–10), *Xinzeng Xing'an huilan* 新增刑案匯覽 (1886; vol. 11), and *Xing'an huilan xubian* 刑案匯覽續編 (1887; vols. 12–15).

Late Imperial Chinese Geomantic Texts, Ritual Manuals, and Fengshui Map Collections

Anonymous, *Dili yuanxu* 地理原序. Baoning, Sichuan. Undated Manuscript, c. 1875–1912.

Anonymous, *Xiao'er guansha tujie* 小兒關煞圖解. 1889 Rpt. Baise: Baise keyin faxingsuo, Undated Reprint.

Bu Yingtian 卜應天 and Xie Zhidao 謝志道, ed. *Chongding jiaozheng kui ban jujie xiaosha jing jietu dili jueyao xuexin fu* 重訂校正魁板句解消砂經節圖地理訣要雪心賦. 1602 Rpt. In *Yuwai hanji zhenben wenku* 域外漢籍珍本文庫, *di 5 ji*, Vol. 19. Chongqing: Xihua shifan daxue chubanshe, 2015, 4–64.

Gao Jiannan 高見南. *Xiangzhai jingzuan* 相宅經纂. Fuzhou: Weigen caotang, 1844.

Gūwalgiya (Gua'erjia) Lianrui 爪爾佳廉瑞. *Qingnang dili jiyao* 青囊地理集要. Beijing: Beijing: Liulichang Wenbaotang, 1907.

Huang Zongsheng 黃宗聖. *Xincan houxu baizhong jing* 新參後續百中經. Chanshan: Fuwentang, d.u./c.1840.

Kong Wenxing 孔聞星 (Che Ying 徹瑩). *Huitu Dili zhizhi yuanzhen* 繪圖地理直指原真. 1692 Rpt. Shanghai: Shanghai jiaojing shanfang yinhang, c. 1875–1912.

Miao Xiyong 繆希雍, ed. *Nanjie ershisi pian* 難解二十四篇, in Mao Jin 毛晉, *Jindai mishu* 津逮秘書, *di si ji* 第四集. c. 1630 Rpt. Shanghai: Boguzhai, 1922. Note: This text is typically appended onto Miao Xiyong's (1546–1627) *Complementary Commentary on the Burial Classic* (*Zangjing yi* 葬經翼) and thus by tradition has been ascribed to him.

Ouyang Chun 歐陽純. *Fengshui ershu xingqi leize* 風水二書形氣類則. Liuyang: Nanshan Ouyang shuyuan, 1831.

Qintianjian louke ke 欽天監漏刻科, ed. *Qintianjian Dili xingshi qieyao bianlun* 欽天監地理醒世切要辨論. Composed 1740; Re-issued 1746; Hong Kong: Xinyi tang youxian gongsi, 2013.

Song Gengping 宋賡平. *Kuangxue xinyao xinbian* 礦學心要新編. Location Unknown: Shuxi Guangshi shan fang, 1902.

Wang Daoheng 王道亨 and Wang Fangzhi 王方智. *Luojing toujie* 羅經透解. Taiyuan: Sihetang, 1824.

Wang Junrong 王君榮. *Xinke huitu yangzhai shishu jicheng* 新刻繪圖陽宅十書集成. 1590 Rpt. Location Unknown: Saoye shanfang, 1882.

Xiong Dian 熊殿 (Zaofu sanren 造福散人), Xiong Weiyao 熊維堯 (Preface), and Yuan Shizhen 袁世振 (Epilogue). *Buju misui* 卜居秘髓. Nanjing: Jinshantang, 1595.

Yamashita Kazumasa 山下和正, ed. *Chūgoku mokuhan fūsui chizushū: kohangi yori saisatsu* 中国木版風水地図集: 古版木より再刷. Tokyo: Yamashita Kazumasa Kenchiku Kenkyūjo, 2010.

Yunlu 允祿, ed. *Qinding xieji bianfang shu* 欽定協紀辨方書. Composed 1739. Beijing: Wu yingdian, 1741.

Zhao Jiufeng 趙九峰 (Zhao Tingdong 趙廷棟). *Dili wujue* 地理五訣. 1786 Rpt. Location Unknown: Chongwentang, 1841.

Zhou Nan 周南 and Lü Lin 呂臨. *Anju jinjing* 安居金鏡. Hangzhou: Shounantang, 1780.

Max Planck Institute for the History of Science (Berlin) Gazetteer Database Sources

Chen, Shih-Pei, Calvin Yeh, Qun Che, and Sean Wang. *LoGaRT: Local Gazetteers Research Tools*. Berlin: Max Planck Institute for the History of Science, 2017.

(In chronological order of publication)

Da Ming Yitong zhi 大明一統志 (1461)
Kangxi Guangping fu zhi 康熙廣平府志 (1677)
Kangxi Shaoxing fu zhi 康熙紹興府志 (1683)
Kangxi Zhejiang tong zhi 康熙浙江通志 (1684)
Kangxi Pengxi xian zhi 康熙蓬溪縣志 (1713)
Yongzheng Sichuan tong zhi 雍正四川通志 (1733)
Qianlong Hengzhou zhi 乾隆橫州志 (1746/1899)
Qianlong Suining xian zhi 乾隆遂寧縣志 (1747)
Qianlong Changsha fu zhi 乾隆長沙府志 (1747)
Qianlong Hezhou zhi 乾隆合州志 (1748)
Qianlong Rong xian zhi 乾隆榮縣志 (1756)
Qianlong Danling xian zhi 乾隆丹棱縣志 (1761)
Qianlong Longchang xian zhi 乾隆隆昌縣志 (1764)
Qianlong Cangxi xian zhi 乾隆蒼溪縣志 (1783)
Jiaqing An xian zhi 嘉慶安縣志 (1812)
Jiaqing Shifang xian zhi 嘉慶什邡縣志 (1813)
Jiaqing Peng xian zhi 嘉慶澎縣志 (1813)
Jiaqing Sichuan tong zhi 嘉慶四川通志 (1816)
Daoguang Baoning fu zhi 道光保寧府志 (1821)
Daoguang Nanjiang xian zhi 道光南江縣志 (1827)

Daoguang Fugou xian zhi 道光扶溝縣志 (1833)
Daoguang Yunmeng xian zhi lüe 道光雲夢縣志略 (1840)
Daoguang Chongqing fu zhi 道光重慶府志 (1843)
Daoguang Nanbu xian zhi 道光南部縣志 (1849)
Xianfeng Langzhong xian zhi 咸豐閬中縣志 (1851)
Tongzhi Pi xian zhi 同治郫縣志 (1870)
Tongzhi Chengdu xian zhi 同治成都縣志 (1873)
Tongzhi Huili zhou zhi 同治會理州志 (1874)
Guangxu Qingfu xian zhi 光緒慶符縣志 (1876)
Guangxu Ninghe xian zhi 光緒寧河縣志 (1880)
Guangxu Yulin zhou zhi 光緒鬱林州志 (1894)
Guangxu Xuzhou fu zhi 光緒敘州府志 (1895)
Guangxu Zhecheng xian zhi 光緒柘城縣志 (1896)
Guangxu Pengxi xian xu zhi 光緒蓬溪縣續志 (1899)
Guangxu Jingyan zhi 光緒井研志 (1900)
Guangxu Yuejuanting quan zhi 光緒越嶲廳全志 (1906)
Minguo Mianzhu xian zhi 民國綿竹縣志 (1920)
Minguo Chongqing xian zhi 民國崇慶縣志 (1926)
Minguo Mianyang xian zhi 民國綿陽縣志 (1932)
Minguo Yuanshi xian zhi 民國元氏縣志 (1933)
Minguo Xinghua xian xiao tong zhi 民國興化縣小通志 (1934)
Minguo Sichuan Qijiang xian xu zhi 民國四川綦江縣續志 (1938)
Minguo Ba xian zhi 民國巴縣志 (1939)
Minguo Chongxiu Guangyuan xian zhi gao 民國重修廣元縣志稿 (1940)
Minguo Sichuan xin dizhi 民國四川新地志 (1946)

Published Primary and Secondary Sources

Adshead, S.A.M. *Province and Politics in Late Imperial China: Viceregal Government in Szechwan, 1898–1911*. London and Malmö: Curzon Press, 1984.

Ahern, Emily M. *The Cult of the Dead in a Chinese Village*. Stanford, CA: Stanford University Press, 1973.

Allee, Mark A. *Law and Local Society in Late Imperial China: Northern Taiwan in the Nineteenth Century*. Stanford, CA: Stanford University Press, 1994.

Anon. "The Szechuan Telegraph Lines: Chungking, Nov. 7." *The Japan Weekly Mail*, November 19, 1898: 516.

Asen, Daniel S. *Death in Beijing: Murder and Forensic Science in Republican China*. Cambridge: Cambridge University Press, 2016.

Atwill, David G. *The Chinese Sultanate: Islam, Ethnicity, and the Panthay Rebellion in Southwest China, 1856–1873*. Stanford, CA: Stanford University Press, 2005.

Baark, Erick. *Lightning Wires: The Telegraph and China's Technological Modernization, 1860–1890*. Vol. 1. Westport, CT: Greenwood Publishing Group, 1997.

Bai, Bin. "Daoism in Graves." In *Modern Chinese Religion*. Vol. 1. Edited by Pierre Marsone and John Lagerwey. Leiden: Brill, 2015, 548–600.

Barbalet, Jack. *Confucianism and the Chinese Self: Re-Examining Max Weber's China*. London: Palgrave Macmillan, 2017.

Bello, David A. *Across Forest, Steppe, and Mountain: Environment, Identity, and Empire in Qing China's Borderlands*. Cambridge: Cambridge University Press, 2016.

Bernhardt, Kathryn, and Philip C. C. Huang. *Civil Law in Qing and Republican China*. Stanford, CA: Stanford University Press, 1994.

Bian, He. *Know Your Remedies: Pharmacy and Culture in Early Modern China*. Princeton: Princeton University Press, 2020.

Bijie xian difang zhi bianzuan weiyuan hui 畢節縣地方志編纂委員會, ed. *Bijie xian zhi* 畢節縣志. Guiyang: Guizhou renmin chubanshe, 1996.

Bird, Isabella Lucy. *The Yangtze Valley and Beyond: An Account of Journeys in China, Chiefly in the Province of Sze Chuan and Among the Man-tze of the Somo Territory*. London: John Murray, 1899.

Bodde, Derk. "'Chinese Laws of Nature': A Reconsideration." *Harvard Journal of Asian Studies* 39 (1979): 139–55.

Bodde, Derk, and Clarence Morris. *Law in Imperial China: Exemplified by 190 Ch'ing Dynasty Cases*. Cambridge, MA: Harvard University Press, 1967.

Bokenkamp, Stephen R. *Ancestors and Anxiety: Daoism and the Birth of Rebirth in China*. Berkeley: University of California Press, 2007.

Bourgon, Jerome. "Uncivil Dialogue: Law and Custom Did Not Merge into Civil Law under the Qing." *Late Imperial China* 23.1 (2002): 50–90.

Bray, Francesca. *Technology and Gender: Fabrics of Power in Late Imperial China*. Berkeley: University of California Press, 1997.

———. "Introduction: Science and Confucian Statecraft in East Asia." In *Science and Confucian Statecraft in East Asia*, edited by Francesca Bray and Jongtae Lim. Leiden: Brill, 2019, 1–28.

Brokaw, Cynthia J. *Commerce in Culture: The Sibao Book Trade in the Qing and Republican Periods*. Cambridge, MA: Harvard University Asia Center, 2007.

Brook, Timothy. *Praying for Power: Buddhism and the Formation of Gentry Society in Late-Ming China*. Cambridge, MA: Harvard University Press and the Harvard-Yenching Institute, 1993.

———. *The Confusions of Pleasure: Commerce and Culture in Ming China*. Berkeley: University of California Press, 1999.

———. "Weber's Religion of China." In *Max Weber's Economic Ethic of the World Religions: An Analysis*, edited by Thomas C. Ertman. Cambridge: Cambridge University Press, 2017, 87–108.

Brook, Timothy, Jérôme Bourgon, and Gregory Blue. *Death by a Thousand Cuts*. Cambridge, MA: Harvard University Press, 2008.

Brown, Tristan G. "The Veins of the Earth: Property, Environment, and Cosmology in Nanbu County." PhD diss., Columbia University, 2017.

———. "The Deeds of the Dead in the Courts of the Living: Graves in Qing Law." *Late Imperial China* 39.2 (2018): 109–55.

———. "A Mountain of Saints and Sages: Muslims in the Landscape of Popular Religion in Late Imperial China." *T'oung Pao* 105.3–4 (2019): 437–91.

———. "The Muslims of 'All Under Heaven': Islam on the Ground in Late Imperial China." *Archives de sciences sociales des religions* 1 (2021): 79–106.

Bruun, Ole. *Fengshui in China: Geomantic Divination Between State Orthodoxy and Popular Religion*. Honolulu: University of Hawaiʻi Press, 2003.

———. *An Introduction to Feng Shui*. Cambridge: Cambridge University Press, 2012.

Buck, John Lossing. *Chinese Farm Economy: A Study of 2866 Farms in Seventeen Localities and Seven Provinces in China*. Chicago: University of Chicago Press, 1930.

Buoye, Thomas M. *Manslaughter, Markets, and Moral Economy: Violent Disputes Over Property Rights in Eighteenth-Century China*. Cambridge: Cambridge University Press, 2000.

Burton-Rose, Daniel. "Terrestrial Rewards as Divine Recompense: The Self-Fashioned Piety of the Peng Lineage of Suzhou, 1650s–1870s." PhD diss., Princeton University, 2016.

———. "Wenchang Buildings in Late Imperial China: A Consideration of the Visual Record in Late Imperial Local Gazetteers." *Journal of the European Association for Chinese Studies* 4.1 (2023): 59–84.

Cai Dongzhou 蔡東洲. "Langzhong Chenshi zupu kaolun" 閬中陳氏族譜考論. *Wenxian* 文獻 3 (1997): 134–51.

———. *Qingdai Nanbu xianya dang'an yanjiu* 清代南部縣衙檔案研究. Beijing: Zhonghua shuju, 2012.

Cai Dongzhou 蔡東洲 and Zhang Liang 張亮. "Nanbu dang'an zhong youguan Songdai Langzhou Chenshi jiazu mu dang'an yanjiu" 南部檔案中有關宋代閬州陳氏家族墓檔案研究. *Zhonghua wenhua luntan* 中華文化論壇 4 (2014): 98–104.

Cao Qiangxin 曹強新. "Qingdai xingbu jianyu yuanliu kaoxi" 清代刑部監獄源流考析. *Fanzui yu gaizao yanjiu* 犯罪與改造研究 10 (2018): 71–74.

Cao Xueqin 曹雪芹 and Gao E. *Honglou meng* 紅樓夢. Beijing: Renmin wenxue chubanshe, 2005.

Carroll, Peter J. *Between Heaven and Modernity: Reconstructing Suzhou, 1895–1937*. Stanford, CA: Stanford University Press, 2006.

Carlitz, Katherine. "Genre and Justice in Late Qing China: Wu Woyao's Strange Case of Nine Murders and Its Antecedents." In *Writing and Law in Late Imperial China: Crime, Conflict, and Judgment*, edited by Robert E. Hegel and Katherine Carlitz. Seattle: University of Washington Press, 2007.

Cassel, Pär. *Grounds of Judgment: Extraterritoriality and Imperial Power in Nineteenth-Century China and Japan*. New York: Oxford University Press, 2012.

Chang, Chung-li. *The Chinese Gentry: Studies on Their Role in Nineteenth-Century Chinese Society*. Seattle: University of Washington Press, 1955.

Chang, Ping-Ying. "Chinese Hereditary Mathematician Families of the Astronomical Bureau, 1620–1850," PhD diss., City University of New York, 2015.

———. *The Chinese Astronomical Bureau, 1620–1850: Lineages, Bureaucracy, and Technical Expertise*. New York: Routledge, 2022.

Chao Yuanfang 巢元方. *Zhubing yuanhou lun* 諸病源候論, edited by Lu Zhaolin 魯兆麟. c. 610 CE Rpt. Shenyang: Liaoning kexue jishu chubanshe, 1997.

Chen, Gilbert. "Living in This World: A Social History of Buddhist Monks and Nuns in Nineteenth-Century Western China." PhD diss., Washington University, 2019.

Chen, Hailian. *Zinc for Coin and Brass: Bureaucrats, Merchants, Artisans, and Mining Laborers in Qing China, ca. 1680s–1830s*. Leiden: Brill, 2019.

———. "Creating Intellectual Space for West-East and East-East Knowledge Transfer: Global Mining Literacy and the Evolution of Textbooks on Mining in Late Qing China, 1860–1911."

In *Accessing Technical Education in Modern Japan*, edited by Erich Pauer and Regine Mathias. Amsterdam: Amsterdam University Press, 2022, 37–69.

Ch'en, Jerome. *The Highlanders of Central China: A History, 1895–1937*. New York: M. E. Sharpe, 1992.

Chen Jinguo 陳進國. *Xinyang, yishi yu xiangtu shehui: fengshui de lishi renleixue tansuo*. 信仰、儀式與鄉土社會：風水的歷史人類學探索. Beijing: Zhongguo shehui kexue chubanshe, 2005.

Chen, Li. "Legal Specialists and Judicial Administration in Late Imperial China, 1651–1911." *Late Imperial China* 33.1 (2012): 1–54.

———. *Chinese Law in Imperial Eyes: Sovereignty, Justice, and Transcultural Politics*. New York: Columbia University Press, 2016.

Chen, Song-Chuan. "The Power of Ancestors: Tombs and Death Practices in Late Qing China's Foreign Relations, 1845–1914." *Past and Present* 239.1 (2018): 113–42.

Chen Tingxian 陳廷賢. *Sichuan Nanbu xian Chenshi zupu* 四川南部縣陳氏族譜. (Preface Dated 1858). Langzhong Municipal Library, Local Rare Books Collection.

Chia, Lucille. "Text and *Tu* in Context: Reading the Illustrated Page in Chinese Blockprinted Books." *Bulletin de l'École française d'Extrême-Orient* (2002): 241–76.

China Ministry of Communications, ed., *List of Post Offices*. 11th issue. Shanghai: Department of the Directorate General of Posts, 1923.

Chiu Pengsheng 邱澎生. *Dang jingji yushang falü: Ming-Qing Zhongguo de shichang bianhua* 當經濟遇上法律：明清中國的市場變化. Taipei: Lianjing chuban shiye gufen youxian gongsi, 2018.

Chiu Pengsheng 邱澎生 and Chen Hsi-yuan 陳熙遠, eds. *Ming-Qing falü yunzuo zhong de quanli yu wenhua* 明清法律運作中的權力與文化. Taipei: Academia Sinica and Linking Publishing Company, 2009.

Ch'ü, T'ung-tsu. *Local Government in China Under the Ch'ing*. Cambridge, MA: Harvard University Press, 1962.

Clément, Sophie, Pierre Clément, and Shin Yong Hak. *Architecture du paysage en Asie orientale: du "fengshui" comme modèle conceptuel et comme pratique d'harmonisation bâti-paysage*. Paris: L'Institut Français d'Architecture, 1982.

Clunas, Craig. *Fruitful Sites: Garden Culture in Ming Dynasty China*. London: Reaktion Books, 1996.

———. *Pictures and Visuality in Early Modern China*. London: Reaktion Books, 1997.

Coggins, Chris. "When the Land Is Excellent: Village Fengshui Forests and the Nature of Lineage, Polity, and Vitality in Southern China." In *Religion and Ecological Sustainability in China*, edited by James Miller, Dan Smyer Yu, and Peter Van der Veer. London: Routledge, 2014, 97–126.

Cohen, Myron L. "Souls and Salvation: Conflicting Themes in Chinese Popular Religion." In *Death Ritual in Late Imperial and Modern China*, edited by James Watson and Evelyn Rawski. Berkeley: University of California Press, 1988, 180–202.

———. "Lineage Organization in North China." *The Journal of Asian Studies* 49 (1990): 509–34.

———. "Family Management and Family Division in Contemporary Rural China." *The China Quarterly* 130 (1992): 357–77.

Collins, William Frederick. *Mineral Enterprise in China*. London: Heinemann, 1918.
Constant, Frédéric. "Thinking with Models: The Construction of Legal Cases as Reflected in Late Qing Local Archives." *T'oung Pao* 107.3-4 (2021): 417–73.
Cronon, William. *Changes in the Land: Indians, Colonists, and the Ecology of New England*. New York: Hill and Wang, 1983.
Crook, David, and Isabel Crook. *Revolution in a Chinese Village: Ten Mile Inn*. 1959 Rpt. London: Routledge, 2006.
Crook, Isabel, and Christina Kelley Gilmartin, *Prosperity's Predicament: Identity, Reform, and Resistance in Rural Wartime China*. Lanham, MD: Rowman & Littlefield, 2013.
Cruikshank, Julie. "'Are Glaciers 'Good to Think With?' Recognising Indigenous Environmental Knowledge." *Anthropological Forum* 22 (2012): 239–50.
Dai, Yingcong. *The Sichuan Frontier and Tibet: Imperial Strategy in the Early Qing*. Seattle: University of Washington Press, 2009.
———. *The White Lotus War: Rebellion and Suppression in Late Imperial China*. Seattle: University of Washington Press, 2019.
Daigle, Jean-Guy. "Challenging the Imperial Order: The Precarious Status of Local Christians in Late-Qing Sichuan." *European Journal of East Asian Studies* (2005): 1–29.
Dardess, John W. "A Ming Landscape: Settlement, Land Use, Labor, and Estheticism in T'ai-ho County, Kiangsi." *Harvard Journal of Asiatic Studies* 49.2 (1989): 295–364.
Davis, Edward. *Society and the Supernatural in Song China*. Honolulu: University of Hawai'i Press, 2001.
Day, Jenny Huangfu. *Qing Travelers to the Far West: Diplomacy and the Information Order in Late Imperial China*. Cambridge: Cambridge University Pres, 2018.
De Groot, Jan Jakob Maria. *The Religious System of China: Its Ancient Forms, Evolution, History and Present Aspect, Manners, Customs and Social Institutions Connected Therewith*, vols. 2 and 3. Leiden: Brill, 1894 and 1897.
Dean, Kenneth. *Lord of the Three in One: The Spread of a Cult in Southeast China*. Princeton: Princeton University Press, 1998.
Deng, Kent. "China's Population Expansion and Its Causes during the Qing Period, 1644–1911." *Economic History Working Paper Series* (219/2015). The London School of Economics and Political Science, London.
Dennis, Joseph R. *Writing, Publishing, and Reading Local Gazetteers in Imperial China, 1100–1700*. Cambridge, MA: Harvard University Asia Center, 2015.
Dong, Madeleine Yue. *Republican Beijing: The City and Its Histories*. Berkeley: University of California Press, 2003.
Doolittle, Justus. *Social Life of the Chinese, with Some Account of Their Religious, Governmental, Educational, and Business Customs and Opinions*, vols. 1 and 2. New York: Harper & Brothers, 1865.
Driscoll, Mark W. *The Whites Are the Enemies of Heaven: Climate Caucasianism and Asian Ecological Protection*. Durham, NC: Duke University Press, 2020.
Duara, Prasenjit. *Culture, Power, and the State: Rural North China, 1900–1942*. Stanford, CA: Stanford University Press, 1988.
Ebrey, Patricia. "Cremation in Sung China." *American Historical Review* 95 (1990): 406–28.

———. "Sung Neo-Confucian Views on Geomancy." In *Meeting of Minds: Intellectual and Religious Interaction in East Asian Traditions of Thought*, edited by Irene Bloom and Joshua A. Fogel. New York: Columbia University Press, 1997, 75–107.

Encyclopædia Britannica, 3rd ed., vol. 7. Edinburgh, 1797.

Edgerton-Tarpley, Kathryn. *Tears from Iron: Cultural Responses to Famine in Nineteenth Century China*. Berkeley: University of California Press, 2008.

Eckfeld, Tonia. *Imperial Tombs in Tang China, 618–907: The Politics of Paradise*. New York: Routledge, 2005.

Eitel, Ernest John. *Feng-shui, or the Rudiments of Natural Science in China*. 1873 Rpt. Hong Kong: Lane, Crawford & Company, 1878.

Elleman, Bruce A., and S.C.M. Paine. *Modern China: Continuity and Change, 1644 to the Present*. Lanham, MD: Rowman & Littlefield Publishers, 2019.

Ellickson, Robert C. "The Costs of Complex Land Titles: Two Examples from China." *Property Rights Conference Journal* 1 (2012): 281–302.

Elliott, Mark C. *The Manchu Way: The Eight Banners and Ethnic Identity in Late Imperial China*. Stanford, CA: Stanford University Press, 2001.

Elman, Benjamin. "Political, Social, and Cultural Reproduction via Civil Service Examinations in Late Imperial China." *The Journal of Asian Studies* 50 (1991): 7–28.

———. *A Cultural History of Civil Examinations in Late Imperial China*. Berkeley: University of California Press, 2000.

———. "Naval Warfare and the Refraction of China's Self-Strengthening Reforms into Scientific and Technological Failure, 1865–1895." *Modern Asian Studies* 38.2 (2004): 283–326.

———. "Early Modern or Late Imperial? The Crisis of Classical Philology in Eighteenth Century China." In *World Philology*, edited by Sheldon Pollock, Benjamin Elman, and Ku-min Kevin Chang. Cambridge, MA: Harvard University Press, 2015, 225–44.

———. *On Their Own Terms: Science in China, 1550–1900*. Cambridge, MA: Harvard University Press, 2009.

———. *Civil Examinations and Meritocracy in Late Imperial China*. Cambridge, MA: Harvard University Press, 2013.

Elvin, Mark. "On Water Control and Management during the Ming and Ch'ing Periods: A Review Article," *Ch'ing-shih wen-t'i* 3 (1975): 82–103.

———. "Female Virtue and the State in China." *Past & Present* 104.1 (1984): 111–52.

———. "How Did the Cracks Open? The Origins of the Subversion of Chin's Late-Traditional Culture by the West." *Thesis Eleven* 57.1 (1999): 1–16.

———. *The Retreat of the Elephants: An Environmental History of China*. New Haven, CT: Yale University Press, 2008.

Entenmann, Robert. "Migration and Settlement in Sichuan, 1644–1796." PhD diss., Harvard University, 1982.

Esherick, Joseph W. *The Origins of the Boxer Uprising*. Berkeley: University of California Press, 1988.

Fairbank, John King, and Kwang-Ching Liu, eds. *The Cambridge History of China: Late C'ing, 1800–1911*, part 2, vol. 2. Cambridge: Cambridge University Press, 1980.

Fang, Qiang. "Hot Potatoes: Chinese Complaint Systems from Early Times to the Late Qing (1898)." *The Journal of Asian Studies* 68.4 (2009): 1105–35.

Faure, Bernard, and Nobumi Iyanaga, eds. *Onmyōdō, The Way of Yin and Yang*. Special issue of the *Cahiers d'Extrême-Asie* (2013).

Faure, David. *The Structure of Chinese Rural Society: Lineage and Village in the Eastern New Territories, Hong Kong*. New York: Oxford University Press, 1986.

———. "Between House and Home: The Family in South China." In *House, Home, Family: Living and Being Chinese*, edited by Ronald G. Knapp and Kai-yin Lo. Honolulu: University of Hawaiʻi Press, 2005, 281–94.

———. *China and Capitalism: A History of Business Enterprise in Modern China*. Hong Kong: Hong Kong University Press, 2006.

———. *Emperor and Ancestor: State and Lineage in South China*. Stanford, CA: Stanford University Press, 2007.

Faure, David, and Helen Siu, eds. *Down to the Earth: The Territorial Bond in South China*. Stanford, CA: Stanford University Press, 1995.

Faure, David, and Ho Ts'ui-p'ing, eds. *Chieftains into Ancestors: Imperial Expansion and Indigenous Society in Southwest China*. Vancouver: University of British Columbia Press, 2013.

Felt, David Jonathan. *Structures of the Earth: Postimperial Metageographies of Early Medieval China*. Cambridge, MA: Harvard University Asia Center, 2021.

Feng, Jiren. *Chinese Architecture and Metaphor: Song Culture in the Yingzao Fashi Building Manual*. Honolulu: University of Hawaiʻi Press, 2012.

Feng Tianyu 馮天瑜, ed. *Wuhan shizhi: Renwuzhi* 武漢市誌: 人物志. Wuhan: Wuhan daxue chubanshe, 1999.

Feng Xianliang 馮賢亮. "Fenying yizhong: Ming-Qing Jiangnan de minzhong shenghuo yu huanjing baohu" 墳塋義塚: 明清江南的民眾生活與環境保護. *Zhongguo shehui lishi pinglun* 中國社會歷史評論 7 (2006): 161–84.

Feuchtwang, Stephan. *An Anthropological Analysis of Chinese Geomancy*. Vientiane, Laos: Vithagna, 1974.

Fischel, William A. *The Economics of Zoning Laws: A Property Rights Approach to American Land Use Controls*. Baltimore, MD: Johns Hopkins University Press, 1987.

Forêt, Philippe. *Mapping Chengde: The Qing Landscape Enterprise*. Honolulu: University of Hawaiʻi Press, 2000.

Freedman, Maurice. *Lineage Organization in Southeastern China*. London: Athlone Press, 1965.

———. *Chinese Lineage and Society: Fukien and Kwangtung*. London: Athlone Press, 1966.

———. *The Study of Chinese Society*. Stanford, CA: Stanford University Press, 1979.

Gerritsen, Anne. *The City of Blue and White: Chinese Porcelain and the Early Modern World*. Cambridge: Cambridge University Press, 2020.

Giersch, Charles Patterson. *Asian Borderlands: The Transformation of Qing China's Yunnan Frontier*. Cambridge, MA: Harvard University Press, 2006.

Gladney, Dru C. *Muslim Chinese: Ethnic Nationalism in the People's Republic*. Cambridge, MA: Harvard Council on East Asian Studies, 1991.

Golas, Peter J. *Science and Civilisation in China*, vol. 5, part 13: *Mining*. Cambridge: Cambridge University Press, 1999.

Gong Yilong 龔義龍. *Zuqun ronghe yu shehui zhenghe: Qingdai Chongqing yimin jiazu yanjiu* 族群融合與社會整合：清代重慶移民家族研究. Beijing: Zhongguo wenshi chubanshe, 2015.

Goossaert, Vincent. "1898: The Beginning of the End for Chinese Religion?" *The Journal of Asian Studies* 65.2 (2006): 1–29.

Gou Deyi 苟德儀. *Qingdai jiceng zuzhi yu xiangcun shehui guanli: Yi Sichuan Nanbu xian wei ge'an de kaocha* 清代基層組織與鄉村社會管理：以四川南部縣為個案的考察. Beijing: Zhonghua shuju, 2020.

———. "The Destruction of Immoral Temples in Qing China." *Journal of Chinese Studies (Special Isssue)*, 2 (2009): 131–53.

Graham, David Crockett. *Folk Religion in Southwest China*. 1961 Rpt.; Washington, DC: Smithsonian Press, 1967.

Greatrex, Roger. "The Illegal Trade in Saltpetre in Southern China in the Eighteenth and Early Nineteenth Centuries." In *Southwest China in a Regional and Global Perspective (c.1600–1911)*, edited by Ulrich Theobald and Jin Cao. Leiden: Brill, 2018, 349–78.

Guo, Weiting. "Social Practice and Judicial Politics in 'Grave Destruction' Cases in Qing Taiwan, 1683–1895." In *Chinese Law: Knowledge, Practice and Transformation, 1530s to 1950s*, edited by Li Chen and Madeleine Zelin. Leiden: Brill, 2015, 84–123.

Guy, R. Kent. *The Emperor's Four Treasuries: Scholars and the State in the Late Ch'ien-lung Era*. Cambridge, MA: Counsil on East Asian Studies, 1987.

———. *Qing Governors and Their Provinces: The Evolution of Territorial Administration in China, 1644–1796*. Seattle: University of Washington Press, 2010.

Halsey, Stephen R. "Sovereignty, Self-Strengthening, and Steamships in Late Imperial China." *Journal of Asian History*, 48.1 (2014): 81–111.

———. *Quest for Power: European Imperialism and the Making of Chinese Statecraft*. Cambridge, MA: Harvard University Press, 2015.

Hansen, Valerie. *Negotiating Daily Life in Traditional China*. New Haven, CT: Yale University Press, 1995.

Hanson, Marta. "The Golden Mirror in the Imperial Court of the Qianlong Emperor, 1739–1742." *Early Science and Medicine* 8.2 (2003): 111–47.

Haraway, Donna. "Situated Knowledges: The Science Question in Feminism and the Privilege of Partial Perspective." *Feminist Studies* 14 (1988): 575–99.

Harrison, Henrietta. *The Man Awakened from Dreams: One Man's Life in a North China Village, 1857–1842*. Stanford, CA: Stanford University Press, 2005.

Harrison, Peter. "'Science' and 'Religion': Constructing the Boundaries." *The Journal of Religion* 86 (2006): 81–106.

Hayes, James. "Geomancy and the Village." In *The Rural Communities of Hong Kong: Studies and Themes*, edited by James Hayes. New York: Oxford University Press, 1983, 146–52.

———. "Specialists and Written Materials in the Village World." In *Popular Culture in Late Imperial China*, edited by David Johnson, Andrew J. Nathan, and Evelyn S. Rawski. Berkeley: University of California Press, 1985, 75–111.

He Wei 何偉. "Jindai Jiangnan kanyuye yanjiu" 近代江南堪輿業研究. PhD diss., Soochow University, 2015.

———. "Jindai kanyu tankuang lilun chutan" 近代堪輿探礦理論初探. *Zhongguo kuangye daxue xuebao* 中國礦業大學學報 5 (2015): 69–73.

Henderson, John B. *The Development and Decline of Chinese Cosmology*. New York: Columbia University Press, 1984.

Hong Jianrong 洪健榮. "Dang [*fengshui*] chengwei [*huoshui*]—Qingdai Taiwan shehui de fengshui jiufen" 當「風水」成為「禍水」——清代臺灣社會的風水糾紛, Part Two. *Tainan wenhua* 臺南文化 62 (2008): 1–46.

Hornibrook, Jeff. *A Great Undertaking: Mechanization and Social Change in a Late Imperial Chinese Coalmining Community*. Albany, NY: Suny Press, 2015.

Hostetler, Laura. *Qing Colonial Enterprise: Ethnography and Cartography in Early Modern China*. Chicago: University of Chicago Press, 2001.

Hu Daojing 胡道靜, ed. *Zangwai daoshu* 藏外道書. Chengdu: Ba-shu shushe, 1992.

Hu Kaiquan 胡開全 and Su Donglai 蘇東來, eds. *Chengdu Longquanyi bainian qiyue wenshu* 成都龍泉驛百年契約文書. Chengdu: Ba-shu shushe, 2012.

Huang, Chin-Shing. *Confucianism and Sacred Space: The Confucian Temple from Imperial Times to Today*. Translated by Jonathan Chin. New York: Columbia University Press, 2020.

Huang, Fei. *Reshaping the Frontier Landscape: Dongchuan in Eighteenth-Century Southwest China*. Leiden: Brill, 2018.

Huang, Philip C. C. *The Peasant Economy and Social Change in North China*. Stanford, CA: Stanford University Press, 1985.

———. *Civil Justice in China: Representation and Practice in the Qing*. Stanford, CA: Stanford University Press, 1996.

———. *Code, Custom, and Legal Practice in China: The Qing and the Republic Compared*. Stanford, CA: Stanford University Press, 2001.

Huang, Philip C. C., and Kathryn Bernhardt, eds. *Research from Archival Case Records: Law, Society and Culture in China*. Leiden: Brill, 2014.

Huang Shangjun 黃尚軍, You Li 遊黎, and Li Guotai 李國太, eds. *Chuan dongbei Qingdai mubei jicheng* 川東北清代墓碑集成. Chengdu: Sichuan minzu chubanshe, 2016.

Huang Yinong 黃一農. "Tongshu: Zhongguo chuantong tianwen yu shehui de jiaorong" 通書——中國傳統天文與社會的交融. *Hanxue yanjiu* 漢學研究 14 (1996): 159–86.

———. *Shehui tianwenxue shi shijiang* 社會天文學史十講. Shanghai: Fudan daxue chubanshe, 2004.

Hung, Ho-fung. *Protest with Chinese Characteristics: Demonstrations, Riots, and Petitions in the Mid-Qing Dynasty*. New York: Columbia University Press, 2011.

Hutson, James. *Mythical and Practical in Szechwan*. Shanghai: The National Review Office, 1915.

Hymes, Robert. *Way and Byway: Taoism, Local Religion, and Models of Divinity in Sung and Modern China*. Berkeley: University of California Press, 2002.

———. "Truth, Falsity, and Pretense in Song China: An Approach through the Anecdotes of Hong Mai." *Zhongguo shixue* 中國史學 15 (2005): 1–26.

Idema, Wilt. *Judge Bao and the Rule of Law: Eight Ballad Stories from the Period 1250–1450*. Hackensack, NJ: World Scientific, 2009.

Isett, Christopher Mills. *State, Peasant, and Merchant in Qing Manchuria, 1644–1862*. Stanford, CA: Stanford University Press, 2007.

Jami, Catherine. *The Emperor's New Mathematics: Western Learning and Imperial Authority during the Kangxi Reign (1662–1722)*. Oxford: Oxford University Press, 2012.

Javers, Quinn. "The Logic of Lies: False Accusation and Legal Culture in Late Qing Sichuan." *Late Imperial China* 35 (2014): 27–55.

Jia Lingli 賈玲利. "Shaanxi Guanzhong diqu nongcun juzhu jianzhu wenhua tantao" 陝西關中地區農村居住建築文化探討. *Sichuan jianzhu kexue yanjiu* 四川建築科學研究 1 (2007): 160–63.

Jiang Haitao 姜海濤. *Qingdai Sichuan diqu fengshui ta shikong fenbu yanjiu* 清代四川地區風水塔時空分布研究. MA thesis, China Southwest University, 2013.

Jiang Huang 蔣鐄, Wu Shengzu 吳繩祖, and Wang Kaizhuo 王開琸, eds. *Jiuyi shanzhi (liangzhong) Yanling zhi* 九疑山志(兩種) 炎陵志. Changsha: Yuelu shushe, 2008.

Jiang, Yonglin. *The Mandate of Heaven and the Great Ming Code*. Seattle: University of Washington Press, 2011.

Jianpeng Deng, Li Chen (trans.), and Yu Wang (trans.) "Classifications of Litigation and Implications for Qing Judicial Practice." In *Chinese Law: Knowledge, Practice and Transformation, 1530s to 1950s*, edited by Li Chen and Madeleine Zelin. Leiden: Brill, 2015, 17–46.

Jin, Cao. "Mint Metal Mining and Minting in Sichuan, 1700–1900: Effects on the Regional Economy and Society." PhD diss., University of Tubingen, 2012.

Jin Shengyang 金生楊, Xie Jiayuan 謝佳元, and Liu Yanwei 劉艷偉. "Qingdai Nanbu xian yanli shoushi chuyi" 清代南部縣鹽厘首事芻議. *Yanye shi yanjiu* 鹽業史研究 3 (2017): 3–9.

Jin Tianzhu 金天柱 and Hai Zhengzhong 海正忠, eds. *Qingzhen shiyi* 清真釋疑. Yinchuan: Ningxia renmin chubanshe, 2002.

Jingmenshi shuili zhi bianzuan weiyuanhui 荊門市水利志編纂委員会, ed., *Jingmenshi shuili zhi* 荊門市水利志. Wuhan: Hubei Jiaoyu chubanshe, 1989.

Johnson, David G. *Spectacle and Sacrifice: The Ritual Foundations of Village Life in North China*. Cambridge, MA: Harvard University Asia Center, 2009.

Jones, Stephen. *In Search of the Folk Daoists in North China*. Farnham, UK: Ashgate Publishing, 2010.

Josephson-Storm, Jason A. *The Myth of Disenchantment*. Chicago: University of Chicago Press, 2017.

Kalinowski, Marc. *Divination et sociéte dans la Chine medievale: Etude des manuscrits de Dunhuang de la Bibliotheque nationale de France et de la British Library*. Paris: Bibliothèque nationale de France, 2003.

Katz, Paul R. "Divine Justice in Late Imperial China: A Preliminary Study of Indictment Rituals." In *Religion and Chinese Society: Volume II, Taoism and Local Religion in Modern China*, edited by John Lagerwey. Hong Kong and Paris: Chinese University of Hong Kong and École française d'Extrême-Orient, 2004, 869–902.

———. *Divine Justice: Religion and the Development of Chinese Legal Culture*. London: Routledge, 2008.

———. *Religion in China and Its Modern Fate*. Waltham, MA: Brandeis University Press, 2014.

Katz, Paul R., and Vincent Goossaert. *The Fifty Years that Changed Chinese Religion, 1898–1949*. Ann Arbor, MI: Association for Asian Studies, 2021.

Keliher, Macabe. *The Board of Rites and the Making of Qing China*. Berkeley: University of California Press, 2019.

Kieschnick, John. *The Impact of Buddhism on Chinese Material Culture*. Princeton: Princeton University Press, 2003.

Kim, Nanny. "River Control, Merchant Philanthropy, and Environmental Change in Nineteenth-Century China." *Journal of the Economic and Social History of the Orient* 52.4 (2009): 660–94.

———. *Mountain Rivers, Mountain Roads: Transport in Southwest China, 1700–1850*. Leiden: Brill, 2019.

Kinzley, Judd C. *Natural Resources and the New Frontier: Constructing Modern China's Borderlands*. Chicago: University of Chicago Press, 2018.

Kiong, Tong Chee. *Chinese Death Rituals in Singapore*. London: RoutledgeCurzon, 2004.

Kleeman, Terry F. *A God's Own Tale: The Book of Transformations of Wenchang, the Divine Lord of Zitong*. Albany: State University of New York Press, 1994.

Knapp, Ronald G. "Siting and Situating a Dwelling: Fengshui, House-Building Rituals, and Amulets." In *House, Home, Family: Living and Being Chinese*, edited by Ronald G. Knapp and Kai-yin Lo. Honolulu: University of Hawai'i Press, 2005, 99–138.

———. *Chinese Bridges: Living Architecture from China's Past*. Clarendon, VT: Tuttle Publishing, 2008

———. *Chinese Houses: The Architectural Heritage of a Nation*. Clarendon, VT: Tuttle Publishing, 2012.

Ko, Dorothy. *The Social Life of Inkstones: Artisans and Scholars in Early Qing China*. Seattle: University of Washington Press, 2017.

Köll, Elisabeth. *Railroads and the Transformation of China*. Cambridge, MA: Harvard University Press, 2019.

Kory, Stephan N. "Presence in Variety: De-Trivializing Female Diviners in Medieval China." *Nan Nü* 18.1 (2016): 3–48.

Kotyk, Jeffrey. "Buddhist Astrology and Astral Magic in the Tang Dynasty." PhD diss., Leiden University, 2017.

Kuhn, Philip A. *Rebellion and Its Enemies in Late Imperial China: Militization and Social Structure, 1796–1864*. Cambridge, MA: Harvard University Press, 1980.

Kurtz, Joachim. *The Discovery of Chinese Logic*. Leiden: Brill, 2011.

Lagerwey, John. *China: A Religious State*. Hong Kong: Hong Kong University Press, 2010.

———. *Paradigm Shifts in Early and Modern Chinese Religion: A History*. Leiden: Brill, 2018.

Lam, Tong. *A Passion for Facts: Social Surveys and the Construction of the Chinese Nation-State, 1900–1949*. Berkeley: University of California Press, 2010.

Lan Yong 藍勇. "Qingchu Sichuan huhuan yu huanjing fuyuan wenti" 清初四川虎患與環境復原問題. *Zhongguo lishi dili luncong* 中國歷史地理論叢 3 (1994): 203–10.

Langzhongshi lishi wenhua mingcheng yanjiuhui 閬中市歷史文化名城研究會, ed., *Mingcheng yanjiu* 名城研究, vol. 14. Langzhong: Gazetteer Office of Langzhong Municipality, 2012.

Latour, Bruno, and Catherine Porter (trans.). *We Have Never Been Modern*. Cambridge, MA: Harvard University Press, 1993.

Lawson, Joseph. *A Frontier Made Lawless: Violence in Upland Southwest China, 1800–1956*. Vancouver: University of British Columbia Press, 2017.

Lavelle, Peter B. *The Profits of Nature: Colonial Development and the Quest for Resources in Nineteenth Century China*. New York: Columbia University Press, 2020.

Lee, Keekok. *The Philosophical Foundations of Chinese Medicine: Philosophy, Methodology, Science*. Lanham, MD: Lexington Books, 2017.

Li Baoping 李保平, Deng Ziping 鄧子平, Han Xiaobai 韓小白, eds. *Kailuan meikuang: Dang'an shiliaoji* 開灤煤礦：檔案史料集, vol. 1. Shijiazhuang: Hebei jiaoyu chubanshe, 2012.

Li Chaozheng 李朝正, ed. *Qingdai Sichuan jinshi zhenglüe* 清代四川進士征略. Chengdu: Sichuan daxue chubanshe, 1986.

Li Chenghua 李成華, ed. *Sichuansheng Cangxi xian hexi fenghuanggong hedong yangyueshan Lishi zongpu* 四川省蒼溪縣河西鳳凰宮河東陽岳山李氏宗譜. Cangxi: Cangxixian xingzi diannao wenyinbu, 2002.

Li Di 李杕. *Quan huo ji* 拳禍記. 1905 Rpt.; Shanghai: Tushan wan yinshuguan, 1923.

Li Dingxin 李定信. *Siku quanshu: Kanyulei dianji yanjiu* 四庫全書：堪輿類典籍研究. Shanghai: Shanghai guji chubanshe, 2007.

Li, Fan, and Michael Lackner. "Contradictory Forms of Knowledge? Divination and Western Knowledge in Late Qing and Early Republican China." In *Coping with the Future: Theories and Practices of Divination in East Asia*, edited by Michael Lackner. Leiden: Brill, 2017, 461–85.

Li, Huaiyin. *The Making of the Modern Chinese State, 1600–1950*. London: Routledge, 2020.

Li Jie 李婕. "Qingdai wanqi Sichuan beibu diqu xiangtu jianzhu jiqi wenhua neihan: Yi Nanbu xian Songjia dayuan weili" 清代晚期四川北部地區鄉土建築及其文化內涵：以南部縣宋家大院為例. *Xihua daxue xuebao* 西華大學學報 1 (2016): 1–6.

Li, Kan. "On the Road to a Modern City: New Transportation Technology and Urban Transformation of Tianjin, 1860–1937." PhD diss., University of Minnesota, 2020.

Li, Lillian M. *Fighting Famine in North China: State, Market, and Environmental Decline, 1690s–1990s*. Stanford, CA: Stanford University Press, 2007.

Li, Mengbi. "The Evolution of the *zhaobi*: Physical Stability and the Creation of Architectural Meaning." *The Journal of Architecture* 25 (2020): 45–64.

Li Pingyi 李平毅. "Zigong chuantong juluo jingguan zhong de fengshui wenhua jiexi" 自貢傳統聚落景觀中的風水文化解析. *Zhonghua wenhua luntan* 中華文化論壇 7 (2014): 121–25.

Li Runqiang 李潤強. "Qingdai jinshi de shikong fenbu yanjiu" 清代進士的時空分布研究. *Xibei shida xuebao* 西北師大學報 1 (2005): 62–69.

Li Siyi 李思逸. *Tielu xiandaixing: Wan Qing zhi Minguo de shikong tiyan yu wenhua xiangxiang* 鐵路現代性：晚清至民國的時空體驗與文化想像. Taipei: Shibao wenhua chuban gongsi, 2020.

Li Wenfu 李文福 and Li Yongqi 李永奇, eds. *Langzhong guji* 閬中古蹟. Beijing: Zhongyang wenxian chubanshe, 2009.

Li Xiaofang 李曉方 and Wen Xiaoxing 溫小興. "Ming Qing shiqi Gannan kejia diqu de fengshui xinyang yu zhengfu kongzhi" 明清時期贛南客家地區的風水信仰與政府控制. *Shehui kexue* 社會科學 1 (2007): 108–14.

Li, Yu. "Social Change During the Ming-Qing Transition and the Decline of Sichuan Classical Learning in the Early Qing." *Late Imperial China* 19.1 (1998): 26–55.

Li Zan 里贊. *Wan Qing zhouxian susong zhong de shenduan wenti: Cezhong Sichuan Nanbu xian de shijian* 晚清州縣訴訟中的審斷問題：側重四川南部縣的實踐. Beijing: Falü chubanshe, 2010.

Li Zhe 李哲. *Zhongguo chuantong shehui fenshan de falü kaocha: Yi Qingdai wei zhongxin* 中國傳統社會墳山的法律考察——以清代為中心. Beijing: Zhongguo zhengfa daxue chubanshe, 2017.

Li Zhenghong 李錚虹. *Sichuan nongye jinrong yu diquan yidong zhi guanxi* 四川農業金融與地權異動之關係. In *Minguo ershi niandai Zhongguo dalu wenti ziliao* 民國二十年代中國大陸問題資料, vol. 89. Taipei: Chengwen chubanshe, 1977.

Li Zhitian (Lai Chi Tim) 黎志添. "Zheng Guanying 'xiandao' yu 'jiushi' de sixiang he shijian: Jianping qi dui Qingmo Minchu daojiao fazhan de yingxiang ji yiyi" 鄭觀應'仙道'與'救世'的思想和實踐：兼評其對清末民初道教發展的影響及意義. *Zhongguo wenhua yanjiusuo xuebao* 中國文化研究所學報 67 (2018): 151–202.

Liang, Linxia. *Delivering Justice in Qing China: Civil Trials in the Magistrate's Court*. Oxford: Oxford University Press, 2007.

Liang Yong 梁勇. *Yimin, guojia yu difang quanshi: Yi Qingdai Ba xian weili* 移民、國家與地方權勢：以清代巴縣為例. Beijing: Zhonghua shuju, 2014.

Liangshan Yizu zizhizhou bowuguan 涼山彝族自治州博物館 and Liangshan Yizu zizhizhou wenwu guanlisuo 涼山彝族自治州文物管理所, eds. *Liangshan lishi beike zhuping* 涼山歷史碑刻注評. Beijing: Wenwu chubanshe, 2011.

Lin, Man-Houng. *China Upside Down: Currency, Society, and Ideologies, 1808–1856*. Cambridge, MA: Harvard University Council on East Asian Studies, 2006.

Lin, Wei-ping. "Boiling Oil to Purify Houses (*zhuyou jingwu* 煮油淨屋): A Dialogue between Religious Studies and Anthropology." In *Exorcism in Daoism: A Berlin Symposium*, edited by Florian C. Reiter. Wiesbaden, Germany: Harrassowitz, 2011, 151–69.

Lipman, Jonathan. *Familiar Strangers: A History of Muslims in Northwest China*. Seattle: University of Washington Press, 1998.

Little, Archibald John. *Through the Yang-tse Gorges*. London: Sampson Low, Marston, Searle & Rivington, Ltd., 1888.

Liu Bingxue 劉冰雪. "Qingdai fengshui zhengsong yanjiu: Yi fenzang jiufen weili" 清代風水爭訟研究：以墳葬糾紛為例. *Zhengfa luntan* 政法論壇 4 (2012): 18–29.

Liu Xiancheng 劉先澄. *Langyuan bianlian jijin* 閬苑區聯集錦. Yinchuan: Ningxia renmin chubanshe, 2010.

Liu, Yanchi. *The Essential Book of Traditional Chinese Medicine*. New York: Columbia University Press, 1988.

Liu, Yonghua. "The World of Rituals: Masters of Ceremonies (*Lisheng*), Ancestral Cults, Community Compacts, and Local Temples in Late Imperial Sibao, Fujian." PhD diss., McGill University, 2003.

———. *Confucian Rituals and Chinese Villagers: Ritual Change and Social Transformation in a Southeastern Chinese Community, 1368–1949*. Leiden: Brill, 2013.

Liu Yunsheng 劉雲生. *Zigong yanye qiyue yuhui jishi* 自貢鹽業契約語彙輯釋. Beijing: Falü chubanshe, 2014.

Loewe, Michael. "The Pivot of the Four Quarters: A Preliminary Enquiry into the Origins and Character of the Ancient Chinese City by Paul Wheatley." *Modern Asian Studies* 2 (1973): 288–91.

———. *Dong Zhongshu, a "Confucian" Heritage and the Chunqiu Fanlu*. Leiden: Brill, 2011.

Long Denggao 龍登高. *Zhongguo chuantong diquan zhidu jiqi bianqian* 中國傳統地權制度及其變遷. Beijing: Zhongguo shehui kexue chubanshe, 2018.

Long Xianzhao 龍顯昭, ed. *Ba-shu fojiao beiwen jicheng* 巴蜀佛教碑文集成. Chengdu: Ba-shu shushe, 2004.

Long Xianzhao 龍顯昭 and Huang Haide 黃海德, eds. *Ba-shu daojiao beiwen jicheng* 巴蜀道教碑文集成. Chengdu: Sichuan daxue chubanshe, 1997.

Lü, Lingfeng. "Eclipses and the Victory of European Astronomy in China." *East Asian Science, Technology, and Medicine* 27 (2007): 127–45.

Lu Zijian 魯子健. *Qingdai Sichuan caizheng shiliao* 清代四川財政史料, vol. 2. Chengdu: Sichuansheng shehui kexueyuan chubanshe, 1988.

Luo Dajing 羅大經. *Helin yulu* 鶴林玉露. Preface 1248. Beijing: Zhonghua shuju, 1983.

Luo Guanzhong 羅貫中. *Sanguo yanyi* 三國演義. Beijing: Renmin wenxue chubanshe, 2005.

Lynn, Richard John. *The Classic of Changes: A New Translation of the I Ching as Interpreted by Wang Bi.* New York: Columbia University Press, 2004.

Macauley, Melissa Ann. *Social Power and Legal Culture: Litigation Masters in Late Imperial China.* Stanford, CA: Stanford University Press, 1997.

MacCormack, Geoffrey. *The Spirit of Traditional Chinese Law.* Atlanta: University of Georgia Press, 1996.

Marks, Robert B. *Tigers, Rice, Silk, and Silt: Environment and Economy in Late Imperial South China.* Cambridge: Cambridge University Press, 1998.

———. *China: An Environmental History (Second Edition).* Lanham, MD: Rowman & Littlefield, 2017.

Marsh, Robert M. "Weber's Misunderstanding of Traditional Chinese Law." *American Journal of Sociology* 106 (2000): 281–302.

Mazumdar, Sucheta. "Rights in People, Rights in Land: Concepts of Customary Property in Late Imperial China." *Extrême-Orient Extrême-Occident* (2001): 89–107.

McDermott, Joseph. *The Making of a New Rural Order in South China: Vol. 1, Village, Land, and Lineage in Huizhou, 900–1600.* Cambridge: Cambridge University Press, 2014.

McMahon, Daniel. "Geomancy and Walled Fortifications in Late Eighteenth Century China." *Journal of Military History* 76.2 (2012): 373–93.

Meyer, Jeffrey F. "'Feng-Shui' of the Chinese City." *History of Religions* 18.2 (1978): 138–55.

Meyer-Fong, Tobie. *What Remains: Coming to Terms with Civil War in 19th Century China.* Stanford, CA: Stanford University Press, 2013.

Miller, Ian M. *Fir and Empire: The Transformation of Forests in Early Modern China.* Seattle: University of Washington Press, 2020.

Miyazaki, Ichisada. *China's Examination Hell: The Civil Service Examinations of Imperial China.* New Haven, CT: Yale University Press, 1981.

Moll-Murata, Christine. *State and Crafts in the Qing Dynasty (1644–1911).* Amsterdam: Amsterdam University Press, 2018.

Mosca, Matthew. *From Frontier Policy to Foreign Policy: The Question of India and the Transformation of Geopolitics in Qing China.* Stanford, CA: Stanford University Press, 2013.

Mugerwa, Swidiq, Moses Moywaywa Nyangito, Huria Nderitu John, and Chris Bakuneta. "Farmers' Ethno-Ecological Knowledge of the Termite Problem in Semi-Arid Nakasongola." *African Journal of Agricultural Research* 6.13 (2011): 3183–91.

Mühlhahn, Klaus. *Criminal Justice in China: A History.* Cambridge, MA: Harvard University Press, 2009.

Murata, Sachiko. *The First Islamic Classic in Chinese: Wang Daiyu's Real Commentary on the True Teaching.* Albany: State University of New York Press, 2017.

Nanjing guomin zhengfu sifa xingzheng bu 南京國民政府司法行政部. *Minshi xiguan diaocha baogao lu* 民事習慣調查報告錄, vol. 1, ed. By Hu Xusheng 胡旭晟, Xia Xinhua 夏新華, and Li Jiaofa 李交發. Beijing: Zhongguo zhengfa daxue chubanshe, 2000.

Naquin, Susan. "The Transmission of White Lotus Sectarianism in Late Imperial China." In *Popular Culture in Late Imperial China*, edited by David Johnson, Andrew J. Nathan, and Evelyn S. Rawski, 255–91. Berkeley: University of California Press, 1985.

———. *Peking: Temples and City Life, 1400–1900*. Berkeley: University of California Press, 2000.

Nappi, Carla. *The Monkey and the Inkpot: Natural History and Its Transformations in Early Modern China*. Cambridge, MA: Harvard University Press, 2010.

Nedostup, Rebecca. *Superstitious Regimes: Religion and the Politics of Chinese Modernity*. Cambridge, MA: Harvard University Asia Center, 2009.

Needham, Joseph. *Science and Civilisation in China*, vol. 1: *Introductory Orientations*. Cambridge: Cambridge University Press, 1954.

———. *Science and Civilisation in China*, vol. 2: *History of Scientific Thought*. Cambridge: Cambridge University Press, 1956.

Ocko, Jonathan. "I'll Take It All the Way to Beijing: Capital Appeals in the Qing." *The Journal of Asian Studies* 47 (1988): 291–315.

Olles, Volker. *Ritual Words: Daoist Liturgy and the Confucian Liumen Tradition in Sichuan Province*. Wiesbaden, Germany: Harrassowitz Verlag, 2013.

Osborne, Anne. "The Local Politics of Land Reclamation in the Lower Yangzi Highlands." *Late Imperial China* 15 (1994): 1–46.

Overmyer, Daniel L. *Local Religion in North China in the Twentieth Century: The Structure and Organization of Community Rituals and Beliefs*. Leiden: Brill, 2009.

Ownby, David. *Brotherhoods and Secret Societies in Early and Mid-Qing China: The Formation of a Tradition*. Stanford, CA: Stanford University Press, 1996.

———. "Recent Chinese Scholarship on the History of Chinese Secret Societies." *Late Imperial China* 22.1 (2001): 139–58.

Pan Sheng 潘晟. *Zhishi, lisu yu zhengzhi: Songdai dilishu de zhishi shehui shitan* 知識、禮俗與政治：宋代地理術的知識社會史探. Nanjing: Jiangsu renmin chubanshe, 2018.

Pankenier, David W. "Characteristics of Field Allocation (*fenye*) Astrology in Early China." *Current Studies in Archaeoastronomy: Conversations Across Space and Time* (2005): 499–513.

———. *Astrology and Cosmology in Early China: Conforming Earth to Heaven*. Cambridge: Cambridge University Press, 2013.

Park, Nancy E. "Corruption in Eighteenth-Century China." *The Journal of Asian Studies* 56.4 (1997): 967–1005.

Parker, Edward Harper. *Up the Yang-tse*. Shanghai: Kelly & Walsh, 1899.

Paton, Michael. *Five Classics of Fengshui: Chinese Spiritual Geography in Historical and Environmental Perspective*. Leiden: Brill, 2013.

Pegg, Richard A. *Cartographic Traditions in East Asian Maps*. Honolulu: University of Hawai'i Press, 2014.

Peerenboom, Randall. "Law and Religion in Early China." In *Religion, Law, and Tradition: Comparative Studies in Religious Law*, edited by Andrew Huxley. New York: Routledge, 2002, 84–107.

Peng Jiusong 彭久松. "Zigong yanye qiyue kaoshi (part two)" 自貢鹽業契約考釋 (二). *Yanye shi yanjiu* 鹽業史研究 1 (1988): 33–40.

Peng Nansheng 彭南生. "Lun yangwu huodong zhong 'fengshui' guan de yingxiang" 論洋務活動中'風水'觀的影響. *Gansu shehui kexue* 甘肅社會科學 6 (2004): 91–94.

Peng Yuxin 彭雨新. *Qingdai tudi kaiken shi* 清代土地開墾史. Beijing: Nongye chubanshe, 1990.

Perdue, Peter C. "Water Control in the Dongting Lake Region during the Ming and Qing Periods." *The Journal of Asian Studies* 41.4 (1982): 747–65.

———. *Exhausting the Earth: State and Peasant in Hunan, 1500–1850.* Cambridge, MA: Harvard University Asia Center, 1987.

Perry, Elizabeth J. *Rebels and Revolutionaries in North China, 1845–1945.* Stanford, CA: Stanford University Press, 1980.

Pomeranz, Kenneth. *The Making of a Hinterland: State, Society, and Economy in Inland North China, 1853–1937.* Berkeley: University of California Press, 1993.

———. "Land Markets in Late Imperial and Republican China." *Continuity and Change* 23.01 (2008): 101–50.

———. *The Great Divergence: China, Europe, and the Making of the Modern World Economy.* Princeton: Princeton University Press, 2009.

Porkert, Manfred. *The Theoretical Foundations of Chinese Medicine: Systems of Correspondence.* Cambridge, MA: MIT Press, 1974.

Pregadio, Fabrizio, ed. *The Encyclopedia of Taoism*. Vol. 1. Abingdon, UK: Routledge, 2008.

Qi Qizhang 戚其章. *Zhongguo jindai shehui sichaoshi* 中國近代社會思潮史. Jinan: Shandong jiaoyu chubanshe, 1994.

Qi Shouhua 祁守華. "Chuanyan yu meitan" 川鹽與煤炭. *Yanyeshi yanjiu* 鹽業史研究 3 (1988): 52–54.

Qi Shouhua 祁守華 and Zhong Xiaozhong 鐘曉鐘, eds. *Zhongguo difang zhi meitan shiliao xuanji* 中國地方志煤炭史料選輯. Beijing: Meitan gongye chubanshe, 1990.

Qiu Jingjia 邱靖嘉. *Tiandi zhijian: Tianwen fenye de lishixue yanjiu* 天地之間：天文分野的歷史學研究. Beijing: Zhonghua shuju, 2020.

Ransmeier, Johanna. *Sold People: Traffickers and Family Life in North China.* Cambridge, MA.: Harvard University Press, 2017.

Raz, Gil. "Daoist Sacred Geography." In *Early Chinese Religion*. Vol. 2, *The Period of Division*, edited by John Lagerwey and Pengzhi Lü. Leiden: Brill, 2009.

Reed, Bradly. *Talons and Teeth: County Clerks and Runners in the Qing Dynasty.* Stanford, CA: Stanford University Press, 2000.

Ricci, Matteo, and Louis J. Gallagher (trans.). *China in the Sixteenth Century: The Journals of Matthew Ricci, 1583–1610.* New York: Random House, 1953.

Ritvo, Harriet. *The Dawn of Green: Manchester, Thirlmere, and Modern Environmentalism.* Chicago, IL: University of Chicago Press, 2009.

Rowe, William T. *Hankow: Conflict and Community in a Chinese City, 1796–1895.* Vol. 2. Stanford, CA: Stanford University Press, 1989.

———. *Saving the World: Chen Hongmou and Elite Consciousness in Eighteenth-Century China.* Stanford, CA: Stanford University Press, 2002.

Ruf, Gregory A. *Cadres and Kin: Making a Socialist Village in West China, 1921–1991.* Stanford, CA: Stanford University Press, 2000.
Ruitenbeek, Klaas. *Carpentry and Building in Late Imperial China: A Study of the Fifteenth-Century Carpenter's Manual Lu Ban Jing.* Leiden: Brill, 1996.
Ruskola, Teemu. *Legal Orientalism.* Cambridge, MA: Harvard University Press, 2013.
Said, Edward W. *Orientalism.* New York: Pantheon Books, 1978.
Schäfer, Dagmar. *The Crafting of the 10,000 Things: Knowledge and Technology in Seventeenth-Century China.* Chicago: University of Chicago Press, 2011.
Schlesinger, Jonathan. *A World Trimmed with Fur: Wild Things, Pristine Places, and the Natural Fringes of Qing Rule.* Stanford, CA: Stanford University Press, 2017.
Schlosberg, David, and David Carruthers. "Indigenous Struggles, Environmental Justice, and Community Capabilities." *Global Environmental Politics* 10.4 (2010): 12–35.
Schneewind, Sarah. *A Tale of Two Melons: Emperor and Subject in Ming China.* Indianapolis: Hackett Publishing, 2006.
———. *Shrines to Living Men in the Ming Political Cosmos.* Leiden: Brill, 2020.
Schurmann, Franz. "Traditional Property Concepts in China." *The Journal of Asian Studies* 15.4 (1956): 507–16.
Scott, James C. *Weapons of the Weak: Everyday Forms of Peasant Resistance.* New Haven, CT: Yale University Press, 1985.
Segawa Masahisa 瀨川昌久. *Zokufu : Kanan Kanzoku no shukyō, fūsui, ijū* 族譜：華南漢族の宗族・風水・移住. Tokyo: Fūkyōsha, 1996.
Sela, Ori. *China's Philological Turn: Scholars, Textualism, and the Dao in the Eighteenth Century.* New York: Columbia University Press, 2018.
Seow, Victor. *Carbon Technocracy: Energy Regimes in Modern East Asia.* Chicago: University of Chicago Press, 2021.
Shaw, Norman. *Chinese Forest Trees and Timber Supply.* London: T. F. Unwin, 1914.
Shi Hansheng 石瀚昇, ed. *Shishi zongpu* 石氏宗譜. Guangyuan: Shishi qingminghui: 2005.
Shiga Shūzō 滋賀秀三. *Shindai Chūgoku no hō to saiban* 清代中國の法と裁判. Tokyo: Sōbunsha, 1984.
———. "Qingdai susong zhidu zhi minshi fayuan de gaikuoxing kaocha: Qing-li-fa" 清代訴訟制度之民事法源的概括性考察：情、理、法. In Wang Yaxin 王亞新 and Liang Zhiping 梁治平, eds., *Ming-Qing shiqi de minshi shenpan yu minjian qiyue* 明清時期的民事審判與民間契約. Beijing: Falü chubanshe, 1998, 19–53.
Shih, James C. *Chinese Rural Society in Transition: A Case Study of the Lake Tai Area, 1368–1800.* Berkeley, CA: University of California Press, 1992.
Shryock, John. *The Temples of Anking and Their Cults: A Study of Modern Chinese Religion.* Paris: Librairie Orientaliste Paul Geuthner, 1931.
Sichuansheng wenwu guanliju 四川省文物管理局, ed. *Sichuan wenwuzhi* 四川文物志, vol. 1. Chengdu: Ba-shu shushe, 2005.
Sivin, Nathan. *Traditional Chinese Medicine in Contemporary China.* Ann Arbor: University of Michigan Center for Chinese Studies, 1987.
———. "Taoism and Science." In *Medicine, Philosophy and Religion in Ancient China: Researches and Reflections*, edited by Nathan Sivin. Aldershot, UK: Variorum, 1995, 1–71.

———. "Science and Medicine in Imperial China—The State of the Field." *The Journal of Asian Studies* 47.1 (1988): 41–90.

Skinner, G. William. "Marketing and Social Structure in Rural China, Part I." *The Journal of Asian Studies* 24.01 (1964): 3–43.

———. "Sichuan's Population in the Nineteenth Century: Lessons from Disaggregated Data." *Late Imperial China* 8.1 (1987): 38–39.

Skinner, William, and Hugh Baker, eds. *The City in Late Imperial China*. Stanford, CA: Stanford University Press, 1977.

Smith, Arthur H. *Chinese Characteristics*. New York: Fleming H. Revell, 1894.

Smith, George, and Alexander Elder, eds. "Feng-shui." *The Cornhill Magazine* 29 (1874): 337–48.

Smith, Richard J. "A Note on Qing Dynasty Calendars." *Late Imperial China* 9.1 (1988): 123–45.

———. *Fortune-Tellers and Philosophers: Divination in Traditional Chinese Society*. Boulder, CO: Westview Press, 1991.

———. "Li Hung-chang's Use of Foreign Military Talent: The Formative Period, 1862–1874." In *Li Hung-chang and China's Early Modernization*. Edited by Samuel Chu and Kwang-Ching Liu, 119–44. Armonk, NY: M. E. Sharpe Press, 1994.

———. *Fathoming the Cosmos and Ordering the World: The Yijing (I ching, or Classic of Changes) and Its Evolution in China*. Charlottesville: University of Virginia Press, 2008.

———. *The I Ching: A Biography*. Princeton: Princeton University Press, 2012.

———. *Mapping China and Managing the World: Culture, Cartography, and Cosmology in Late Imperial Times*. London: Routledge, 2013.

———. *The Qing Dynasty and Traditional Chinese Culture*. Lanham, MD: Rowman and Littlefield, 2015.

———. "The Legacy of Daybooks in Late Imperial and Modern China." In *Books of Fate and Popular Culture in Early China*, edited by Donald Harper and Marc Kalinowski. Leiden: Brill, 2017.

———. "The Transnational Travels of Geomancy in Premodern East Asia, c. 1600–c. 1900: PARTS I and II." *Transnational Asia*, 2.1, May 9, 2019: 1–112 (Part I); 1–130 (Part II).

Snyder-Reinke, Jeffrey. "Afterlives of the Dead: Uncovering Graves and Mishandling Corpses in Nineteenth-Century China." *Frontiers of History in China* 11 (2016): 1–20.

———. "Cradle to Grave: Baby Towers and the Politics of Infant Burial in Qing China." In *The Chinese Deathscape: Grave Reform in Modern China*, edited by Thomas S. Mullaney. Stanford, CA: Stanford University Press, 2019. https://chinesedeathscape.supdigital.org/read/cradle-to-grave

———. *Dry Spells: State Rain Making and Local Governance in Late Imperial China*. Cambridge, MA: Harvard University Asia Center, 2009.

Sommer, Matthew. *Polyandry and Wife-Selling in Qing Dynasty China: Survival Strategies and Judicial Interventions*. Berkeley: University of California Press, 2015.

Stapleton, Kristin. *Civilizing Chengdu: Chinese Urban Reform, 1895–1937*. Cambridge, MA: Harvard University Asia Center, 2000.

Statman, Alexander. *A Global Enlightenment: France, China, and the Idea of Progress*. Chicago: University of Chicago Press, 2023.

Steinhardt, Nancy S. *Chinese Architecture: A History*. Princeton: Princeton University Press, 2019.
Su, Te-cheng, and Hui-min Lai. "Resplendent Innovations: Fire Gilding Techniques at the Qing Court." In *Making the Palace Machine Work*, edited by Martina Siebert, Kai Jun Chen, and Dorothy Ko. Amsterdam: Amsterdam University Press, 2021, 157–86.
Suleski, Ronald. *Daily Life for the Common People of China, 1850–1950: Understanding Chaoben Culture*. Leiden: Brill, 2018.
Sun Li 孫藜. *Wan Qing dianbao jiqi chuanbo guannian* 晚清電報及其傳播觀念. Shanghai: Shanghai shiji chuban jituan, 2007.
Sutton, Donald S. "Death Rites and Chinese Culture: Standardization and Variation in Ming and Qing Times." *Modern China* 33.1 (2007): 125–53.
Sweeten, Alan Richard. *China's Old Churches: The History, Architecture, and Legacy of Catholic Sacred Structures in Beijing, Tianjin, and Hebei Province*. Leiden: Brill, 2019.
Szonyi, Michael. "The Illusion of Standardizing the Gods: The Cult of the Five Emperors in Late Imperial China." *The Journal of Asian Studies* (1997): 113–35.
———. *Practicing Kinship: Lineage and Descent in Late Imperial China*. Stanford, CA: Stanford University Press, 2002.
———. *The Art of Being Governed*. Princeton: Princeton University Press, 2017.
Tang, Cindy Q., Yongchuan Yang, Masahiko Ohsawa, Arata Momohara, Jingze Mu, and Kevin Robertson. "Survival of a Tertiary Relict Species, *Liriodendron chinense* (*Magnoliaceae*), in Southern China, with Special Reference to Village *fengshui* Forests." *American Journal of Botany* 100.10 (2013): 2112–19.
Taussig, Michael. *The Devil and Commodity Fetishism in South America*. Chapel Hill: University of North Carolina Press, 1980.
———. "Viscerality, Faith, and Skepticism: Another Theory of Magic." In *In Near Ruins: Cultural Theory at The End of the Century*, edited by Nicholas B. Dirks. Minneapolis: University of Minnesota Press, 1998, 221–56.
Teng, Emma. *Taiwan's Imagined Geography: Chinese Colonial Travel Writing and Pictures, 1683–1895*. Cambridge, MA: Harvard University Asia Center, 2006.
Teng, Ssu-yü, and John King Fairbank. *China's Response to the West: A Documentary Survey, 1839–1923*. Cambridge, MA: Harvard University Press, 1979.
Ter Haar, Barend. *The White Lotus Teachings in Chinese Religious History*. Leiden: Brill, 1992.
Theiss, Janet. *Disgraceful Matters: The Politics of Chastity in Eighteenth-Century China*. Berkeley: University of California Press, 2005.
Tu Qinghong 塗慶紅. "Lüe lun Qingdai Sichuan yanshang dui difang jiaoyu de zhichi" 略論清代四川鹽商對地方教育的支持. *Chengdu shifan daxue xuebao* 成都師範大學學報 12 (2019): 7–12.
Twitchett, Denis. "Law and Religion in East Asia." In *Encyclopedia of Religion*, vol. 8, edited by Mircea Eliade. New York: Macmillan, 1987, 469–72.
Tz'u Sung (Ci Song). *The Washing Away of Wrongs: Forensic Medicine in Thirteenth-Century China*, translated by Brian E. McKnight. Ann Arbor: University of Michigan Press, 1981.
Verellen, Franciscus. "'Evidential Miracles in Support of Taoism': The Inversion of a Buddhist Apologetic Tradition in Late Tang China." *T'oung Pao* (1992): 217–63.
Vermeer, Eduard B. "Population and Ecology along the Frontier in Qing China." In *Sediments of Time: Environment and Society in Chinese History*, edited by Mark Elvin and Ts'ui-jung Liu. Cambridge: Cambridge University Press, 1998, 235–79.

Vogel, Hans Ulrich. "'That Which Soaks and Descends Becomes Salty': The Concept of Nature in Traditional Chinese Salt Production." In *Concepts of Nature: A Chinese-European Cross-Cultural Perspective*, edited by Han Ulrich Vogel and Günter Dux, 469–515. Leiden: Brill, 2010.

Von Glahn, Richard. *An Economic History of China: From Antiquity to the Nineteenth Century*. Cambridge: Cambridge University Press, 2016.

Von Richthofen, Ferdinand. *Baron Richthofen's Letters, 1870–1872*. Shanghai: North China Herald Office, 1872.

Vries, Peer. *State, Economy and the Great Divergence: Great Britain and China, 1680s–1850s* London: Bloomsbury Publishing, 2015.

Waley-Cohen, Joanna. "Politics and the Supernatural in Mid-Qing Legal Culture." *Modern China* 19.3 (1993): 330–53.

Walravens, Hartmut. "*Vorhersagen von Sonnen- und Mondfinsternissen in Mandjurischer und Chinesischer Sprache.*" *Monumenta Serica* 35 (1981): 431–84.

Wang Daolü 王道履. *Nanbu xian xiangtu zhi* 南部縣鄉土志 (1906 Rpt.). In Yao Leye 姚樂野, ed., *Sichuan daxue tushuguan guancang zhenxi Sichuan difangzhi congkan* 四川大學圖書館館藏珍稀四川地方志叢刊, vol. 3. Chengdu: Ba-shu shushe, 2009, 325–415.

Wang, Di 王笛. *Kuachu fengbi de shijie: Changjiang shangyou quyu shehui yanjiu, 1644–1911* 跨出封閉的世界: 長江上游區域社會研究, 1644–1911. Beijing: Zhonghua shuju, 2001.

———. *Street Culture in Chengdu: Public Space, Urban Commoners, and Local Politics, 1870–1930*. Stanford, CA: Stanford University Press, 2003.

———. *The Teahouse: Small Business, Everyday Culture, and Public Politics in Chengdu, 1900–1950*. Stanford, CA: Stanford University Press, 2008.

Wang, Fei-Hsien. *Pirates and Publishers: A Social History of Copyright in Modern China*. Princeton: Princeton University Press, 2019.

Wang Hongbin 王宏斌. *Wan Qing haifang: Sixiang yu zhidu yanjiu* 晚清海防: 思想與製度研究. Beijing: Shangwu yinshuguan, 2005.

Wang, Robin R. *Yinyang: The Way of Heaven and Earth in Chinese Thought and Culture*. Cambridge: Cambridge University Press, 2012.

Wang, Xing. *Physiognomy in Ming China: Fortune and the Body*. Leiden: Brill, 2020.

Watson, James L. "Funeral Specialists in Cantonese Society: Pollution, Performance, and Social Hierarchy." In *Death Ritual in Late Imperial and Modern China*, edited by James Watson and Evelyn Rawski, 109–34. Berkeley: University of California Press, 1988.

———. "Standardizing the Gods: The Promotion of T'ien Hou ('Empress of Heaven') along the South China Coast, 960–1960." In *Popular Culture in Late Imperial China*, edited by David Johnson, Andrew J. Nathan, and Evelyn S. Rawski, 292–324. Berkeley: University of California Press, 1985.

Weber, Max, and Hans H. Gerth (trans.). *The Religion of China: Confucianism and Taoism*. 1951 Rpt.; New York: MacMillan Publishing Company, 1964

Wei, Fan. "Village Fengshui Principles." In *Chinese Landscapes: The Village as Place*, edited by Ronald G. Knapp, 35–45. Honolulu: University of Hawai'i Press, 1992.

Wei, Shuge. "Circuits of Power: China's Quest for Cable Telegraph Rights, 1912–1945." *Journal of Chinese History* 3.1 (2019): 113–35.

Wei Shunguang 魏順光. "Qingdai zhongqi fenchan zhengsong wenti yanjiu: Jiyu Ba xian dang'an wei zhongxin de kaocha"清代中期墳產爭訟問題研究：基於巴縣檔案為中心的考察 PhD diss., Southwest University of Political Science and Law, 2011.

———. "Qingdai zhongqi de 'jiefen zisong' xianxiang yanjiu: Jiyu Ba xian dang'an wei zhongxin de kaocha" 清代中期的'藉墳滋訟'現象研究：基於巴縣檔案為中心的考察. *Qiusuo* 求索 4 (2014): 159–163.

Weller, Robert P. *Unities and Diversities in Chinese Religion.* London: Macmillan Press, 1987.

———. *Discovering Nature: Globalization and Environmental Culture in China and Taiwan.* Cambridge: Cambridge University Press, 2006.

Wesley-Smith, Peter. "Identity, Land, Feng Shui and the Law in Traditional Hong Kong." *Australian Journal of Law and Society* 10 (1994): 213–39.

Wheatley, Paul. *The Pivot of the Four Quarters: A Preliminary Enquiry into the Origins and Character of the Ancient Chinese City.* Chicago: Aldine Publishing Company, 1971.

White Jr., Lynn. "The Historical Roots of Our Ecologic Crisis." *Science* 155 (1967): 1203–7.

Whiteman, Stephen. "Kangxi's Auspicious Empire: Rhetorics of Geographic Integration in the Early Qing." In *Chinese History in Geographical Perspective*, edited by Yongtao Du and Jeff Kyong-McClain. Lanham, MD: Lexington Books, 2013, 33–54.

———. *Where Dragon Veins Meet: The Kangxi Emperor and His Estate at Rehe.* Seattle: University of Washington Press, 2019.

Will, Pierre-Étienne. *Bureaucracy and Famine in Eighteenth Century China.* Translated by Elborg Forster. Stanford, CA: Stanford University Press, 1990.

———. "Officials and Money in Late Imperial China: State Finances, Private Expectations, and the Problem of Corruption in a Changing Environment." In *Corrupt Histories*, edited by Emmanuel Kreike and William Chester Jordan. Rochester, NY: University of Rochester Press, 2004): 29–95.

———. "Developing Forensic Knowledge through Cases in the Qing Dynasty." In *Thinking with Cases: Specialist Knowledge in Chinese Cultural History*, edited by Charlotte Furth, Judith T. Zeitlin, and Ping-chen Hsiung, 62–100. Honolulu: University of Hawai'i Press, 2007.

———. *Handbooks and Anthologies for Officials in Imperial China: A Descriptive and Critical Bibliography.* Leiden: Brill, 2020.

Will, Pierre-Étienne, and R. Bin Wong. *Nourish the People: The State Civilian Granary System in China, 1650–1850.* Ann Arbor: Michigan Monographs in East Asian Studies, 1991.

Witte Jr., John. *The Reformation of Rights: Law, Religion, and Early Modern Calvinism.* Cambridge: Cambridge University Press, 2008.

Wong, Evy, Loh Li Cheng, Chuah Siew Boon, Wong Su Ee, and Julie Chong. *Chinese Auspicious Culture.* Singapore: Asiapac Books, 2012.

Woodside, Alexander. "Some Mid-Qing Theorists of Popular Schools: Their Innovations, Inhibitions, and Attitudes toward the Poor." *Modern China* 9.1 (1983): 3–35.

Wooldridge, Chuck. *City of Virtues: Nanjing in an Age of Utopian Visions.* Seattle: University of Washington Press, 2015.

Wright, Arthur. "The Cosmology of the Chinese City." In *The City in Late Imperial China*, edited by William Skinner. Stanford, CA: Stanford University Press, 1977, 33–73.

Wright, Mary Clabaugh. *The Last Stand of Chinese Conservatism: The T'ung-Chih Restoration (1862–1874)*. Stanford, CA: Stanford University Press, 1957.

Wu, Albert. "Superstition and Statecraft in Late Qing China: Towards a Global History." *Past & Present* 255.1 (2021): 279–316.

Wu, Huiyi. "Guns, Maize, and Europeans: Early Modern Globalization in Local Gazetteers." In *Knowing the Empire in Early Modern China and Spain*, edited by Mackenzie Cooley and Huiyi Wu. Forthcoming.

Wu Jianxin 吳建新 and Zhong Haiyan 衷海燕. "Ming-Qing Guangdongren de fengshuiguan: Difang liyi yu shehui jiufen" 明清廣東人的風水觀：地方利益與社會糾紛. *Xueshu yanjiu* 學術研究 2 (2007): 98–106.

Wu Peilin 吳佩林. "Qingmo xinzheng shiqi guanzhi hunshu zhi tuixing: Yi Sichuan wei li" 清末新政時期官制婚書之推行—— 以四川為例. *Lishi yanjiu* 歷史研究 5 (2011): 78–96.

———. "Jin sanshi nian lai guonei dui Qingdai zhouxian susong dang'an de zhengli yu yanjiu" 近三十年來國內對清代州縣訴訟檔案的整理與研究. *Beida falü pinglun* 北大法律評論 (2011): 258–272.

———. *Qingdai xianyu minshi jiufen yu falü zhixu kaocha* 清代縣域民事糾紛與法律秩序考察. Beijing: Zhonghua shuju, 2013.

———. "Qingdai difang shehui de susong shitai" 清代地方社會的訴訟實態. *Qingshi yanjiu* 清史研究 4 (2013): 29–40.

Wu Peilin 吳佩林 and Deng Yong 鄧勇. "Qingdai Sichuan Nanbu xian jingyanye gailun—Yi Qingdai Nanbu xian yamen dang'an wei zhongxin de kaocha" 清代四川南部縣井鹽業概論—— 以清代四川南部縣衙門檔案為中心的考察. *Yanyeshi yanjiu* 鹽業史研究 1 (2008): 40–52.

Wu Peilin 吳佩林 and Wan Haiqiao 萬海蕎. "Qingdai zhouxianguan renqi 'sannian yiren' shuo zhiyi: Jiyu Sichuan Nanbu xian zhixian de shizheng fenxi" 清代州縣官任期"三年一任"說質疑—— 基於四川南部縣知縣的實證分析. *Qinghua daxue xuebao* 清華大學學報 3 (2018): 63–72.

Wu, Shellen. "Mining the Way to Wealth and Power: Late Qing Reform of Mining Law (1895–1911)." *The International History Review* 34.3 (2012): 581–99.

———. *Empires of Coal: Fueling China's Entry into the Modern World Order, 1860–1920*. Stanford, CA: Stanford University Press, 2015.

Wu Yicheng 武奕成 and Shen Weiwei 沈瑋瑋. "Shilun Qingdai yilai de kuangye huanjing baohu" 試論清以來的礦業環境保護. *Lanzhou xuekan* 蘭州學刊 1 (2011): 120–27.

Wu, Yi-Li. *Reproducing Women: Medicine, Metaphor, and Childbirth in Late Imperial China*. Berkeley: University of California Press, 2010.

Wu Youru 吳友如, ed. *Dianshizhai huabao: Dake tang ban* 點石齋畫報：大可堂版, vol. 14. Shanghai: Shanghai huabao chubanshe, 2001.

Wyman, Judith. "The Ambiguities of Chinese Antiforeignism: Chongqing, 1870–1900." *Late Imperial China* 18 (1997): 86–122.

Xia Xiangrong 夏湘蓉, Li Zhongjun 李仲均, and Wang Genyuan 王根元. *Zhongguo gudai kuangye kaifashi* 中國古代礦業開發史. Beijing: Dizhi chubanshe, 1980.

Xihua shifan daxue quyu wenhua yanjiu zhongxin 西華師範大學區域文化研究中心 and Nanbu xian difangzhi bangongshi 南部縣地方志辦公室, eds. *Tongzhi zengxiu Nanbu xian zhi* 同治增修南部縣志. 1870 Rpt. Chengdu: Ba-shu shushe, 2014.

Xu, Yinong. *The Chinese City in Space and Time: The Development of Urban Form in Suzhou*. Honolulu: University of Hawai'i Press, 2000.

Xu, Yue. "Sichuan's Promotion of Education and Activities of Felling Temple Trees in the Late Qing Dynasty." *Frontiers of History in China* 3.3 (2008): 406–31.

Yang, C. K. *Religion in Chinese Society: A Study of Contemporary Social Functions of Religion and Some of Their Historical Factors*. Berkeley: University of California Press, 1970.

Yang, Claire Yi. "Death Ritual in the Tang Dynasty (618–907): A Study of Cultural Standardization and Variation in Medieval China." PhD diss., University of California at Berkeley, 2019.

Yang Guanqiong 楊冠瓊. *Dangdai Zhongguo xingzheng guanli moshi yange yanjiu* 當代中國行政管理模式沿革研究. Beijing: Beijing shifan daxue chubanshe, 1999.

Yang Jialuo 楊家駱, ed. *Yangwu yundong wenxian huibian* 洋務運動文獻彙編, vol. 6. Taipei: Shijie shuju, 1963.

Yang, Yuda. "Silver Mines in Frontier Zones: Chinese Mining Communities along the Southwestern Borders of the Qing Empire." In *Mining, Monies, and Culture in Early Modern Societies*, edited by Nanny Kim and Keiko Nagase-Reimer. Leiden: Brill, 2013, 87–114.

Yee, Cordell D. K. "Traditional Chinese Cartography and the Myth of Westernization." In *History of Cartography*, vol. 2, Book 2: *Cartography in the Traditional East and Southeast Asian Societies*, edited by J. B. Harley and David Woodward, 170–202. Chicago: University of Chicago Press, 1994.

Yeoh, Brenda S. A. *Contesting Space in Colonial Singapore: Power Relations and the Urban Built Environment*. Singapore: National University of Singapore Press, 2013.

Ying Liangeng 應廉耕. *Sichuansheng zudian zhidu* 四川省租佃制度. Chongqing: Sichuansheng nongcun jingji diaocha weiyuanhui, 1941.

Yoon, Hong-key, ed. *P'ungsu: A Study of Geomancy in Korea*. Albany: State University of New York Press, 2017.

Yoon, Wook. "Dash Expectations: Limitation of the Telegraphic Service in the Late Qing." *Modern Asian Studies* 3 (2015): 832–57.

You, Ziying. *Folk Literati, Contested Tradition, and Heritage in Contemporary China: Incense Is Kept Burning*. Bloomington: Indiana University Press, 2020.

Young, Ernest P. *Ecclesiastical Colony: China's Catholic Church and the French Religious Protectorate*. New York: Oxford University Press, 2013.

Yu, Xin. "Publishing at the Grassroots: Print Culture and Rural Society in Early Modern China." PhD diss., Washington University in St. Louis, 2022.

Yu Zhengsong 余正松 and Zheng Jiewen 鄭傑文. "Qingdai Nanbu xian dang'an ji qi jiazhi" 清代南部縣檔案及其價值. *Wenxian* 文獻 1 (2008): 85–92.

Yu Zhengui 余振貴 and Lei Xiaojing 雷曉靜, eds. *Zhongguo huizu jinshilu* 中國回族金石錄. Yinchuan: Ningxia renmin chubanshe, 2001.

Yuan Hui 袁慧 and Jin Shengyang 金生楊. "Qingdai Nanbu xian yanliju de shezhi" 清代南部縣鹽厘局的設置. *Zhongguo yan wenhua* 中國鹽文化 11 (2018): 35–44.

Yuan Shushan 袁樹珊. *Zhongguo lidai buren zhuan* 中國歷代卜人傳. Shanghai: Runde shuju, 1948.

Yuan Yongbin 袁用賓. *Nanbu xian yutu kao* 南部縣輿圖考 (1853; 1869; 1896 Rpt.). In *Sichuan daxue tushuguan guancang zhenxi Sichuan difangzhi congkan* 四川大學圖書館館藏珍稀四川地方志叢刊, Vol. 3, ed. Yao Leye 姚樂野. Chengdu: Ba-shu shushe, 2009, 217–324. Note:

from its first printing in 1853 onward, Nanbu's atlas was accorded with different titles. A copy of the 1853 edition, titled *Xianjing fenfang tushuo* 縣境分方圖說, is held at the Shanghai Library, a copy of the 1869 edition, titled *Nanbu xian yutu shuo* 南部縣輿圖說, is held at National Library of China. And this 2009 version reprints the 1896 edition.

Yue Qintao 岳欽韜. *Yi Shanghai wei zhongxin: Hu Ning, Hu Hang Yong tielu yu jindai Changjiang sanjiaozhou diqu shehui bianqian* 以上海為中心：滬寧、滬杭甬鐵路與近代長江三角洲地區社會變遷. Beijing: Zhongguo shehui kexue chubanshe, 2016.

Zelin, Madeleine. *The Magistrate's Tael: Rationalizing Fiscal Reform in Eighteenth-Century Ch'ing China*. Berkeley: University of California Press, 1984.

———. "The Rights of Tenants in Mid-Qing Sichuan: A Study of Land-Related Lawsuits in the Ba xian Archives." *The Journal of Asian Studies* 45.3 (1986): 499–526.

———. *The Merchants of Zigong: Industrial Entrepreneurship in Early Modern China*. New York: Columbia University Press, 2005.

———. "Eastern Sichuan Coal Mines in the Late Qing." In *Empire, Nation and Beyond: Chinese History in Late Imperial and Modern Times, A Festschrift in honor of Frederic Wakeman, Jr.*, edited by Joseph W. Esherick, Wen-hsin Yeh, and Madeleine Zelin, 105–22. Berkeley: University of California Press, 2006.

Zeng Xiaomei 曾曉梅 and Wu Mingran 吳明冉, eds. *Qiangzu shike wenxian jicheng* 羌族石刻文獻集成, vol. 3. Chengdu: Ba-shu shushe, 2017.

Zgonjanin, Sanja. "Quoting the Bible: The Use of Religious References in Judicial Decision-Making." *City University of New York Law Review* 9 (2005): 31–91.

Zhang Aihua 張愛華. *Wenhua ruan quanli shiye xia de jiapu yanjiu: Yi Ming-Qing Anhui Jing xian Zhushi xilie jiapu wei yangben* 文化軟權力視野下的家譜研究：以明清安徽涇縣朱氏系列家譜為樣本. Tianjin: Tianjin renmin chubanshe, 2020.

Zhang Chuanyong 張傳勇. "Qingdai de 'tingsang bude shijin' lun tanxi: Jianji Qingdai guojia zhili 'tingsang buzang' wenti de duice" 清代的"停喪不得仕進"論探析—— 兼及清代國家治理"停喪不葬"問題的對策. *Zhongguo shehui lishi pinglun* 中國社會歷史評論 1 (2009): 281–98.

Zhang Fang 張昉. "Yi Yueyang Zhangguyingcun weili tantao Xiangbei minju fengshui" 以岳陽張谷英村為例探討湘北民居風水. *Shanxi jianzhu* 山西建築 12 (2006): 19–29.

Zhang, Lawrence. "The Legacy of Success: Office Purchase and State-Elite Relationship in Qing China." *Harvard Journal of Asiatic Studies* 73:2 (2013): 259–97.

Zhang, Meng. *Timber and Forestry in Qing China: Sustaining the Market*. Seattle: University of Washington Press, 2021.

Zhang Peiguo 張佩國. *Linquan, fenshan yu miaochan* 林權、墳山與廟產. Beijing: Zhongguo shehui kexue chubanshe, 2014.

Zhang, Taisu. "Cultural Paradigms in Property Institutions." *Yale Journal of International Law* 41 (2016): 347–413.

———. *The Laws and Economics of Confucianism: Kinship and Property in Preindustrial China and England*. Cambridge: Cambridge University Press, 2017.

———. "Moral Economies in Early Modern Land Markets: History and Theory." *Law & Contemporary Problems*. 80 (2017): 107–33.

Zhang, Ting. *Circulating the Code: Print Media and Legal Knowledge in Qing China*. Seattle: University of Washington Press, 2020.

Zhang, Xiaowei. *The Politics of Rights and the 1911 Revolution*. Stanford, CA: Stanford University Press, 2018.

Zhang, Xiaoye. "Legitimate, but Illegal: Case Studies of Civil Justice in the Ming and Qing Dynasties." *Etudes chinoises* 28 (2009): 73–94.

Zhang Xuejun 張學君 and Ran Guangrong 冉光榮. *Ming-Qing Sichuan jingyan shigao* 明清四川井鹽史稿. Chengdu: Sichuan renmin chubanshe, 1984.

Zhang Yanmei 張艷梅. "Qingdai Sichuan hanzai shikong fenbu yanjiu" 清代四川旱災時空分布研究. MA Thesis, China Southwest University, 2008.

Zhang, Zhibin, and Paul U. Unschuld. *Dictionary of the Ben cao gang mu, vol. 1: Chinese Historical Illness Terminology*. Berkeley: University of California Press, 2014.

Zhao Jishi 趙吉士. *Jiyuan ji suoji* 寄園寄所寄. 1695 Rpt. Hefei: Huangshan shushe, 2008.

Zhao Weini 趙妮妮. *Shenduan yu jinxu: Yi wan Qing Nanbu xian hunyin lei anjian wei zhongxin* 審斷與矜卹：以晚清南部縣婚姻類案件為中心. Beijing: Falü chubanshe, 2013.

Zhong Haiyan 衷海燕. "'Kuangmai' yu 'longmai' zhi zheng: Qingdai guanyu Duanyan kaicai de 'fengshui' lunshuo" "礦脈"與"龍脈"之爭：清代關於端硯開采的"風水"論說. *Huanan nongye daxue xuebao* 華南農業大學學報 4 (2007): 101–5.

Zhongguo renmin daxue qingshi yanjiusuo 中國人民大學清史研究所, ed. *Qingdai de kuangye* 清代的礦業 (vols. 1 and 2). Beijing: Zhonghua shuju, 1983.

Zhou, Guangyuan. "Beneath the Law: Chinese Local Legal Culture During the Qing Dynasty." PhD diss., University of California, Los Angeles, 1995.

Zhou, Yongming. *Historicizing Online Politics: Telegraphy, the Internet, and Political Participation in China*. Stanford, CA: Stanford University Press, 2006.

Zhu Cishou 祝慈壽. *Zhongguo gongye jishu shi* 中國工業技術史. Chongqing: Xinhua shudian jingxiao, 1995.

Zhu Huimin 朱慧敏. "Ming-Qing Huizhou jiapu xiangzhuan chutan" 明清徽州家譜像傳初探. *Ningxia daxue xuebao (Renwen shehui kexueban)* 寧夏大學學報 (人文會科學版) 39 (2017): 45–50.

Zhu Xi, and Joseph Adler (trans.). *The Original Meaning of the Yijing: Commentary on the Scripture of Change*. New York: Columbia University Press 2020.

Zigongshi dang'anguan 自貢市檔案館, Beijing jingji xueyuan 北京經濟學院, and Sichuan daxue 四川大學, eds. *Zigong yanye qiyue dang'an xuanji (1732–1949)* 自貢鹽業契約檔案選輯 (1732–1949). Beijing: Zhongguo shehui kexue chubanshe, 1985.

INDEX

academies, 71–72, 80, 111–12, 120–21, 139–41, 157; Aofeng Academy, 122, 197–99; Brocade Screen Academy, 116–17, 120; Hanlin Academy, 186, 205, 263n54
agriculture, 12–13, 37, 46, 50, 56–57, 61, 67, 87, 99, 151–52, 179, 185, 205; destruction of graves for, 20, 29, 52; threats to from mining, 159–65, 169–72, 175, 187. *See also* food costs; irrigation
Ahern, Emily, 2
amulets, 71
An County 安縣 (Sichuan), 170
Analects 論語, 27
ancestral halls, 23, 32–33, 54, 65–66, 79–80, 83, 96–99, 101, 149–150, 153–54
ancestral veneration, 23, 29–30, 34, 41–43, 184, 247n16
architecture, 1, 70–71, 81, 229. *See also* houses
astrology, 63, 128, 130–34, 147, 228, 264n72, 269n158. *See also* cosmology; field allocation system (*fenye* 分野); lodges (astral)
Astronomical Bureau (*Qintianjian* 欽天監), 6, 22, 51, 109, 130, 132, 261n4, 265n94
Autumn Assizes, 50, 52, 74, 102, 107

Ba County 巴縣 (Sichuan), xxiii, 12–13, 15–16, 30, 113, 170, 172–74, 180, 196, 200, 212. *See also* Chongqing Prefecture (Sichuan)
Ba County Qing Archive (defined), 15

bandits, 73, 102, 119, 149, 152, 204–5, 211, 269n163
Bao, Judge (999–1062), 50
Baoning Prefecture 保寧府 (Sichuan), 65–66, 83, 94, 102–4, 114–6, 119–120, 123–5, 130, 166. *See also* Langzhong County 閬中縣 (Sichuan)
bazi 八字. *See* birthtimes
Beijing, 50–51, 81, 114, 136, 144, 155, 162–63, 205, 228
Bird, Isabella (1831–1904), 112
birthtimes, 25–27, 101, 227
Board of Punishments (*Xingbu* 刑部), 40, 50–52, 54, 203, 51, 245n90, 245n90
Boards of the Imperial Government, 50–51, 245n90
Bodde, Derk (1909–2003), 3
Book for Magistrates 牧令書, 67
Boxer Uprising (1899–1901), 114, 203, 205, 211, 213, 220, 222, 281n85
Brandt, Maximilian August Scipio von (1835–1920), 210
Bray, Francesca, 62
bribery, 54, 67, 154, 197, 284n18. *See also* corruption
bridges, 10, 45–46, 66, 81, 96, 98–101, 111, 125, 143, 149–154
Brief Account of Taming the Rapids 平灘紀略, *A*, 179–80
Brook, Timothy, 186
Bruun, Ole, 2, 8
Buck, John Lossing (1890–1975), 33
Buddhism, 121, 125–26, 130, 181, 185. *See also* temples, Buddhist

buildings, auspicious positioning and construction time of, 134, 136, 139, 144–47. *See also* houses, auspicious positioning and construction time of

buildings, geomantic veins of 76, 79–80. *See also* dragon veins; earth veins

burial: auspicious timing of, 22, 25, 27, 136, 165 (*see also* graves, auspicious location of); delayed, 22, 27, 157; illicit (*daozang* 盜葬), 21, 29–31, 40–41, 54, 58–59, 251n84; reburial, 16, 27, 29–30, 40, 168, 208. *See also* graves

Cai Dongzhou 蔡東洲, 101

Cai Shengyuan 蔡升元 (1652–1722), 137. *See also* civil examinations (*"zhuangyuan"* 狀元)

calendar: imperial, xxi–xxii, 12, 119, 133, 165, 224, 228, 248n30, 283n6, 283n15; Western-learning and, 132–33, 266n106

capital punishment. *See* Autumn Assizes

capitalism, 19, 138, 220

Carroll, Peter, 70

cartography. *See* maps

cemeteries. *See* burial; graves

Censorate (*Duchayuan* 都察院), 50–51, 54

Changming 常明 (Tunggiya clan; d. 1817), 119

Changshou County 長壽縣 (Sichuan), 173–4

Chen, Hailian, 271–72

Chen Hongmou 陳宏謀 (1696–1771), 159–60

Chen Jinguo 陳進國, 2

Chen Qizhang 陳其璋 (d.u.), 207

Chen Yi 陳彝 (c. 1827–96), 201–3, 209

Chengdu County 成都縣/Prefecture 府 (Sichuan), xxiii, 18, 111, 114, 123, 130, 132, 134–35, 139, 174–75, 185–89, 203

children, harm to, 23–27, 33, 35, 39–40, 55, 76, 219. *See also* earth veins, harm to

Chinese Forest Trees and Timber Supply, 217

Chongqing Department 崇慶州 (Sichuan), 218–9

Chongqing Prefecture 重慶府 (Sichuan), 12, 15, 30, 45, 123, 170, 172, 185, 195–96, 203. *See also* Ba County 巴縣 (Sichuan)

Chongqing Uprising of 1886, 196, 210

Chuanzhu 川主 ("Lord of Sichuan" or "River God"), 152, 182

churches, 2, 72, 196, 198, 220. *See also* Jesuits; missionaries

city walls. *See* walls, city

civil examinations, 1, 11, 16, 24, 108–57, 187, 214–15, 220–21, 226; geomantic influences on success of, 120–26, 139–41, 146, 150, 153, 166, 171, 178, 198–99, 267n127; *jinshi* 進士, 108–9, 113–14, 127, 138, 147, 153, 200; levels of, 112–13; *zhuangyuan* 狀元, 96, 125, 129, 137, 267

class differences and tensions, 62, 83, 88, 101, 154, 173, 198, 208. *See also* gentry; rural society

Classic for Hitting the Mark Every Time 百中經, 25

Classic of Changes 易經, 6, 85, 154, 227, 229

clay. *See under* mining

coal. *See under* mining

Coggins, Chris, 8

Cohen, Myron, 206, 249n51

Collected Statutes of the Great Qing 大清會典, 22–23, 127–28. *See also Imperially Endorsed Collected Statutes and Precedents of the Great Qing* 欽定大清會典事例; *Imperially Endorsed Collected Statutes and Regulations of the Great Qing* 欽定大清會典則例

Collins, William Frederick (1882–1956), 160

Comments on Cases from Sichuan Litigation 蜀訟批案, 30. *See also* Wang Dingzhu 王定柱 (c. 1761–1830)

Communism, 48, 56, 80, 229

compass, geomantic (*luopan* 羅盤), 7, 64, 85–92, 132–33, 188, 228, 243n38, 266n106. *See also* earthly branches; heavenly stems; trigrams

Compendium of Ritual Words 法言會纂, 182

Compilation of Writings on the Statecraft of Our August Dynasty 皇清經世文編, 99

Complete Book Concerning Happiness and Benevolence, A 福惠全書, 66

Complete Collection of Effective Prescriptions for Women 婦人大全良方, 260

Complete Conspectus of the Grand Canal in Shandong 山東運河備覽, 64
Confucianism, 4, 6, 10, 27, 112, 116, 126–27, 142, 157, 182, 194, 197–98, 205, 226, 268n141, 276n128. *See also* temples, Confucian
conservation, 8, 168, 171, 176, 229, 283n5. *See also* industrialization
Conspectus of Penal Cases, The 刑案匯覽, 52–53, 125
contracts, 9, 22, 24, 34, 36–37, 39, 44–48, 58–59, 146, 165, 167–68, 172, 219, 227, 251n92, 282n122
copper. *See under* mining
Correct Doctrines of Fengshui (Composed by the Qing Astronomical Bureau) 欽天監風水正論, xxv, 6–7, 17, 73, 107, 109, 144
corruption, 53–54, 93, 125, 158, 174, 197, 215, 227. *See also* bribery
cosmology, 4, 6–7, 50, 55–56, 101, 105, 131–32, 139, 156, 165, 227–29, 265n92. *See also* astrology; field allocation system (*fenye* 分野); lodges (astral)
Court of Judicial Review (*Dalisi* 大理寺), 50–51
cremation, 10, 20, 246n1
Crook, David (1910–2000) and Isabel (b. 1915), 48

Dai Sanxi 戴三錫 (1758–1830), 53
dams, 10, 108. *See also* floods; rivers; water, role of in fengshui
Dan-Xin Archive, 16, 68
Daning County 大寧縣 (Sichuan), 170
Daoguang Emperor/reign, 15, 27, 53, 80, 93, 182
Daoism, 23, 115–16, 125–26, 128, 130–31, 135, 179, 181–85, 188–89, 225. *See also* shrines, Daoist
deities, 3, 109, 119, 182, 204; astral, 50, 81, 121–2, 127, 131; city gods, 53–54, 198; 218–19; and mining, 181–84. *See also* Chuanzhu 川主; Erlang 二郎; Kui star 魁星; Ling star 欞星 (Wenqu 文曲); religion; Wenchang 文昌
Deliberative Assembly (Nanbu), 216–17

Department of Punishments, 245n90. *See also* yamen
Department of Works, 42, 61, 69, 76, 79, 245n90, 256n19. *See also* yamen
development, 5–8, 14, 150–1; in the late Qing, 225, 283n5; in Nanbu, 213–18; in the Qing calendar, 164–5. *See also* industrialization
Dianshizhai Pictorial 點石齋畫報, 203, 205, 207
distilleries, 10, 148
Distinguishing Correct Earthly Principles 地理辨正, 167
divination, 7, 13, 22, 25, 59, 63, 91, 99, 127–28, 130, 166, 229
doctors. *See* medicine
Draft History of the Qing 清史稿, 162
dragon veins, 7, 44–45, 65–66, 69, 87–90, 96–99, 115, 119, 150–51, 153–54, 161, 167, 172, 176, 182, 198, 243n45; harm to 8, 146–48, 160, 163–64, 167, 186, 195; of Sichuan, 184–90, 195, 205. *See also* buildings, geomantic veins of; earth veins
Dream of Red Mansions 紅樓夢, A, 55, 147
Driscoll, Mark, 200
drought, 13, 25, 116, 120, 151, 153, 161, 184–86, 196, 199, 203–5, 216–19. *See also* famine; food costs; rain
Duara, Prasenjit, 219

earth veins, 7–8, 54–55, 64, 85, 109, 150–51, 167, 205, 243n42, 243n45; harm to, 8, 10, 23–24, 28, 31, 33, 39–40, 52, 55, 87, 122, 145, 168, 173, 187, 201–2, 205, 219. *See also* buildings, geomantic veins of; children, harm to; dragon veins
earthly branches, 7, 64, 85, 243n38. *See also* compass, geomantic (*luopan* 羅盤)
earthly principles (*dili* 地理), 25, 31, 62–63, 87, 95–96
Earthly Principles for Guiding the Original Truth 地理直指原真, 85–86, 92, 137–38. *See also* Kong Wenxing 孔聞星 (1620–1705)
eclipses, 130–32, 227–28, 265n92, 265n94
Edgerton-Tarpley, Kathryn, 187

Elman, Benjamin, 62, 108–9, 194, 228
Elvin, Mark, 160, 205
Erlang 二郎, 97–100. *See also* Chuanzhu 川主
examinations. *See* civil examinations

famine, 161, 185–88, 200, 205. *See also* drought; food costs
Fan Zengxiang 樊增祥 (1846–1931), 84–87, 145–46, 202
Faure, David, 23, 190
Feuchtwang, Stephan, 2
fiction and novels. *See Dream of Red Mansions, A; Romance of the Three Kingdoms; Plum in the Golden Vase, The*
field allocation system (*fenye* 分野), 131–32, 228. *See also* astrology; cosmology; lodges (astral)
fire, risk of, 72, 139, 144–45, 156
five agents, 6, 62, 72, 139–40, 142, 165–66
Five Secrets of Earthly Principles 地理五訣, 27, 72
floods, 13, 25, 120, 151, 180, 196, 199, 204, 208, 220. *See also* dams; rivers
food costs, 159, 162, 185, 203. *See also* drought; famine
Freedman, Maurice, 2, 261n9
fuel, 162, 166, 170, 175, 186, 206, 215, 218, 221, 274n960; costs and shortages, 33, 36, 57, 217, 273n75. *See also* mining, coal; timber
Fujian Province, 21
furnaces, 33, 144, 147, 156, 166, 197, 199

Gansu Province, xxiii, 83, 115, 132, 181, 203
Gaoguan Temple 高觀寺, 103–5
Gate of Heavenly Peace (*Tian'anmen* 天安門), 50–51
genealogies, 2, 16–18, 41–44, 49, 54, 59, 61, 64–67, 70, 95–103, 106, 218, 260n111
General's Arrow (*jiangjun jian* 將軍箭), 25–27. *See also* children, harm to; horoscopes; wind disease (*fengbing* 風病)
gengtie 庚貼. *See* horoscopes
gentry, 68, 87, 100, 111–15, 119–21, 124, 127, 143, 146–54, 156, 171–72, 176–78, 187–91, 196–200, 214–16; definition of, 18, 108, 112–13, 214–15; ranks of, 112–13. *See also* rural society, elites in
geomancers, xxi, 5, 7, 20, 22, 31, 48, 58, 63, 84–85, 107, 120, 128, 134–35, 137–41, 150, 153, 166, 179, 181, 247n8, 272n48–49; and legal testimony, 75–76, 79, 168; unskilled, 24–25, 29, 31, 59
Gladney, Dru, 117–18
gods. *See* deities; religion
Golas, Peter (1937–2019), 160, 183
gold. *See under* mining
Golden Mirror for Peaceful Living 安居金鏡, 72–73, 76, 79, 81, 144, 148
Golden Mirror of Medical Orthodoxy, Imperially Commissioned 御纂醫宗金鑑, 106
Goossaert, Vincent, 215
government: circuit, 4, 30, 51, 69, 249n44; county, 4, 9–11, 15–16, 22, 30–31, 51, 69, 95, 104–5, 128, 156, 225 (*see also* yamen); prefecture, 4, 51, 69, 123, 128; provincial, 30–31, 51, 69, 164, 225
Graham, David (1884–1961), 135
grave trees, 20, 34, 47, 246n4. *See also* trees, fengshui
graves, 1, 5–7, 9–11, 15–18, 20–60, 83–94; ancient (adoption of), 20, 41–44, 49, 58, 66, 101; as financial investments, 20, 36, 48, 55, 84–85; auspicious positioning of, 22–25, 27, 40, 43, 85–87, 94, 99, 102, 137–38, 166–67 (*see also* burial, auspicious timing of); communal lineage cemeteries, 29–34, 37–39, 42–43, 47–49, 54, 58; destruction or desecration of, 20–22, 29, 37, 40–41, 49–50, 52–53, 173, 199–202, 204–5, 207–11, 243, 246n4, 278n30, 280n84, 281n85; forbidden land surrounding, 21, 168; geomantic harm to, 159–60, 162–68, 172–75, 178, 184, 201–2, 204, 207, 210, 212; maintenance of, 20, 23, 39, 43, 45, 54, 58, 65, 173; protection of, 22, 29, 37, 40, 48–50, 55, 65, 84, 87, 95, 102; sale of, 20–21, 30, 44–48, 58, 172, 208; "selling the land but keeping the graves," 21, 30, 45–46, 84, 173; unmarked, 33, 42. *See also* burial; tombstones

INDEX   319

Great Qing Code 大清律例, 3, 21–23, 40, 162, 225, 228; 1730 amendment to, 45–46; 1768 amendment to, 44; 1788 amendment to, 29, 31; 1817 amendments to, 21, 45
Great Qing Gate (*Da Qingmen* 大清門), 51
groves, fengshui. *See* trees, fengshui
Guangdong Province, xxiii, 16, 177, 185
Guangping County 廣平縣/Prefecture 府 (Zhili), 129
Guangshan County 光山縣 (Henan), 21
Guangxu Emperor/reign, 15, 130, 167, 195–96, 199
Guangyuan County 廣元縣 (Sichuan), xxiii, 170, 276n150
Guo, Weiting, 22
Guo Songtao 郭嵩燾 (1818–91), 208

Halberd Gate (*Jimen* 戟門), 141–42
Halsey, Stephen, 194
Han people, 24, 71, 111–12, 115–18, 134, 220, 263n42
handbooks, judicial, 10, 14, 16, 18, 65–67, 106
Hangzhou Prefecture 杭州府, 72, 126, 157, 207
Hanson, Marta, 107
harm to fengshui, 1, 4, 8, 19, 30, 35, 45, 53, 59, 77, 79, 144–46, 151, 153, 163–64, 175, 180–81, 193, 195, 200, 204–5, 216, 218, 220–21. *See also* dragon veins, harm to; earth veins, harm to; graves, geomantic harm to; mining, geomantic objections to and bans on; railways; telegraphs; trees, cutting of
heavenly stems, 7, 64, 85, 243n38. *See also* compass, geomantic (*luopan*)
Hebei Province. *See* Zhili Province
Hengzhou 橫州 (Guangxi), 126, 148
Hiyande (Xiande 憲德, Silut clan; d. 1740), 57
homicide, 50, 52–54, 244n67
Hong Xiuquan 洪秀全 (1814–64), 109, 111. *See also* Taiping Civil War
Honglou meng 紅樓夢. *See Dream of Red Mansions, A*
horoscopes, 25–27, 227, 248n30

houses, 70–83; auspicious positioning and construction time of, 70–72, 99, 144, 165 (*see also* buildings, auspicious positioning and construction time of); central halls, 75–77, 83; courtyards and lightwells, 71, 73, 79–83; "illnesses" (*zhaibing* 宅病) of, 73, 80; roofs, 71–72, 81, 257n48
Huang, Fei, 111
Huang Chao 黃巢 (835–84), 109
Huang, Philip, 3
Huang Liuhong 黃六鴻 (d.u.; 1651 *juren* 舉人), 66, 99–100
Huayi Corporation 華益公司, 212
Huili Department 會理州 (Sichuan), xxiii, 262
Huizhou 徽州 (region), 15, 65–66, 71, 97, 99, 139
Hunan Province, xxiii, 15, 27, 72, 181, 203
hungry ghosts, 59

imperialism, 14, 17–18, 192–93, 201
Imperially Endorsed Almanac for Time Selection 欽定選擇通書, 6
Imperially Endorsed Collected Statutes and Precedents of the Great Qing 欽定大清會典事例, 163, 246n1, 272n36, 278n30, 280n84
Imperially Endorsed Collected Statutes and Regulations of the Great Qing 欽定大清會典則例, 162, 266n112, 271n25
Imperially Endorsed Treatise on Harmonizing Times and Distinguishing Directions 欽定協紀辨方書, 6, 12, 17, 107, 134, 164, 224
India, 201
industrialization, 14, 17, 19, 160, 193–95, 208–9, 220, 226–27, 283n5. *See also* conservation; infrastructure; railways; telegraphs; Western nations, science and technology of
infrastructure, 19, 109, 150, 152, 157, 179, 180, 185, 192–193, 245n90. *See also* industrialization; railways; telegraphs
inscriptions, 24, 33, 35, 42–44, 49, 59, 103, 115–16, 118–20, 122, 134, 138, 152–54, 178, 218
Instructions for Magistrates Published by Imperial Order 欽頒州縣事宜, 67

irrigation, 10–11, 13, 15, 17, 21, 28, 38, 67, 99, 149, 163, 188, 224. *See also* agriculture; water, role of in fengshui

Islam, 112, 114–19, 125. *See also* mosques

Jami, Catherine, 228
Japan, 2, 70, 193–94, 213
Jesuits, 78, 105, 130, 132, 227–28. *See also* churches; missionaries
Jialing River 嘉陵江. *See under* rivers
Jiang, Yonglin, 4
Jiang Dahong 蔣大鴻 (c. 1620–1714), 54, 167
Jiangbei subprefecture 江北廳 (Sichuan), 31
Jiangnan 江南 (region), 16, 71, 95, 106, 111, 113–14, 127, 137, 157, 226
Jiangsu Province, 113, 200, 203
Jiangxi Province, 15, 186
Jiaqing Emperor/reign, 53, 119
Jin Ping Mei 金瓶梅. *See Plum in the Golden Vase, The*
Jinan Prefecture 濟南府 (Shandong), 164
Jing'an 景安 (Niohuru clan; d. 1822), 51
Jingdezhen 景德鎮 (Jiangxi), 186
Jingyan County 井研縣 (Sichuan), 93
jinshi 進士. *See under* civil examinations
Judgments of Fanshan 樊山判讀, 84, 202
judicial punishments: conscription, 54; corporal, 20, 29, 41, 53, 143, 173, 203, 246n2, 246n4, 251n84, 254n133; execution, 22, 52, 102, 107, 281n85; exile, 22, 46, 52, 54, 102, 107; imprisonment, 72, 284n18; penal servitude, 52, 203, 254n133

Kai County 開縣 (Sichuan), 50, 52
Kangxi Emperor/reign, 132, 163
Katz, Paul, 4, 22, 215
Khoja Abd Allāh (d. 1689), 115–19, 196
kinship groups, 2, 32, 36, 44, 54, 66, 87, 102–3, 178, 211, 217
Knapp, Ronald, 71
Ko, Dorothy, 177–78
Köll, Elisabeth, 208

Kong Wenxing 孔聞星 (1620–1705), 85, 137–38
Kui star 魁星, 121–22, 125, 153; relationship to Kui lodge 奎宿 (written with the *da* radical), 264n72
Kuirun 奎潤 (Aisin-Gioro clan; 1829–90), 207

Lagerwey, John, 183
land: legal disputes over, 15, 20–22, 27, 30, 36–48, 52–53, 57–59, 65, 67, 69, 79, 88–91, 97, 159; "right of first refusal" of kin, 45–47, 55; ritual fields (*jitian* 祭田), 46, 55, 106, 178, 253n113; sale of, 20–1, 30, 37–39, 44–48, 55–56, 187; untaxed, 22, 48
language of power, fengshui as, 45, 71, 87, 176
Langzhong County 閬中縣 (Sichuan), 104, 114, 116, 118, 124–25
Lavelle, Peter, 194
law codes. *See Collected Statutes of the Great Qing* 大清會典; *Great Qing Code* 大清律例; *Substatutes of the Board of Punishments in Current Use* 刑部現行則例
lead. *See under* mining
Leshan County 樂山縣 (Sichuan), xxiii, 174–75
Li Benzhong 李本忠 (c. 1760–1840), 179–81, 185, 190
Li Hongzhang 李鴻章 (1823–1901), 193–94, 200–1, 203–4, 208–11, 220, 277n2
Li Rong 李榕 (1819–90), 94
Li Weijun 李維鈞 (d. 1727), 163
Li Xuejin 黎學錦 (1776–1838), 119–20
Li Zhengyong 李徵庸 (1848–1902), 212
life force (*qi*), 6–8, 33, 39, 62, 85, 94, 133, 140, 147, 150, 153, 166–67, 169, 184, 198–99, 202, 205
lightwell (*tianjing* 天井), 81–83
Ling star 櫺星 (Wenqu 文曲), 127, 139, 141–42, 197, 264n72
litigation masters (*songshi* 訟師), 29, 31, 63, 228
Little, Archibald John (1838–1908), 123
Liu Mingchuan 劉銘傳 (1836–96), 208
Liu Xianting 劉獻廷 (1648–95), 166–67
Liu Xihong 劉錫鴻 (d. 1891), 204–5, 209

INDEX   321

Liu Yuan 劉沅 (1767–1855), 182
lodges (astral), 131–33, 147, 264n72, 266n106, 269n158. *See also* astrology; cosmology; field allocation system (*fenye* 分野)

Macauley, Melissa, 9
Manchu: rule, 107, 114, 127, 130, 192, 220, 226; language, 254
manuals (geomantic), 5, 10, 17, 25–27, 54, 62–64, 70–75, 79–81, 85, 102, 106–7, 122, 132, 147–49, 156, 224, 228, 253n128; and civil examinations, 137–42 (*see also* civil examinations)
maps: genealogical, 64–67, 95–103, 106; judicial, 24, 35–39, 42–43, 45, 59, 61, 63–70, 76–77, 79, 83–94, 105–7, 175, 177–78, 188–89; *kaifang jili* 開方計里 cartographic method, 64; of graves, 83–94, 256n19, 256n25; of residences, 70–83; of temples and shrines, 103–5; regulations for, 65–70; Western cartographic method, 105
marriage, 101–2, 107, 165, 227
medicine, 6, 11, 63, 102, 106, 110, 115, 129, 134–35, 228, 248n28, 261n8, 266n112
memorial to the throne: palace (*zouzhe* 奏摺), 53–54, 57, 119, 163–64, 185, 193, 195, 201–5, 207–9, 220–1; routine (*tiben* 題本), 50–53, 254n131
Meyer, Jeffrey, 155
Mianzhu County 綿竹縣 (Sichuan), xxiii, 168
Miao Tong 繆彤 (1627–97), 137. *See also* civil examinations ("*zhuangyuan*" 狀元)
migration, 12, 32, 49, 57, 88, 97, 103, 152, 226
military, 3, 14, 22, 65, 115, 134, 158, 193–94, 196, 198, 201, 213
militias, 13, 61, 88, 95, 100, 103, 105, 156, 187
millstones, 17, 33, 73, 88–93, 96, 143–45
Min River 岷江. *See under* rivers
Mineral Enterprise in China, 160
Ming dynasty (1368–1644), 2, 10, 17, 32, 72, 87, 95–97, 105–6, 111, 127–30, 132, 139, 163, 169, 176, 181, 226–27
Ming-Qing transition, 12, 101, 114

mining, 8, 10–11, 18, 158–92, 194–95; clay, 186–89; coal, 10, 33, 52, 123, 158–59, 162–63, 168–176, 205, 208, 271n28, 273n66; copper, 158, 161–62, 176; geomantic insurance for, 161, 179–84; geomantic objections to and bans on, 158–65, 167–78, 187–89, 209, 212–14, 270n1, 271n28; gold, 158, 162, 176–79, 183, 213–14, 216, 274n102; lead, 159; regulation of, 161–65, 211–12; salt, 113, 158–59, 166–68, 171, 174, 196–99, 272nn48–49, 278n29 (*see also* Salt Tax Bureau); saltpeter, 272n43; late Qing schools of, 195, 212; stone quarrying, 40, 83, 96, 158–59, 162, 164, 177–79, 196, 214; tin, 159; zinc, 159
missionaries, 2, 17, 72, 78, 169, 188, 196–98, 200, 203. *See also* churches; Jesuits
monasteries and nunneries, 97–100, 103, 126, 137, 173, 181, 216, 250n74.
Morris, Clarence (1903–1985), 3
mosques, 111, 117–18, 143, 202, 262n35, 263n38. *See also* Islam
Mount Aofeng 鼇峰山, 121–22, 197–99, 214
mountains, role of in fengshui, 7–8, 34, 67, 85–87, 89–92, 101, 115–18, 133, 150, 153, 159–60, 162, 166, 168–69, 176, 178, 183, 188, 204–6. *See also* water, role of in fengshui
Muslims. *See* Islam; mosques

Nanbu County 南部縣 (Sichuan), xxiii, 12–16, 18, 30, 33, 36, 58, 69, 113–14, 166, 170, 176–77, 196–97, 199–200, 213–219, 221, 228. *See also* Deliberative Assembly (Nanbu)
Nanbu County Qing Archive (defined), 14–17
Nanjiang County 南江縣 (Sichuan), 114, 125, 203
Nappi, Carla, 62
Naquin, Susan, 162
New Culture Movement, 70, 229
New Edition of the Core Essentials on Mining Studies 礦學心要新編, 195
New Policies (1901–11), 213–18
Ninghe County 寧河縣 (Zhili), 117
noise pollution, 144, 146–48, 225. *See also* pollution

noodle shops, 146–48
North China Plain 34, 81, 126, 206, 221, 281n85
nuns. *See* monasteries and nunneries

oil presses, 143–49, 151, 156, 202
Olles, Volker, 182
On the Origins and Symptoms of Various Illnesses 諸病源候論, 260
opera performance, 34, 134, 181
Opium Wars (1839–42; 1856–60), 158, 192, 200
Orientalism, 2–3, 5, 62. *See also* Western nations
Osborne, Anne, 58–59
Ouyang Chun 歐陽純 (d.u.), 27–28, 72

pagodas (*ta* 塔), 10, 17–18, 120–26, 139–41, 196, 198–99, 202
pavilions (*ge* 閣), 120–22, 124–25, 139–42, 197
Peng County 彭縣 (Sichuan), 126
People's Republic of China, 229
Personal Views on Learning Government 學治臆說, 67
Pi County 郫縣 (Sichuan), 58
Plum in the Golden Vase, The 金瓶梅, 136–7
pollution, 8. *See also* conservation; noise pollution
Pomeranz, Kenneth, 33
ponds (*chi* 池): lotus, 80; *pan* 泮, 80, 141–42, 268n140
population growth, 12, 17, 21, 29, 36, 46, 56–58, 114, 127, 159, 175, 182
Preface to the Origins of Earthly Principles 地理原序, 102, 137–38
prisons, 50–52, 72, 144
property disputes. *See* land, legal disputes over

Qianlong Emperor/reign, 73, 106–7, 130–32, 146, 162–64, 199, 271n28
Qianwei County 犍為縣 (Sichuan), 174–75
Qingfu County 慶符縣 (Sichuan), xxiii, 146–47
quarrying. *See* mining, stone quarrying

railways, 2, 8, 193–94, 200–1, 204–11, 218, 221–22; Beijing–Hankou (Lu-Han 盧漢, later Jing-Han 京漢) Railway, 218; Tianjin–Pukou (Jin-Pu 津浦) Railway, 208; Shanghai–Nanjing (Hu-Ning 滬寧) Railway, 221; Tianjin–Tongzhou (Jin-Tong 津通) Railway, 207
rain, 31, 71, 81, 153, 185; summoning, 116, 126, 186, 216–17, 219. *See also* drought
Raz, Gil, 183
Records of the Unity of the Great Ming 大明一統志, 101
Reed, Bradley, 11
religion, 2–5, 8, 56, 108, 116, 128, 191, 219, 221, 223. *See also* ancestral veneration; Buddhism; Confucianism; Daoism; deities; Islam; monasteries and nunneries; mosques; sacrificial offerings; shrines; temples
Ren Wenyi 任文翌 (d.u.; 1745 *jinshi* 進士), 108
Renhe County 仁和縣 (Zhejiang), 113
Republican era (1912–49), 219, 229
Republican Revolution (1911–12), 219
Richthofen, Ferdinand von (1833–1905), 169, 185–86
ritual, 3–4, 11, 18, 22, 34, 126, 128, 130–31, 134–36, 179–84, 186, 219, 225, 227–28; *jiao* 醮 30, 34, 184. *See also* ancestral veneration; graves, maintenance of; opera performances; sacrificial offerings
ritual fields (*jitian* 祭田). *See under* land
rivers, 65, 71, 89, 96, 99, 123, 149–53, 169–70, 176–77, 179–80, 186–89, 204, 220; Jialing, xxiii, 12, 120, 125, 153, 170, 176–77, 213; Min, xxiii, 186–89; Yangzi, xxiii, 12, 123, 179–81, 185. *See also* dams; floods; water, role of in fengshui
Romance of the Three Kingdoms 三國演義, 63
Rong County 榮縣 (Sichuan), xxiii, 152, 171
Rowe, William, 126
rural society, 1–2, 16, 21, 23, 28–29, 41, 44, 56, 59, 65, 87–88, 92–93, 100, 103–4, 110, 135, 150, 155–57, 202–4, 213–17, 219; elites in, 13, 28, 83–84 (*see also* gentry); poor in, 21, 41, 45

INDEX 323

sacrificial offerings, 29–30, 34, 41–43, 47, 134, 160n115, 181, 183–84, 215, 246n6, 252n106
salt. *See under* mining
Salt Tax Bureau, 197–99. *See also* mining, salt
Sanguo yanyi 三國演義. *See Romance of the Three Kingdoms*
Santai County 三台縣 (Sichuan), 126
Schäfer, Dagmar, 62–63
Schneewind, Sarah, 155
schools of fengshui (Form and Compass), 6–7, 66, 107, 150
schools, 5, 10, 70, 111–13, 118, 120–21, 139–42, 147–50, 199, 213, 215–16, 222, 263n50, 268n140; mining, 195, 212. *See also* temple schools; Yin-yang schools
screen walls. *See* walls, screen (*zhaobi* 照壁)
Secret Essentials of Judging Houses 卜居秘髓, 71–73, 139–40
security units. *See* surveillance units
Segawa Masahisa 瀬川昌久, 2
Self-Strengthening Movement (1861–95), 193–94
sha 煞 (death-inducing malady), 25, 33, 198. *See also* children, harm to
Shaanxi Province, xxiii, 17, 71, 81, 83–84, 93, 126, 132, 203
Shandong Province, xxiii, 17, 114, 136, 164
Shanxi Province, xxiii, 71, 81, 181, 203
Shaw, Norman (1878–1955), 217
Shen Baozhen 沈葆楨 (1820–79), 201
Shen Shixiu 申士秀 (d.u.; 1763 *jinshi* 進士), 146–48
Shifang County 什邡縣 (Sichuan), 151–52, 253n125
Shixianli 時憲曆 or *Shixianshu* 時憲書. *See* calendar, imperial
shrines, 24, 36, 47, 50, 77, 80, 83, 99–100, 103–5, 120, 141–42, 198, 218, 221; Daoist, 103, 117, 125, 144, 157, 159; Islamic, 114–19, 125. *See also* temples
Shunchang County 順昌縣 (Fujian), 21
Sino-Japanese War of 1894–95, 194
Six Boards of the Imperial Government. *See* Boards of the Imperial Government

Skinner, William (1925–2008), 164
Smith, Arthur H. (1845–1932), 169
smuggling, 168, 197
Snow Heart Rhapsody 雪心賦, 144, 148
Sommer, Matthew, 9
Song dynasty (960–1279), 10, 95–97, 101–2, 111, 121, 226
Song Gengping 宋賡平 (d. c. 1910), 195, 209, 222
Song Yingxing 宋應星 (1587–1666), 169
souls, 23
Stapleton, Kristin, 196
Steinhardt, Nancy, 70
Substatutes of the Board of Punishments in Current Use 刑部現行則例, 162
Sufism. *See* Islam.
Suining County 遂寧縣 (Sichuan), xxiii, 108
superstition (*mixin* 迷信), 218, 229. *See also* Western nations, misconceptions of fengshui in
surveillance units (*baojia* 保甲), 61, 68, 88, 95, 99–100, 102–3, 113, 168, 187
Szonyi, Michael, 97

Taiping Civil War (1850–64), 12, 95, 109, 113–14, 164, 193, 200, 220, 244n61, 246n1
Taiwan, 3, 16, 22, 41, 223, 229
Tang dynasty (618–907), 96–97, 102, 126, 141, 214
Taussig, Michael, 56–57
taxation, 9, 12–13, 39, 44, 58, 159, 162–64, 171, 185, 190, 196–200, 212, 220, 244n61, 245n90; granary books, 253n122, 253n124; of graves, 48–50; *lijin* 釐金 tax, 171, 185, 197, 278n29; Salt Tax Bureau, 197–99 (*see also* mining, salt)
telegraphs, 2, 8, 193–94, 200–4, 208, 220–22, 279n48
temple schools, 127, 139, 141–42, 156, 265n80
temples, 4, 32–37, 53–54, 61, 69–71, 88–90, 95–100, 103–6, 116–18, 126, 150, 152, 154, 172–74, 180, 202, 214–19; Buddhist, 35–36, 103, 115, 121, 125, 144, 157, 159; Confucian, 80–81, 122, 127, 142, 156, 198; Islamic, 116–17. *See also* shrines

Ten Books on Residences 陽宅十書, 79
termites, 25, 27–28
Thorough Explanation of the Geomantic Compass, A 羅經透解, 130, 132–33, 266n106
Thousand Steps Corridor, 50–51
Three Essentials of Residences 陽宅三要, 72
Three Judicial Offices, 50–55
Tian Wenjing 田文鏡 (1662–1733), 67
Tianjin Massacre (1870), 193
timber, 12, 32–34, 36, 49, 53, 103–4, 111, 126, 159, 168–70, 205–6, 217, 221–22. See also trees
tin. See under mining
tombstones, 33, 35, 43, 138, 250n62, 270n176. See also graves
Tongchuan Prefecture 潼川府 (Sichuan), 123
Tongzhi Emperor/reign, 40, 118, 189
towers (ta 塔). See pagodas (ta 塔)
Treatise on State Building Standards 營造法式, 70
Treaty of Tianjin (1858), 200, 210
trees: ancient, 32, 66, 96–97, 103, 122, 125; cutting of, 8, 10, 21, 24, 33–37, 40, 45, 53–54, 97, 103–4, 165, 172, 204–5, 213–221, 250n63 (see also timber); cypress, 91, 103–4, 122, 125, 217, 219; fengshui, 20–21, 25, 32–39, 44–45, 47–49, 53–54, 59, 213, 215, 217, 219, 249n52; inauspicious, 73; planting of, 32–33, 35–36, 39, 73, 97, 216–18; sale of, 36, 49, 53–54, 218, 281n104; theft of, 59, 263n42
trigrams, 64, 72, 85. See also compass, geomantic (luopan 羅盤)
Tulbingga (Tu'erbing'a 圖爾炳阿, Tunggiya clan; d. 1765), 57

veins. See buildings, geomantic veins of; dragon veins; earth veins

walls: city, 120, 125–26, 139, 144, 147–48, 188, 219, 225; screen (zhaobi 照壁), 127, 139–42
Wang Dacai 王達材 (b. 1829), 58
Wang Dingzhu 王定柱 (c. 1761–1830), 30–31, 45–47, 68
Wang Huizu 汪輝祖 (1731–1807), 67
Wang Qingxi 汪清溪 (d.u.), 69
Wang Yunhui 王允輝 (d.u.; 1801 jinshi 進士), 164

water, role of in fengshui, 1, 7–8, 10, 27, 44, 64, 80, 89, 93, 98–99, 108, 137–40, 144, 149–52, 159–60, 165, 184, 229. See also dams; mountains, role of in fengshui; rivers
Water Dragon Classic 水龍經, 54
water exits (shuikou 水口), 122, 151
Watson, James, 224
Weber, Max (1864–1920), 2
Wei Shunguang 魏順光, 15, 31
Weller, Robert, 2
Wenchang 文昌, 109, 117, 121–22, 139–41, 261n10, 264n72
Wenqu star 文曲星. See Ling star 櫺星
Wesley-Smith, Peter, 154
Western Hills (Xishan 西山), 162, 271n28
Western nations: calendar of, 132–33, 266n106; confrontations with, 2, 192, 203, 209, 213, 220; imperialism of, 14, 17–18, 192–93, 201; legal systems of, 4, 211; misconceptions of fengshui in, 1–2, 8, 61–62, 160, 194–95, 200, 210; Orientalism in, 2–3, 5, 62; science and technology of, 105, 130, 194–95, 202, 204, 209, 216, 220, 227–28, 265n94 (see also railways; telegraphs)
Wheatley, Paul (1921–99), 155
White, Lynn (1907–87), 8
White Lotus Rebellion (1796–1804), 13, 95, 97, 102, 119, 123, 135, 157, 244n61, 261n10
White Tiger, 50
Will, Pierre-Étienne, 67, 187
wind disease (fengbing 風病), 39–40. See also children, harm to; General's Arrow (jiangjun jian 將軍箭)
women, 73–75, 95, 102, 243n41, 260nn114–15
Wooldridge, Chuck, 200
Wright, Arthur (1913–76), 155
Wu, Huiyi, 132
Wu, Shellen, 211
Wu Peilin 吳佩林, 14
Wu Tang 吳堂 (1813–76), 184–85, 187–90

Xi'an Prefecture 西安府 (Shaanxi), 114
Xiande 憲德 (Silut clan; d. 1740). See Hiyande

Xianfeng Emperor/reign, 58, 135, 158, 160, 177, 182, 191
Xie Zhidao 謝志道 (d.u.), 144, 147–48
Xinghua County 興化縣 (Jiangsu), 203
Xinjiang, 132
Xiong Weiyao 熊維堯 (d.u.), 72
Xiong Zuozhou 熊佐周 (1892–1958), 166–67
Xu Dong 徐棟 (1793–1865), 67, 256n25
Xu Shilin 徐士林 (1684–1741), 67–68, 93
Xu Yue 徐躍, 215
Xu Zhixiang 徐致祥 (1838–99), 205

yamen, 24, 31, 48–50, 52, 67, 69, 79, 93, 127–132, 136, 146, 176, 197, 217, 228; departments of, 15. *See also* Department of Punishments; Department of Works; government, county
Yang, C. K. 楊慶堃 (Chuan-kwang/Qingkun; 1911–99), 157
Yang Yunsong 楊筠松 (c. 834–900), 138
yangzhai 陽宅. *See* houses
Yangzi River 揚子江. *See under* rivers
Yao Lide 姚立德 (1718–83), 64
Yi (Lolo) people, 112, 263n42
Yilong County 儀隴縣 (Sichuan), 53
Yin-yang cosmology, 6–7, 22–23, 63–64, 70, 226–28, 265n92
Yin-yang masters, 128, 134–36, 145
Yin-yang officers, 12, 68, 79, 109, 127–39, 141, 156, 226
Yin-yang schools, 7, 79, 127–29

Yinzhai 陰宅. *See* graves
Yixuan 奕譞 (Aisin-Gioro clan; 1840–91), 210–11
Yongzheng Emperor/reign, 14, 65–66, 142, 163
Yu Yue 俞樾 (1821–1907), 205
Yuan Shikai 袁世凱 (1859–1916), 211, 280n84, 281n85
Yuan Shizhen 袁世振 (d. 1631), 72
Yunmeng County 雲夢縣 (Hubei), 141
Yunnan Province, 111

Zelin, Madeleine, 158
Zeng Guofan 曾國藩 (1811–72), 11, 109, 111
Zhang, Lawrence, 127
Zhang Tingyu 張廷玉 (1672–1755), 163–64
Zhang Zhongyou 張仲友 (d.u.), 76–77, 79
Zhao Guolin 趙國麟 (1673–1751), 163
Zhao Jishi 趙吉士 (1628–1706), 15
Zhao Tingdong 趙廷棟 (d.u.), 27, 73
Zhejiang Province, xxiii, 113, 137, 157
Zheng Guanying 鄭觀應 (1842–1922), 194, 209, 280n76
Zhili Province, xxiii, 17, 117, 211, 218, 280n84. *See also* Ninghe County 寧河縣 (Zhili)
Zhou Renji 周人驥 (1696–1763), 15
zhuangyuan 狀元. *See under* civil examinations
Zhuge Liang 諸葛亮 (181–234), 62–63
zinc. *See under* mining
Zongli Yamen 總理衙門, 210, 212, 221
Zuo Zongtang 左宗棠 (1812–85), 194

A NOTE ON THE TYPE

This book has been composed in Arno, an Old-style serif typeface in the classic Venetian tradition, designed by Robert Slimbach at Adobe.

GPSR Authorized Representative: Easy Access System Europe - Mustamäe tee 50, 10621 Tallinn, Estonia, gpsr.requests@easproject.com

www.ingramcontent.com/pod-product-compliance
Lightning Source LLC
Jackson TN
JSHW021716241025
93124JS00001B/1